the INSTRUMENT *flight manual*

THIRD EDITION

THE INSTRUMENT RATING

William K. Kershner

Iowa State University Press / Ames, Iowa

Composed by Cushing-Malloy, Ann Arbor, Mich.
Printed in the United States of America

First edition, 1967, four printings; second edition, 1969, seven printings

Third edition, 1977
Second printing, 1977
Third printing, 1979
Fourth printing, 1980
Revised fifth printing, 1982
Sixth printing, 1984
Seventh printing, 1985
Eighth printing, 1986
Ninth printing, 1987
Tenth printing, 1988

International Standard Book No. 0-8138-0839-1
Library of Congress Catalog Card Number: 67-16913

WILLIAM KERSHNER has over four decades of experience in flying 100 types and models of airplanes from
light trainers to jets. As a ground instructor and flight
instructor he has taught many students — using the principles which he stresses in his books. He also holds
the commercial and airline transport pilot certificate,
and is the author of THE STUDENT PILOT'S FLIGHT
MANUAL, THE ADVANCED PILOT'S FLIGHT MAN-
UAL, and the FLIGHT INSTRUCTOR'S MANUAL (Iowa
State University Press).

TO THE MEMORY OF

WILLIAM THOMAS PIPER, SR.

INTRODUCTION

THIS MANUAL is aimed at the pilot interested in obtaining an instrument rating and the already rated pilot wanting to do refresher work.

As it has been written for the general aviation pilot, airspeeds, expected clearances, and altitudes used are based on a high-performance, retractable gear, four-place, general aviation airplane, unless otherwise noted. Jet and other high altitude (above 18,000 feet) IFR operations are best left for coverage in other texts.

One area most neglected by many current texts on instrument flying is that of air traffic control. This is the most complicated (it seems) part of flying IFR. One of the primary aims of the *Instrument Flight Manual* is to cover this subject in somewhat more detail than usual and yet keep it as uncomplicated as possible. Whether this is successful will, of course, be judged by the reader.

One concern that arose during the writing of this manual was that of repeating certain things already covered in *Student Pilot's Flight Manual* and *Advanced Pilot's Flight Manual,* such as flight instrument operation, airplane performance, and other areas. In each case the attempt was made to add to the previous material information that would directly apply to instrument flying. It was felt that the repeating of basics was necessary to have the book complete in itself. The question was whether the reader of this book has immediate access to that basic material (or had forgotten, for instance, how a vertical speed instrument operates), so the information was inserted. In most such cases, new illustrations were drawn to show a different view of the situation, even though the ones used in the first two books would have worked as well.

Needless to say, the charts in this book are *not* to be used for navigation purposes. The Air Traffic Control system is changing fast. Many changes had to be made to the manuscript during the process of writing because of this and finally a cutoff date for changes had to be set. If by the time this gets to the reader's hands some of the particular example airport approaches no longer exist, please consider them as general examples. As far as frequencies are concerned (FSS, Center and tower), even the government publications are hard put to keep up with changes and it takes at least six months for a big revision of a book, so use the AIM or *Airport/Facilities Directory* for such existing numbers. The same goes for Weather information.

In researching this book, I came away from the Center, towers, Flight Service Stations, and Weather Bureau Airport Stations with even more respect, if possible, for their personnel. Many of them are active pilots themselves and know how it is to be "on the other end."

I was fortunate in having knowledgeable people help me on this book; however, any errors are mine. I would like to thank the following:

Appreciation is expressed to Mr. W. W. Parker, Chief of the Memphis Air Route Traffic Center for allowing me to visit the Center at my convenience and particular thanks must go to Roy G. Koeller, ATC Specialist at the Memphis Center, for answering numerous questions and furnishing information on the Center and air traffic control.

John Omohundro, Jr., Chief, Air Traffic Control Tower; Spencer Wise, Tower Watch (George) Supervisor; Ramon Nelson, also Tower Watch Supervisor; and especially James Ayers, ATC Specialist, all of the Nashville tower, must be thanked for the time and effort they put into reviewing the chapters applying to air traffic control. Their suggestions were a great help in that part.

Special recognition must go to Jack LeBarron, top aviation writer and pilot, for sending most useful information (most of which he personally evolved) not generally available to the instrument trainee.

Herbert Price, Watch Supervisor, and E. D. De Shields, ATC Specialist, at the Chattanooga tower, who always had information when needed.

John Lenti and George Bullis, of the Memphis tower, for furnishing information and answering questions.

Willis Singletory, Chief of the Memphis Flight Service Station; and John Edwards, George Roe, and George Rhodes, of that facility, for their help in discussing the functions of the FSS in filing an instrument flight plan.

Berl Henry, M. H. Smith, H. E. Pritchard, and Delbert Robertson of the Chattanooga Weather Bureau Airport Station for sending actual (past date) weather sequences and forecases for use in illustrations.

John Hornaday and Bill Whitmore, of the FAA General Aviation District Office at Nashville, for reviewing the manuscript and offering both practical help and encouragement for its completion. Elmour D. Meriwether, veteran pilot of Nashville, who reviewed the manuscript and made valuable suggestions. Fred C. Stashak, Chief Stress Engineer of Piper Aircraft Corp., for sending information on aircraft stress.

W. D. Thompson, Chief of Flight Test and Aerodynamics of Cessna Aircraft Company, who kindly furnished needed performance data on Cessna airplanes for use in writing this manual. John Kruk, Electrical Design Engineer of Piper Aircraft Corp., for providing much needed electrical background knowledge aimed at the needs of the instrument pilot. Allen W. Hayes, pilot, instructor and engineer, of Ithaca, New York, who cut through manuscript verbosity and corrected errors and misprints and who made valuable suggestions. Norbert A. Weisand, Senior Product Engineer, De-icing Systems Engineering, B. F. Goodrich Co., for furnishing data on that equipment.

Appreciation is expressed to personnel of the Coast and Geodetic Survey of the Environmental Science Services Administration, especially Frank McClung, for assuring that I received the latest charts available at the time the manuscript was in preparation.

I want to thank the avionics and other manufacturers who furnished information on their products. They are listed in the Bibliography.

Col. Leslie McLaurin, manager of the Sewanee airport, who reviewed the manuscript and used it as an instrument refresher and who made suggestions valuable for that area of the manual. Mr. and Mrs. Phillip Werlein, of Sewanee, who helped at a critical stage of the manuscript by giving needed extra working room. Mrs. Mary Lou Chapman, of St. Andrews School in Sewanee, for furnishing working space at the school. The College Entrance Examination Board (Southern Region) here at Sewanee, aided the effort to a large degree by allowing me to use their facilities.

INTRODUCTION TO THE THIRD EDITION

This edition covers in more detail the Center and terminal area radar-computer systems and updates terms, ATC procedures and available COM/NAV equipment.

As was the case of the first two editions, I had advice and suggestions of knowledgeable people but, again, any errors are mine.

At the Memphis Center, I particularly owe thanks to Carl Graves, who patiently explained the Computer system and answered my many questions over a two day period there. Roy Koeller sent me much needed information from the Center before my visit so that I could be prepared to ask reasonable questions.

At the Nashville tower facility, Don Wells devoted much time to showing me the operations and bringing me up to date on the latest procedures from the terminal controller's standpoint, under both radar and non-radar conditions. Others at the Nashville tower who let me "sit in" at the radar consoles and answered questions were Galyon Northcutt, David Pyrdum, Bill Duke and Bill Allen. Ed Stoddart sent information to me before the visit.

The weather chapter of this book was the result of the help of the late Ray McAbee, former FSS specialist who helped me in every way possible. Tennessee aviation will miss him.

The specialists at the Chattanooga Weather Service Office should have gotten tired of my calls and questions but never indicated it, and Hugh Pritchard, Ed Higdon and others were more than generous with information.

C. S. (Chuck) Davis of NARCO AVIONICS sent many photos and data with permission to use them. I would also like to thank Dave Speer of KING RADIO for use of photos and information he sent. Appreciation is also in order for Nat Toulon of Sewanee and Fred Gardner of the FAA at OKC, for their help.

I owe much thanks to William C. Lewis, Jr., CFI-I of Madison, Wisconsin, who reviewed the second edition and made a great number of good suggestions which were incorporated here.

And, of course, most thanks must go to my wife, who encouraged, and typed and retyped as this edition progressed and whose patience was outstanding throughout.

WILLIAM K. KERSHNER

Sewanee, Tennessee

CONTENTS

Part One

AIRPLANE PERFORMANCE AND
BASIC INSTRUMENT FLYING

Chapter 1

THE INSTRUMENT RATING

FOR A LONG TIME NOW, you've sat on the ground and watched other pilots take off into weather that kept you haunting the Weather Service Office at Podunk Greater International Airport or other such well-known places. You squeaked in by the skin of your teeth (the airport went well below VFR minimums shortly after you got in and has been that way for days), and the bitter part about it is that the tops are running only three or four thousand feet. It's CAVU above, and the weather at your destination is very fine VFR — and there you sit. That guy, over at the Flight Service Station counter, is filing IFR and is going — and he doesn't look as though he has any more on the ball than you have. After a few occasions of this nature, you've decided to get that instrument rating. Or maybe your decision came about because one time you were a "grey-faced, pinheaded holeseeker" (Fig. 1-1). Looking back at it, you'll have to confess that you were pinheaded to get in such a predicament, and while you couldn't see your face, it sure felt grey from your side of it. If that hole hadn't showed up when it did, well, that could have put you between a rock and a hard place.

The instrument-rated pilot is still held in some awe by the nonrated people at the airport. Unfortunately, the pilots with this rating don't always try to dispel the awe, but that's only human. Generally speaking, the two *extreme* schools of thought by those considering the instrument rating are: (1) It is a license to fly anywhere, anytime, and weather will no longer be an important consideration; or (2) it will be used only as an emergency method of getting down and may never be needed.

If you belong to the first group, give up any idea of getting an instrument rating. You'll be a menace to the rest of us clear thinkers and very likely have an exciting, but extremely brief, career.

If you are in Group 2, you could be wasting your time and money by getting an instrument rating for use only in an emergency — you may never use it. However, it is good training and would help the other areas of your flying.

Of course, actually *you* don't fall into either of the extremes. You know that there will be times after getting the rating that you'll still be sitting on the ground because of the weather. But you will be able to get out more often than is the case now.

One thing you'll notice as you work on the rating is that all your flying will become more precise. You'll be much more aware of altitude and heading and how power and airspeed combinations affect performance.

FAR 61.65 THE REQUIREMENTS FOR THE INSTRUMENT RATING — AIRPLANE

General. To be eligible for an instrument rating (airplane) or an instrument rating (helicopter), an applicant must:

1. hold at least a current private pilot certificate with an aircraft rating appropriate to the instrument rating sought

2. be able to read, speak, and understand the English language

3. comply with the applicable requirements of this section.

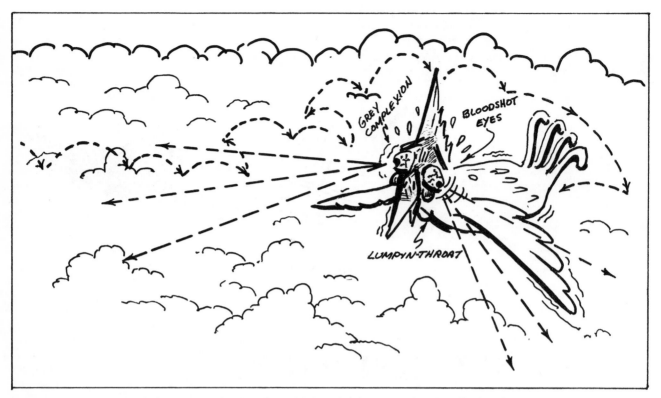

Fig. 1-1. The grey-faced, pinheaded holeseeker has an exciting but often brief career.

Ground instruction. An applicant for the written test for an instrument rating must have received ground instruction, or have logged home study in at least the following areas of aeronautical knowledge appropriate to the rating sought:

1. the regulations of this chapter that apply to flight under IFR conditions, the Airman's Information Manual, and the IFR air traffic system and procedures

2. dead reckoning appropriate to IFR navigation, IFR navigation by radio aids using the VOR, ADF, and ILS systems, and the use of IFR charts and instrument approach plates

3. the procurement and use of aviation weather reports and forecasts, and the elements of forecasting weather trends on the basis of that information and personal observation of weather conditions

4. the safe and efficient operation of airplanes or helicopters, as appropriate, under instrument weather conditions.

Written test. An applicant for an instrument rating must pass a written test appropriate to the instrument rating sought on the subjects in which ground instruction is required.

Flight instruction and skill – airplanes. An applicant for the flight test for an instrument rating (airplane) must present a logbook record certified by an authorized flight instructor showing that he has received instrument flight instruction in an airplane in the following pilot operations, and has been found competent in each of them:

1. control and accurate maneuvering of an airplane solely by reference to instruments

2. IFR navigation by the use of the VOR and ADF

systems, including compliance with air traffic control instructions and procedures

3. instrument approaches to published minimums using the VOR, ADF, and ILS systems (instruction in the use of the ADF and ILS may be received in an instrument ground trainer and instruction in the use of the ILS glide slope may be received in an airborne ILS simulator)

4. cross-country flying in simulated or actual IFR conditions, on Federal airways or as routed by ATC, including one such trip of at least 250 nautical miles, including VOR, ADF, and ILS approaches at different airports

5. simulated emergencies, including the recovery from unusual attitudes, equipment or instrument malfunctions, loss of communications, and engine-out emergencies if a multiengine airplane is used, and missed approach procedures.

Practical test. An applicant for an instrument rating must pass a flight test in an airplane or a helicopter, as appropriate. The test must include instrument flight procedures selected by the inspector or examiner conducting the test to determine the applicant's ability to perform competently the IFR operations on which instruction is required.

Flight experience. An applicant for an instrument rating must have at least the following flight time as a pilot:

1. a total of 125 hours of pilot flight time, of which 50 hours are as pilot in command in cross-country flight in a powered aircraft with other than a student pilot certificate. Each cross-country flight must have a landing at a point more than 50 nautical

miles from the original departure point.

2. forty hours of simulated or actual instrument time, of which not more than 20 hours may be instrument instruction by an authorized instructor in an instrument ground trainer acceptable to the Administrator

3. fifteen hours of instrument flight instruction by an authorized flight instructor, including at least 5 hours in an airplane or a helicopter, as appropriate.

If you are getting the instrument rating "on your own" and not going through a formal program, you'll have to think about a means of simulating instrument conditions in the airplane. One method is the hooded visor which when worn cuts the vision down to that of only straight ahead. It is the most simple and inexpensive arrangement, being put on like a cap, but it restricts side vision to the extent of requiring a great deal of head turning to adjust power, set radios, and check engine instruments. Such quick head turning tends to invite vertigo, a condition in which the pilot *knows* (well, he *thinks* he knows) that the airplane is not doing what the instruments indicate.

While we're on the subject, some guys think that they can grab a hood and go out and practice instruments solo. Not only would that be a bad situation, it's in violation of FAR 91.21 which says, "No person may operate a civil aircraft in simulated instrument flight unless (1) an appropriately rated pilot occupies the other control seat as safety pilot, (2) the safety pilot has adequate vision forward and to each side of the aircraft or a competent observer in the aircraft adequately supplements the vision of the safety pilot.

If you are using a single-engine airplane for your instrument instruction and the instructor determines that the flight can be conducted safely (and you have a private certificate with appropriate category and class ratings), a single throwover control wheel may be used. In earlier times dual control wheels were required for all types of instruction.

Try to work it so that once you start working on the rating, you can go on with it. Don't stretch the program over too long a period. Stretching it out may make it necessary to use a part of each flight as a review. It's also best to be flying as you study for the written — one area helps the other. But get the written out of the way before you have those last few hours of brush-up time prior to the flight test.

During the training period when you are flying cross-country VFR, fly airways as much as feasible. Borrow a Low Altitude Enroute Chart (or you may want to start subscribing to one of the chart services available) and fly as if you were on an IFR flight plan. Of course, if you are flying VFR, you actually will be flying some altitude plus 500 and will be looking out for other airplanes all the time. Also, you'll do no hooded work unless you have an "appropriately rated pilot" in the right seat, but you can smooth out your estimates even flying VFR.

Get as much as possible of your dual instruction in the later stages on actual instruments, filing a flight plan and flying in the clouds (with an appropriately rated instructor, of course). It's a more realistic situation than practicing with the hood, and your confidence will be increased. This doesn't mean that you and the instructor will go out and crack through the worst squall line you can find or fly into the worst icing conditions seen in your area for twenty-nine years, but will choose the type of weather to "practice" in. The Regulations are such that you don't have to have any actual instrument experience in that 40 hours required for the rating but, if you do have some, you'll enjoy more that first flight on actual instruments.

You might talk to some of the Approved Schools in your area. (They are certificated under Part 141 and require less time.) Also, if you plan on getting a commercial certificate, not having an instrument rating can limit you severely, so that's another reason to get cracking.

After you get the rating, don't go busting into IFR with a vengeance. Take it easy and set yourself comparatively high minimum weather conditions. As your experience increases and as you get better equipment, you can gradually lower your minimums to those as published on the Approach Charts. You will also have to keep up your proficiency to a safe level; if you get rusty you'll have to ease back into it again.

Chapter 2

THE FLIGHT AND ENGINE INSTRUMENTS

THE FLIGHT INSTRUMENTS will naturally now be of even greater interest and value than before, and it is extremely important that you understand how they work. Not only must you know how to fly by reference to them but you will have to be aware of what you as a pilot must do to keep them operating properly and be able to recognize signs of impending instrument or system trouble. This chapter will cover the flight instruments and other instruments of the most importance to the instrument pilot. For instance, you know that the gyro horizon (or attitude indicator as it is also termed) is one of the most important of the flight instruments but, to date, have probably paid very little attention to the suction gage which can give warning of possible problems with the vacuum-driven instruments. The ammeter also will be of added importance; an electrical failure while flying under instrument conditions would pose many more problems than would be found if you lost the electrically driven flight instruments and radios during VFR operations. An electrical failure, for instance, *could* cause you to lose the airspeed indicator in icing conditions.

REQUIRED INSTRUMENTS AND EQUIPMENT

Visual Flight Rules (Day)

For flying VFR (day), the airplane is required to have the following instruments and equipment (FAR 91):
1. Airspeed indicator.
2. Altimeter.
3. Magnetic direction indicator.

4. Tachometer for each engine.
5. Oil pressure gage for each engine using pressure system.
6. Temperature gage for each liquid-cooled engine.
7. Oil temperature gage for each air-cooled engine.
8. Manifold pressure gage for each altitude engine.
9. Fuel gage indicating the quantity of fuel in each tank.
10. Landing gear position indicator, if the aircraft has retractable landing gear.
11. If the aircraft is operated for hire over water and beyond power-off gliding distance from shore, a Very pistol and approved flotation gear readily available to each occupant.
12. Approved safety belts for all occupants who have reached their second birthday, and after December 4, 1980, each safety belt must be equipped with an approved metal-to-metal latching device.
13. For small civil airplanes manufactured after July 18, 1978, an approved shoulder harness for each front seat.

Visual Flight Rules (Night)

For VFR flight at night, the following instruments and equipment are required in addition to those specified for VFR day flying:
1. Approved position lights.
2. An approved aviation red or aviation white anticollision light system on all large aircraft and on all small aircraft when required by the aircraft's

airworthiness certificate and on all small aircraft after August 11, 1972. In the event of the failure of any light on the anticollision light system, operations with the aircraft may be continued to a stop where repairs or replacement may be made.

3. If the aircraft is operated for hire, one electric landing light.

4. An adequate source of electrical energy for all installed electrical and radio equipment.

5. One spare set of fuses or three spare fuses of each kind required.

Instrument Flight Rules

For IFR flight, the following instruments and equipment are required in addition to those specified for VFR day and VFR night flying:

1. Two-way radio communications system and navigational equipment appropriate to the ground facilities to be used.

2. Gyroscopic rate of turn indicator

3. Slip-skid indicator.

4. Sensitive altimeter adjustable for barometric pressure.

5. A clock displaying hours, minutes and seconds with a sweep-second pointer or digital presentation.

6. Generator of adequate capacity.

7. Gyroscopic bank and pitch indicator (attitude indicator).

8. Gyroscopic direction indicator (heading indicator or equivalent).

PITOT-STATIC INSTRUMENTS

These are the flight instruments that indicate air pressure or change in pressure and include the airspeed indicator, altimeter, and rate of climb (or vertical speed) indicator.

Airspeed Indicator

The airspeed indicator is an air pressure gage calibrated to read in miles per hour or knots rather than pounds per square foot (psf). The airspeed system is made up of the pitot and static tubes and the airspeed indicator itself. As the airplane moves through the air, the relative wind exerts an impact pressure, or dynamic pressure, in the pitot tube which expands a diaphragm linked to an indicating hand (Fig. 2-1).

In addition to the dynamic pressure, static air pressure also exists in the pitot tube. As shown in Figure 2-1, the diaphragm contains both dynamic *and* static pressures. The static tube allows the static pressure to enter the instrument *case* so that these two static pressures cancel each other as far as the diaphragm is concerned, and it expands only as a function of the dynamic pressure.

Dynamic pressure, sometimes called "q," has the equation $\frac{\rho V^2}{2}$, where ρ (pronounced "rho") is the

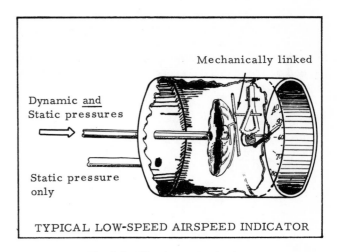

TYPICAL LOW-SPEED AIRSPEED INDICATOR

Fig. 2-1. The airspeed indicator.

air density in slugs per cubic foot, and the V is the *true* velocity of the air in feet per second. A slug is a unit of mass and may be found by dividing the weight of an object by the acceleration of gravity (32.2 feet per second per second). Hence, a 161-pound man would have a mass of 5 slugs; $\frac{161}{32.2} = 5$. A beauty queen weighing 128.8 pounds would have a mass of 4 slugs, which is certainly an unromantic way to think of her.

Realizing that the dynamic pressure is made up of the combination of one-half the density *times* the true speed (squared) of the air particles, you can see that an "indicated" airspeed of 150 knots could either result from high density and comparatively low speed of the air or a lower density and higher true airspeed. The density of the air at sea level is 0.002378 slugs per cubic foot, and at an indicated airspeed of 150 knots at sea level, the dynamic pressure is approximately 76.3 pounds per square foot. The airspeed indicator is a pressure gage, and when 76.3 pounds per square foot of dynamic pressure is reached, it duly indicates 150 knots. The airspeed indicator is calibrated for sea level density, and the 150 knots indicated would also be the true airspeed at *sea level standard conditions* (29.92 inches of mercury pressure and 59° F, or 15° C). The airspeed indicator cannot compensate for density change; it can only indicate the combination of density *and* velocity of the air.

At 10,000 feet, the air density is only about 3/4 of that at sea level; hence, if the plane is indicating 150 knots at that altitude, it is meeting the fewer air particles at a higher speed than was done at sea level in order to get the same dynamic pressure (indicated airspeed). If you are interested in the mathematics of the problem, the following is presented:

Dynamic pressure (psf) $= \frac{\rho V^2}{2}$.

At sea level V = 150 knots = 254 fps.

Dynamic pressure $= \frac{0.002378}{2} \times (254)^2 = 76.3$ psf.

7

At 10,000 feet the standard air density is 0.001756 slugs per cubic foot (see the Appendix). As the airplane is indicating 150 knots at 10,000 feet, the dynamic pressure is also 76.3 psf, and the true airspeed or true relative speed of the air can be found by solving for V as follows:

$$76.3 = \frac{0.001756}{2} \times V^2$$

$$V^2 = \frac{152.6}{0.001765}; \qquad V = \sqrt{\frac{152.6}{0.001756}};$$

$$V = 295 \text{ fps, or } 175 \text{ knots.}$$

You do this type of calculation with your computer (whether you know it or not). You can check the above with your computer (the standard temperature at 10,000 feet is -5° C). You don't work with feet per second, however. You'll note that an indicated (or rather calibrated) airspeed of 150 knots at 10,000 feet density altitude gives a true airspeed of 175 knots (174+).

In the illustration, it was assumed that the airspeed indicator was giving you the exact, straight story; this is not always the case. (It's not even usually the case.) On your computer you are working with calibrated airspeed (C.A.S.), which is the indicated airspeed (I.A.S.) corrected for errors in the airspeed system — (includes errors in the instrument plus errors in the pitot-static system normally called position and/or installation errors). Your airplane may have an airspeed correction table which allows the correcting of indicated to calibrated airspeed. In the majority of cases in practical application for smaller airplanes, airspeed system error is ignored, and indicated airspeed is assumed to equal calibrated airspeed in the cruise range. At low speeds near the stall, however, the difference between I.A.S. and C.A.S. can be 10 mph — or more.

Another term used is *equivalent airspeed* (E.A.S.), and this is C.A.S. corrected for compressibility effects. This is not of consequence below 250 knots and 10,000 feet, so that it's not likely that you would need a compressibility correction table for your present work. Normally your corrective steps will be: I.A.S. to C.A.S. to T.A.S. (true airspeed). If you have no correction card for instrument error, it will be: I.A.S. to T.A.S. If you were operating at altitudes and speeds where compressibility effects existed, note the full number of steps would be: I.A.S. to C.A.S. to E.A.S. to T.A.S. The problem is that the static air in the *pitot* tube is being packed (compressed) and gives a high reading (remember the pitot tube is measuring *both* dynamic *and* static pressures) so that the effect is, as far as the airspeed indication is concerned, that of a higher dynamic pressure than actually exists. In other words, the C.A.S. is higher than it should be, and computing for E.A.S. gives the true picture.

Airspeed Indicator Markings

The FAA requires that the airspeed indicator be marked for various important speeds and speed ranges. Figure 2-2 shows the required markings:

Red Line — Never exceed speed (V_{NE}). This speed should not be exceeded at any time.

Yellow Arc — Caution range. Strong vertical gusts could damage the airplane in this speed range; therefore, it is best to refrain from flying in this speed range when encountering turbulence of any intensity. The caution range starts at the maximum structural cruising speed V_{NO} and ends at the never exceed speed (V_{NE}).

Green Arc — Normal operating range. The airspeed at the lower end of this arc is the flaps-up, gear-up, power-off stall speed at gross weight, V_{S_1}. For most airplanes, the landing gear position (full up or full down) has no effect on stall speed. The upper end of the green arc is the maximum indicated airspeed (V_{NO}) where no structural damage would occur in moderate vertical gust conditions (30 fps).

White Arc — The flap operating range. The lower limit is the power-off stall speed with recommended landing flaps (V_{S_0}) at gross weight (gear extended and cowl flaps closed), and the upper limit is the maximum flap operating speed (full flaps).

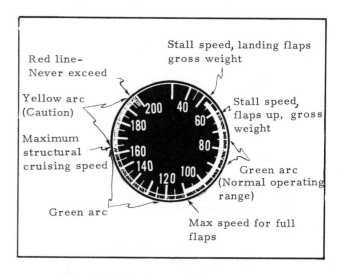

Fig. 2-2. The airspeed indicator markings.

Older airplanes have the airspeed indicator markings as *calibrated* airspeed in mph or knots. Newer airplanes will have the airspeed markings as *indicated* airspeed in *knots*. As a general rule, assume that 1976 model (and later) airspeed indicators will be marked in KIAS, but you should confirm this in the *Pilot's Operating Handbook* or *Airplane Flight Manual*. (More about the POH at the end of this chapter.)

Altimeter

The altimeter is the most important of the three instruments of the pitot-static group as far as instrument flying is concerned. It is an aneroid barometer calibrated to read in feet instead of inches of mercury. Its job is to measure the static pressure (or ambient pressure as it is sometimes called) and register this fact in terms of feet or thousands of feet.

The altimeter has an opening that allows static (outside) pressure to enter the otherwise sealed case. A series of sealed diaphragms or "aneroid wafers" within the case are mechanically linked to the three indicating hands. Since the wafers are sealed, they retain a constant internal "pressure" and expand or contract in response to the changing atmospheric pressure surrounding them in the case. As the aircraft climbs, the atmospheric pressure decreases and the sealed wafers expand; this is duly noted by the indicating hands as an increase in altitude.

Standard sea level pressure is 29.92 inches of mercury, and the standard sea level temperature is 15° C, or 59° F. The altimeter is calibrated for this condition, and any change in local pressure must be corrected by the pilot. This is done by using the setting knob to set the proper barometric pressure (corrected to sea level) in the setting window. For instance, a station at an elevation of 670 feet above sea level has an *actual* barometric pressure reading of 29.45 inches of mercury according to its barometer. Since the pressure drop is 1.06 inches of mercury for the first 1000 feet above sea level (see Appendix), an addition of 0.71 inches to the actual reading of 29.45 will correct the pressure to the sea level value of 30.16 inches of mercury. This, of course, assumes that the pressure drop is standard. This is the normal assumption and accurate enough for *indicated* altitude, which will be discussed shortly.

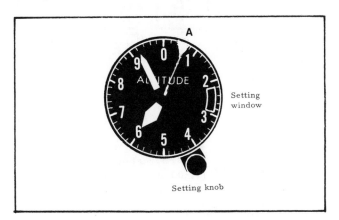

Fig. 2-3. The altimeter. The 10,000 foot hand (A) won't be shown on the exercises following in this book.

There are several altitudes that will be of interest to you:

Indicated Altitude is the altitude read when the altimeter is set to the local barometric pressure corrected to sea level as just mentioned.

True Altitude is the height above sea level.

Absolute Altitude is the height above the terrain.

Pressure Altitude is the altitude read when the altimeter is set to 29.92. This indication shows what your altitude would be if the altimeter setting were

29.92 — that is, if it were a standard pressure day at sea level.

Density Altitude is the pressure altitude computed with temperature. The density altitude is used in performance. If you know your density altitude, air density can be found by tables, and airplane performance calculated. You go through this step every time you use a computer to find the true airspeed. You use the pressure altitude and the outside air temperature (O.A.T,) at that altitude to get the true airspeed. Usually, there's not enough difference in pressure altitude and indicated altitude to make it worthwhile to set up 29.92 in the setting window, so that the usual procedure is to use the *indicated* altitude and O.A.T.

The fact that the computer used pressure altitude and temperature to obtain density altitude in finding true airspeed didn't mean much, as you were only interested in the final result. You may not even have been aware that you were working with density altitude during the process. Most computers also allow you to read the density altitude directly by setting up pressure altitude and temperature. This is handy in figuring the performance of your airplane for a high-altitude and/or high-temperature take-off or landing. The *Pilot's Operating Handbook* gives graphs or figures for take-off and landing performance at the various density altitudes. After finding your density altitude, you can find your predicted performance in the *Pilot's Operating Handbook*. Computers are not always available, and the manufacturer sometimes furnish conversion charts with the *Pilot's Operating Handbook* (Fig. 2-4).

Suppose you are at a pressure altitude of 5000 feet, and the outside air temperature is 90° F. Using the conversion chart, you see that your density altitude is 8000 feet (Fig. 2-4). Looking at the take-off curves for your airplane, you can find your expected performance at that altitude.

You and other pilots fly *indicated altitude*. When you're flying cross-country, you will have no idea of your exact altitude above the terrain (although over level country you can check airport elevations in your area, subtract this from your indicated altitude, and have a barnyard figure). Over mountainous terrain, this won't work, as the contours change too abruptly for you to keep up with them. As you fly, you'll get altimeter settings from various ground stations and keep up to date on pressure changes so your indicated altitude will be correct.

The use of indicated altitude for all planes makes good sense in that all pilots are using sea level as a base point and proper assigned altitude separation results.

Altimeter Errors

Instrument Error — If you set the current barometric pressure (corrected to sea level) for your airport, the altimeter should indicate the field elevation when you're on the ground. FAR 91.170 specifies that airplanes operating in controlled airspace (IFR) must have had each static pressure system and

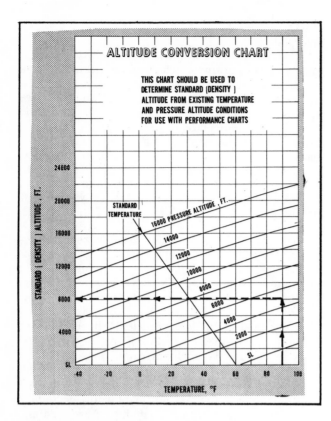

Fig. 2-4. The altitude conversion chart. Move up the 90° line until the 5000-foot pressure altitude is reached; directly across from this point is the standard (density) altitude for that combination (8000 feet). *(Piper Aircraft Corp.)*

each altimeter instrument tested by the manufacturer or an FAA approved repair station within the past 24 calendar months.

Pressure Changes — When you fly from a high-pressure area into a low-pressure area, the altimeter "thinks" you have climbed and will register accordingly — even if you haven't changed altitude. You'll see this and will fly the plane down to the "correct altitude" and will actually be low. (This is a gradual process, and you will be easing down over a period of time to maintain what is the "correct altitude.") When you fly from a low- to a high-pressure area, the altimeter thinks you've let down to a lower altitude and registers too low. A good way to remember (although you can certainly reason it out each time) is: HLH — High to Low, altimeter reads High. LHL — Low to High, altimeter reads Low. (High to low -- look out below!)

You can see that it is worse to fly from a high- to a low-pressure area as far as terrain clearance is concerned. Double-check altimeter settings as you fly IFR en route.

Temperature Errors — The equation of state, which shows the relationship between pressure, density, and temperature of the atmosphere, notes that atmospheric pressure is proportional to the temperature. If the temperature is above normal, the pressure will be higher than normal (constant density). Therefore, if you are flying at a certain indicated altitude and the temperature is higher than

normal, the pressure at your altitude is higher than normal. *The altimeter registers lower* than your *true* altitude. If the temperature is lower, the pressure is lower and the altimeter will register accordingly — *low temperature, altimeter reads high.*

You might remember it this way, using the letters H and L as in pressure change: Temperature High, altimeter reads Low — HL. Temperature Low, altimeter reads High — LH. Or maybe it's easier to remember HALT (High Altimeter because of Low Temperature).

The best thing, however, is to know that higher temperature means higher pressure (and vice versa) at altitude and reason it out from there.

The temperature error is zero at sea level (or at the elevation of the station at which the setting is obtained) and increases with altitude, so that the error could easily be 500 to 600 feet at the 10,000-foot level. In other words, you can have this error at altitude even if the altimeter reads correctly at sea level. Temperature error can be found with a computer as shown in Figure 2-5. For indicated altitude, this error is neglected; but it makes a good question for an instrument rating written exam or flight test, so keep it in mind.

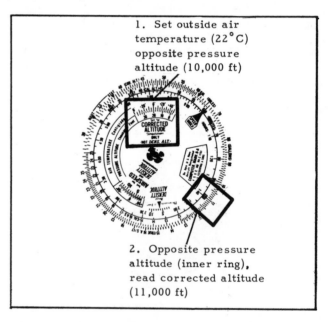

1. Set outside air temperature (22°C) opposite pressure altitude (10,000 ft)

2. Opposite pressure altitude (inner ring), read corrected altitude (11,000 ft)

Fig. 2-5. Correcting the altimeter for temperature errors.

These errors (particularly temperature errors which are normally ignored) affect everybody in that area (though slightly differently for different altitudes), so that the altitude separation is still no problem. Temperature errors could cause problems as far as terrain clearance is concerned, however.

A final altimeter note: For computer work, you are told to use the *pressure altitude* to find the true airspeed. For practical work, use *indicated altitude* (current sea level setting) for true airspeed computations. Remember that the T.A.S. increases about 2 per cent per thousand feet so the most you will be off will be 2 per cent. That is, your sea level altimeter setting could possibly be 28.92 or 30.92 but this is extremely unlikely. So ...

Assume that a total error of no more than 1 per cent will be introduced by use of indicated altitude. For a 200-knot airplane, this means you could be 2 knots off true airspeed. But the instrument error or your error in reading the instrument could be this much.

As you progress in your instrument flying to heavier and more complex equipment, you'll use more sophisticated altitude indicators such as encoding altimeters (used with the transponder) and radar altimeters (which give absolute altitude readings). These will be covered in more detail in later chapters as their use is introduced.

Rate of Climb or Vertical Speed Indicator

Like the altimeter, the vertical speed indicator has a diaphragm. But unlike the altimeter, it measures the *rate of change* of pressure rather than the pressure itself.

The diaphragm has a tube connecting it to the static tube of the airspeed indicator and altimeter (or the tube may just have access to the cabin air pressure in the case of cheaper or lighter installations). This means that the inside of the diaphragm has the same pressure as the static pressure of the air surrounding the airplane. Opening into the otherwise sealed instrument case is a capillary tube, which also is connected to the static system of the airplane.

Figure 2-6 is a schematic diagram of a typical vertical speed indicator. As an example, suppose the airplane is flying at a constant altitude. The pressure within the diaphragm is the same as that of the air surrounding it in the instrument case. The rate of climb is indicated as zero.

The plane is put into a glide or dive. Air pressure inside the diaphragm increases at the same rate as that of the surrounding air (1). However, because of the small size of the capillary tube, the pressure in the instrument case does not change at the same rate (2). In a glide or dive, the diaphragm

expands, the amount of expansion depending on the difference of pressures. As the diaphragm is mechanically linked to a hand (3), the appropriate rate of descent in hundreds (or thousands) of feet per minute is read on the instrument face (4).

In a climb, the pressure in the diaphragm decreases faster than that within the instrument case, and the needle will indicate an appropriate rate of climb.

Because in a climb or dive the pressure in the case is always "behind" the diaphragm pressure in this instrument, a certain amount of lag results. The instrument will still indicate a vertical speed for six to nine seconds after the plane has been leveled off. That's why the vertical speed indicator is not used to maintain altitude. On days when the air is bumpy, this lag is particularly noticeable. The vertical speed indicator is used, therefore, either when a constant rate of ascent or descent is needed or as a check of the plane's climb, dive, or glide rate. The sensitive altimeter is used to maintain a constant altitude, although the vertical speed instrument can show the trend away from a desired altitude — if you realize that the lag is present.

The pointer should read zero while the airplane is on the ground, and any deviation from this can be corrected by turning the adjustment screw on the instrument.

There is a more expensive vertical speed indicator (Instantaneous Vertical Speed Indicator) on the market that does not have lag and is very accurate even in bumpy air. It contains a piston-cylinder arrangement whereby the airplane's vertical acceleration is immediately noted. The pistons are balanced by their own weights and springs. When a change in vertical speed is effected, the pistons are displaced and an immediate change of pressures in the cylinders is created. This pressure is transmitted to the diaphragm, producing an almost instantaneous change in indication. After the acceleration-induced pressure fades, the pistons are no longer displaced, and the diaphragm and capillary tube act as on the old type of indicator (as long as there is no acceleration). The actions of the acceleration elements and the diaphragm-capillary system overlap for smooth action.

It's possible to fly this type of instrument as accurately as an altimeter, but its price is understandably higher than that of the standard vertical speed indicator.

The Pitot-Static System

The three instruments just discussed must have a dependable source of static (outside) air pressure in order to operate accurately. Figure 2-7 shows a schematic diagram of the pitot-static system and the instruments.

The static system shown in Figure 2-7 uses a Y-type vent system to decrease static errors in yaw. The locations of the static vents are carefully

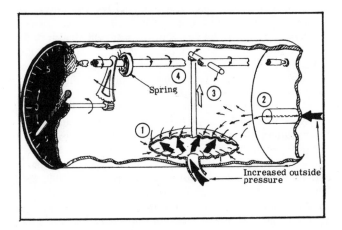

Fig. 2-6. The vertical speed indicator and how it reacts to a descent.

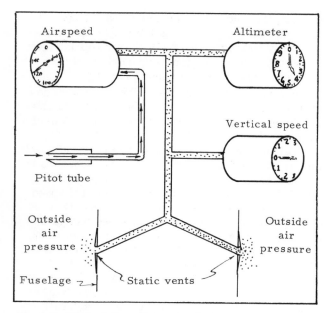

Fig. 2-7. Pitot-static system using flush type Y static vents in fuselage.

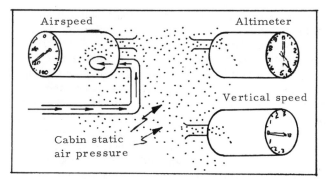

Fig. 2-9. Simple pitot-static system.

chosen to obtain the most accurate static (outside) pressure. The usual location is on each side of the fuselage between the wing and the stabilizer. (You've seen the signs "Keep this vent clean.") This is usually the most accurate system of those used.

Another pitot-static system is shown in Figure 2-8. The static vent is located in the pitot-static tube.

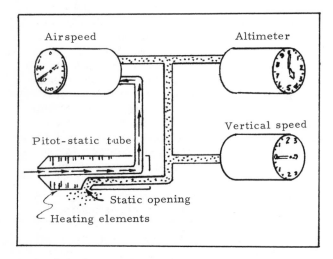

Fig. 2-8. System with static opening in pitot-static tube.

The system shown in Figure 2-8 is not usually as accurate as the Y system, and in addition, the static opening at this location may be more susceptible to icing over if the airplane does *not* have pitot heat. (An airplane that expects to fly in icing had better have pitot heat!) This could mean loss of *all three* of the pitot-static instruments — not just the airspeed as would be the case of the pitot-tube-Y-vent system. (You'd still have static pressure to the airspeed indicator in the Y-vent system but no impact pressure, so it would be out of the running if the

pitot tube iced over. This will be covered in more detail at the end of this section.)

Figure 2-9 shows the most simple pitot-static system as used on older light trainers.

The instruments as shown in Figure 2-9 get the static pressure from the cabin. Because of the effect of the air passing by the cabin, a venturi effect may result, and the static pressure will be lower than the actual outside pressure, which would mean a slightly high airspeed and altimeter indication. Once the airspeed is stabilized, the vertical speed indicator will not be affected because it is a "rate" instrument and would measure *change* of pressure as mentioned earlier.

FAR Part 23 (Airworthiness Standards for Normal, Utility and Acrobatic Category Airplanes) notes that the static air vent system must be such that the opening and closing of windows, airflow variation, and moisture or other foreign matter does not seriously affect its accuracy. (FAR 23.1325)

Pitot-Static System Problems

Pitot System — The big problem you can expect to encounter as far as the *pitot* system is concerned is that of ice closing the pitot tube (pressure inlet). The airspeed will be the only instrument affected in this case, and the prime indication of this occurring is a decrease in airspeed indication with no change in power altitude, or attitude. The application of pitot heat, if available, is the move to make. It's best, however, to apply pitot heat before you enter an area of suspected icing and leave it on until clear. However, the pitot heat is a great current drain, and under some conditions, you may want to use it intermittently.

Static System — The more complex airplanes have an alternate static source that can be used should the primary system get stopped up. This normally consists of a selector which the pilot turns to the "alternate" setting which opens the system to cabin air (nonpressurized cabin). This, then, may have the same inaccuracies described for the system as discussed for the older system shown in Figure 2-9. (But it's a lot better than no static source at all.) Opening windows and vents and using the heater will affect the airspeed indicator readings on the alternate static source selections for many airplanes. With some airplanes the alternate static selection may cause the

12

altimeter to read *lower* than normal at some indicated airspeeds — which would be the opposite you'd expect from "theory." Check the *Pilot's Operating Handbook* for the airplane you are flying. Have at least a general figure for corrections for the airspeed and altimeter using the alternate static source at cruise and expected approach speeds. For larger airplanes a separate co-pilot alternate air source is available.

For airplanes without an alternate source, one means of getting static pressure to the three instruments is to break the glass on the face of the rate of climb as it is considered the least important of the three, *but stop a minute:* Figure 2-10 is another picture of the rate of climb instrument, showing the effects of using it (vertical speed) as a source of static pressure for the system.

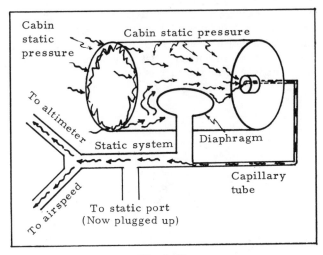

Fig. 2-10.

Note that the only source of static pressure is through the face of the instrument and thence through the capillary tube into the static system. Because the capillary tube is specifically designed to create a lag in pressure changes, the airspeed and altimeter will lag in response as compared to the "true" static pressure changes. The rate of climb will indicate in reverse as can be seen by analyzing Figure 2-10. Compare what would be happening with that discussed in the normal action of the rate of climb. In Figure 2-10, if the airplane climbs, the static pressure in the case surrounding the diaphragm would drop immediately while the pressure in the diaphragm would still be holding up, as the change must "work its way" through the capillary tube. The *diaphragm would expand*, which would give an indication that the plane was *descending!* It might be added that the *rate* would be accurate; the *direction* of vertical speed would be wrong. Of course, if you broke the glass and punched on through to leave a good-sized hole in the diaphragm, the other two instruments wouldn't have any lag (just the errors mentioned previously), but your vertical speed instrument would be kaput.

By breaking the glass in the airspeed or altimeter (easy-does it!), all three instruments will be

about as accurate as they would be with a cabin-alternate source.

The theoretical results of a *suddenly and completely plugged* static system in flight would be:

Airspeed — The airspeed would still be accurate as long as the static pressure trapped in the system was the same as the actual "outside" static pressure. If the airplane descended, the actual static pressure would be greater than that trapped in the system so the airspeed would read *high*. If it climbed, the airspeed would read *low*. You can see this by looking back at Figure 2-1. At the lower altitude, the diaphragm would expand farther than normal for a particular dynamic pressure because *only a part* of the static pressure entering the pitot tube would be cancelled by the now comparatively low static pressure trapped in the case. Naturally, the degree of error would depend on the altitude change. If the trapped static pressure has a pressure of that found at 10,000 feet, the airplane has descended, and the pitot tube is taking in the true dynamic or impact pressure *plus* the static pressure of *sea level*, the result would be an airspeed of awesome values indeed! (You probably would lose the instrument in the descent while you were still 7000 to 8000 feet up.)

Altimeter — The altimeter would read the altitude at which the complete stoppage occurred — and *that's all*. This would be a hairy situation IFR in that you might be easing up — or down — into the next guy's assigned altitude — or you might be easing down to connect with a cloud full of rocks.

The Vertical Speed Instrument — The same thing will happen to it as happened to the altimeter — *nothing*. Easing the nose up or down in cruise by watching the attitude gyro does not result in an indication of rate of climb or descent on the vertical speed instrument. What will more likely happen is that all instruments will lag considerably with altitude change because the system itself will not be perfectly sealed throughout. There would be enough leakage of outside air into the system for the instruments to tend to operate normally, though with so much lag as to make them of questionable value.

If you have reason to believe that the normal static system is plugged you'd better switch to the alternate or *carefully* break the glass of the airspeed indicator. It might be better to *wreck* the rate of climb as mentioned, rather than risk damaging the other two instruments.

MAGNETIC INDICATORS

The magnetic compass is basically a magnet that aligns itself with the lines of the earth's magnetic field — the airplane turns around it.

The magnets in the compass tend to align themselves parallel to the earth's lines of magnetic force. This tendency is more noticeable as the Magnetic North Pole is approached. The compass would theoretically point straight down when directly over the pole. The compass card magnet assembly is mounted so that a low center of gravity fights this dipping tendency. This mounting to fight dip causes certain errors to be introduced into the compass readings as follows:

Northerly Turning Error

In a shallow turn, the compass leads by about 30° when passing through South and lags about 30° when passing through North. On passing East and West headings in the turn, the compass is approximately correct. (The value of 30° is a round figure for U.S. use; it's actually equal to the latitude of the area in which the compass is being used.)

For instance, you are headed South and decide to make a right turn and fly due North. As soon as the right bank is entered, the compass will indicate about 30° of right turn, when actually the nose has hardly started to move. *So, when a turn is started from a heading of South, the compass will indicate an extra fast turn in the direction of bank.* It will then hesitate and move slowly again, so that as the heading of West is passed, it will be approximately correct. The compass will lag as

North is approached, so that you will roll out when the magnetic compass indicates 330° (or "33"). To be more accurate, you should start the roll-out early, the number of degrees of your latitude *plus* the number of degrees you would allow for the roll-out. Thus, at a latitude of 35° N using 5° for roll-out, you would start the roll-out 40° early or, in this case, when 320° is indicated.

Figure 2-11 shows the reactions of the compass to the 180° right turn from a heading of South.

If you had made a left turn from a South heading, the same effects would have been noticed: an immediate indication of turn in the direction of bank, a correct reading at the heading of East, and a compass lag of 30° when headed North.

If you start a turn from a heading of North, the compass will initially register a turn in the opposite direction but will soon race back and be approximately correct as an East or West heading is passed. It will then lead by about 30° as the airplane's nose points to Magnetic South. The initial errors in the turn are not too important. Set up your turn and know what to expect after the turn is started.

Here is a simple rule to cover the effects of

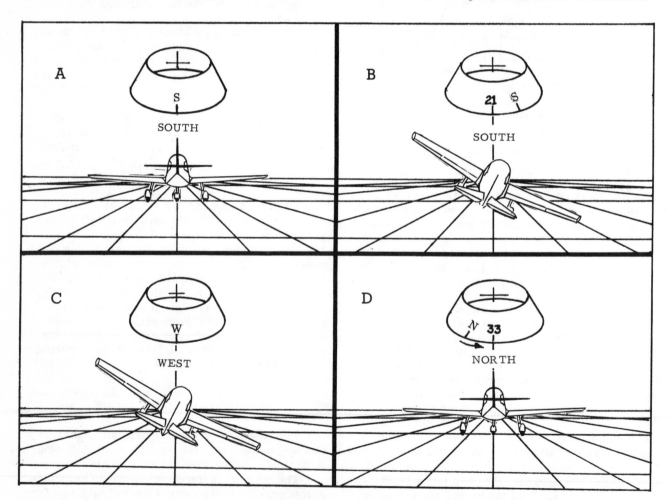

Fig. 2-11. A. When the airplane is flying straight and level headed Magnetic South, the compass is correct (disregarding deviation). B. As soon as the bank is entered, the compass indicates 210°("21"). C. As the nose passes West, the compass is reasonably accurate. D. In this example the airplane has been quickly rolled out when the compass indicated "33". The compass will immediately start to roll to indicate North. For accuracy, turns using the compass as a reference should be held below 20° bank.

14

bank (assuming a shallow bank of 20° or less — if the bank is too steep, the rule won't work).

Northerly turning errors (Northern Hemisphere):

North Heading — Compass *lags* 30° at the start of turn or in the turn.

South Heading — Compass *leads* 30° at the start of the turn or in the turn.

East or West Heading — Compass correct at start of turn or in the turn.

Just remember that North *lags* 30° and South *leads* 30° and this covers the problem. Actually 30° is a round figure; the lead or lag for rolling out of a turn depends on the latitude and angle of bank being used, but 30° is close enough for the work you'll be doing with the magnetic compass and is easy to remember.

Acceleration Errors

Because of its correction for dip, the compass will react to acceleration and deceleration of the airplane. This is most apparent on East or West headings where *acceleration results in a more northerly reading. Deceleration gives a more southerly heading.* Remember the term ANDS (*A*ccelerate and the compass "turns" *N*orth; *D*ecelerate and the compass "turns" *S*outh).

The magnetic compass reads correctly *only* when the airplane is in straight and level unaccelerated flight (and sometimes not even then). In bumpy air, the compass oscillates so that readings are difficult to take and more difficult to hold. The fluid in the case (acid-free white kerosene) is designed to keep the oscillations at a minimum, but the problem is still there.

Variation

The magnetic compass naturally points to the Magnetic North Pole, and this leads to the necessity of correcting for the angle between the Magnetic and Geographic North Poles.

In your earlier VFR flying days you measured the course from a midpoint meridian and this was the "True Course" or the course referred to as the True or Geographic North Pole. To get the magnetic course, the following applied (and still does). *Going from True to Magnetic:*

East Is Least — *Subtract* East variation as shown on the sectional or WAC chart.

West Is Best — *Add* West variation as shown on the sectional or WAC chart.

The variation (15° E or 10° W, etc.) given by the isogonic lines means that the Magnetic North Pole is 15° East or 10° West of the True North Pole — from your position. Naturally, if you happen to be at a point where the two poles are "in line," the variation will be zero. IFR Enroute Charts and Approach Charts are oriented with respect to Magnetic North so variation is already taken care of for that type of flying.

Deviation

The compass has an instrument error due to electrical equipment and the ferrous (iron) metal parts of the plane. This error varies between headings, and a correction card is placed near the compass, showing these errors for each 30°.

The compass is "swung," or corrected, on a compass rose — a large calibrated circle painted on the concrete ramp or taxiway away from metal interference such as hangars. The airplane is taxied onto the rose and corrections are made in the compass with a nonmagnetic screwdriver. The engine should be running and normal radio and electrical equipment on. Attempts are made to balance out the errors — better to have all headings off a small amount than some correct and others badly in error. The corrections are noted on the compass card which is posted at a prominent spot near the compass.

As a review for use for navigation purposes (and for use on the written if necessary) the following steps would apply:

Remember *TVMDC* or *T*rue *V*irgins *M*ake *D*ull *C*ompany; or *T*he *V*ery *M*ean *D*epartment of *C*ommerce (left over from the days when aviation was under the jurisdiction of the Department of Commerce).

1. *T*rue course (or heading) plus or minus *V*ariation gives *M*agnetic course (or heading).

2. *M*agnetic course (or heading) plus or minus *D*eviation gives *C*ompass course (or heading).

The chances are that in your normal flying you've paid little attention to deviation and have been doing fine. But remember, now that you plan on getting that instrument rating, there'll be some pretty good questions on the subject, so it might be a good idea to start thinking about it again.

If you lost all gyro instruments, and had no other method of keeping the wings level during a descent to get out of clouds, the magnetic compass could be used. Set up a heading of South on the mag compass. A deviation from this heading would mean that the wings weren't level and the airplane was turning. You would make corrections as necessary to stay on the South heading. Why South? One reason is that acceleration errors are smallest on North or South headings. Another is that the compass deviations on a South heading are in the *proper direction* and exaggerated. (On a heading of North any bank will cause the compass to swing in the *opposite* direction. This could be confusing for wing leveling purposes.)

The magnetic float compass has many quirks, but once you understand them, it can be a valuable aid. One thing to remember — the mag compass "runs" on its own power and doesn't need electricity or suction to operate. This feature may be important to you some day when your other more expensive direction indicators have failed.

The Remote Indicating Compass

A more sophisticated and expensive type of directional indicator is the remote indicating compass. The transmitter or magnetic "brain" of the assembly is usually located at a position well away from disturbing elements of iron or electrical leaks — often in or near one of the wing tips.

Fig. 2-12. Slaved compass system. (1) The pictorial navigation indicator here is the panel display for the slaved system (which only affects the H/I function). The contrasting colors of the indicators isn't seen here. (2) The slaving control and compensator is panel mounted and the pilot can select either the slaved or free gyro modes. The meter indicates when the system is being slaved. (3) The magnetic slaving transmitter (remoted). (4) The gyro stabilization unit containing the slaving circuitry (remote mounted). (*King Radio Corp.*)

The transmitter is electrically connected through an amplifier to the indicator on the instrument panel. Figure 2-12 shows some components.

This one is connected or synchronized to a gyro for damping the oscillations, in which case they are called magnetic slaved gyro compasses. The magnetic compass is continually correcting the precession of the gyro automatically, instead of the pilot manually resetting a H/I by reference to the float compass during the flight.

Note that the system in Figure 2-12 has a selector by which the system can either be a slaved or a free gyro. (Near the magnetic poles the magnetic compass has large errors, so the free gyro selection is best in those areas.) The slaving rate (when a slaved compass is selected) may be in the order of 2° per minute, and the synchronizing knob (Fig. 2-12) can be used to reset the indicating hand for large deviations such as might exist when the equipment is turned on for the flight.

There are several different designs of this type of compass gyro, and the basic characteristics were covered in a general manner here. You should make it a point to become familiar with the advantages and limitations of this instrument if your plane is, or becomes, so equipped.

GYRO FLIGHT INSTRUMENTS

Principles of Operation

The gyro instruments depend on two main properties of the gyroscope for operation: "rigidity in space" and "precession." Once spinning, the gyroscope resists any effort to tilt its axis (or plane of rotation). The attitude indicator and heading indicator operate on this principle. If a force is exerted to try to change the plane of rotation of a rotating gyro wheel, the gyro resists. If the force is insistent, the gyro reacts as if the force had been exerted at a point 90° around the wheel (in the direction of rotation). Precession is the property used in the operation of the needle of the turn and slip indicator (or needle and ball as you may call it) (Fig. 2-13).

Vacuum-driven Instruments

For the less expensive airplanes, the gyro instruments are usually vacuum-driven, either by an

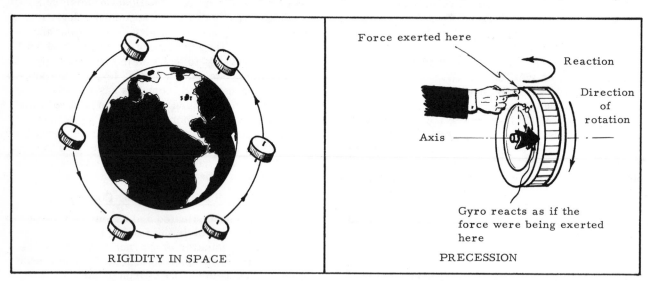

RIGIDITY IN SPACE

Force exerted here

Reaction

Direction of rotation

Axis

Gyro reacts as if the force were being exerted here

PRECESSION

Fig. 2-13. Rigidity in space and precession are the two principles used in the operation of gyro instruments.

16

engine-driven pump or venturi system. A disadvantage of the venturi system is that its efficiency depends on airspeed, and the venturi tube itself causes slight aerodynamic drag. Although a venturi system can be installed on nearly any airplane in a short while, the engine-driven vacuum pump is best for actual instrument operations, since it starts operating as soon as the engine(s) start. Multi-engine airplanes have a vacuum pump on each engine so that the vacuum-driven instruments will still operate in the event of an engine failure. Each pump has the capacity to carry the system. The multiengine airplane will have either a manual or automatic means to select each power source and a means to indicate the power being supplied by each source. The failure of an instrument or energy supply from one source will not interfere with the operations of the other instruments or source (FAR 23.1331).

The gyro instruments usually operate at a suction of 4.0 inches of mercury (29.92 inches of mercury is standard sea level pressure). The 4.0 inches of mercury shows a *relative* difference between the outside air pressure and the air in the vacuum system. The operating limits for the attitude and heading indicators are normally from 3.8 to 4.2 inches of mercury of suction, whereas the turn and slip uses a lower suction of 1.8 to 2.1 inches of mercury. The automatic pilot may use one or more of the panel gyros as its "brain," and the usual requirement is for a higher suction. Although the earlier suction figures probably will apply, check for the normal values for your particular airplane and equipment.

Some airplanes use a pressure-pump system which does the same job.

Errors in the instruments may arise as they age and bearings become worn, or the air filters get clogged with dirt. Low suction means low rpm and a loss in efficiency of operation. One of the greatest enemies of the vacuum-driven gyro instruments is tobacco smoke. The gum resulting from smoking in the cabin over a period of time can cause operational problems.

Electrically-driven Instruments

The electrically-driven gyro instruments got their start when high-performance aircraft such as jets began to operate at very high altitudes. The vacuum-driven instruments lost much of their efficiency in the thin air, and a different source of power was needed.

Below 30,000 feet, either type of gyro performs equally well. It is common practice to use a combination of electrically- and vacuum-driven instruments for safety's sake, should one type of power source fail. A typical gyro instrument group for a single-piloted airplane would probably include a vacuum-driven A/I and H/I and an electric turn and slip. Large airplanes have two complete sets of flight instruments, one set of gyros vacuum-driven and the other set electrically driven.

An advantage of the electric instruments is that the gyro horizon is usually smaller than the vacuum-driven type (which is larger than standard instrument size), leaving more room on the instrument panel for other instruments. Many of the newer electric attitude indicators will not tumble, and aerobatics such as loops, rolls, etc., may be done by reference to the instrument.

Attitude Indicator

The attitude indicator or gyro horizon (or artificial horizon or attitude gyro) operates on the "rigidity in space" principle and is an attitude instrument. The plane of rotation of the gyro wheel is horizontal and maintains this position, with the airplane (and instrument case) being moved about it (Fig. 2-14).

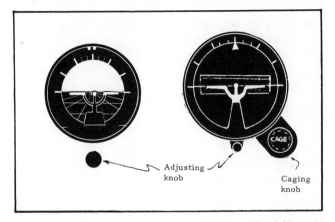

Fig. 2-14. Two types of attitude indicators. The A/I on the right is an older type of instrument.

Attached to the gyro is a face with a contrasting horizon line on it. When the instrument is operating correctly, this line will always represent the actual horizon. A miniature airplane attached to the case moves with respect to this artificial horizon precisely as the real airplane moves with respect to the real horizon. A knob allows you to move the miniature airplane up or down to compensate for small deviations in the horizontal line position.

There are limits of operation on the less expensive vacuum-driven attitude indicators (A/I), and these are, in most cases, 70° of pitch (nose up or down) and 100° of bank. The gyro will "tumble" above these limit stops and will give false information when the gyro is forced from its rotational plane. The instrument also will give false information during the several minutes required for it to return to the normal position after resuming straight and level flight.

"Caging" is done by a knob located on the instrument front. Because it is possible to damage the instrument through repeated tumbling, this caging is a must before you do deliberate acrobatics. The caging knob is useful also for quickly resetting the attitude indicator if it has tumbled. Some attitude indicators have caging knobs; some don't. The caging knobs are often removed by the aircraft manufacturer for various reasons, and if the instrument has tumbled, several minutes of straight and level flight may be required to let it erect itself again. While the chances are slim of getting into an attitude resulting in tumbling of the instrument, you would be wise to do at least one of the following: (1) Install a non-tumbling A/I in your airplane, (2) install a caging knob in the old type A/I if it doesn't have one, or (3) maintain proficiency in flying the turn and slip indicator or turn coordinator.

This instrument allows the pilot to get an immediate picture of the plane's attitude. It can be used to establish a standard-rate turn if necessary, as can be shown.

A good rule of thumb to find the amount of bank needed for a standard-rate turn at various airspeeds is to divide your airspeed (mph) by 10 and add 5 to the answer. For instance, airspeed = 150 mph; $\frac{150}{10}$ = 15; 15 + 5 = 20° bank required. This thumb rule is particularly accurate in the 100 to 200 mph range, as you can see by checking Figure 2-15.

V mph	Degrees Angle of Bank ∅
60	8
70	9.5
80	11
90	12
100	13.5
110	14.6
120	16
130	17
140	18.5
150	20
160	21
170	22
180	23
190	24.5
200	25.5
250	31
300	35.5
600	55
1000	67

Fig. 2-15. The bank required for a standard-rate turn (3° per second).

For airspeed in knots, divide it by 10 and add one-half of the answer. For 130 knots (150 mph), the angle of bank required is: $\frac{130}{10}$ = 13 + (one-half of 13) = 13 + 6.5 = 19.5° of bank required. It was just noted that the angle of bank required for 150 mph is 20°.

Keep an eye on the actions of your attitude gyro day by day as you fly it. If it's wobbling, slow to erect or has excessive errors don't use it for IFR work.

Heading Indicator

The heading indicator H/I functions because of the principle of "rigidity in space" as did the gyro horizon. In this case, however, the plane of rotation is vertical. The heading indicator has a compass card or azimuth scale which is attached to the gyro gimbal and wheel. The wheel and card are "fixed" by the gyro action, and as in the case of the magnetic compass, the airplane turns around them (Fig. 2-16).

The heading indicator has no magnet that causes it to "point" to the Magnetic North Pole and must be set to the heading indicated by the magnetic compass. The heading indicator should be set when the magnetic compass is reading correctly. This is generally done in straight and level flight when the magnetic compass has "settled down."

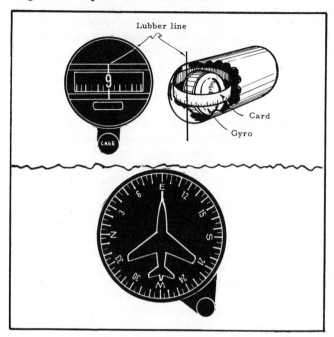

Fig. 2-16. Two types of heading indicator presentations. The older type at the top of the illustration.

The advantage of the heading indicator is that it does not oscillate in rough weather and gives a true reading during turns when the magnetic compass is erratic. A setting knob is used to cage the instrument for acrobatics and to get the proper heading.

A disadvantage of the older types of heading indicators is that they tumble when the limits of 55° nose up or down or 55° bank are exceeded. Although, if you happen to be maneuvering (pitching or rolling) parallel to the plane of rotation of the gyro wheel, this limitation does not apply. For instance, on some heading indicators, the plane of rotation of the gyro is in the 090°-270° line through the card. The airplane could be looped starting at a heading of 090° or 270° without tumbling. Other makes have the plane of rotation on the 180°-360° line through the instrument card, and the rule just cited would be in reverse. You shouldn't be doing such actions, but it's an interesting note.

The heading indicator creeps and must be reset with the magnetic compass about every 15 minutes. (Maximum allowable creep is 3° in 15 minutes.)

More expensive gyros, such as are used by the military and airlines, are connected with a magnetic compass in such a way that this creep is automatically compensated for as was noted earlier.

18

The greatest advantage of the heading indicator is that it allows you to turn directly to a heading without the allowance for lead or lag necessary with a magnetic float compass, but it doesn't have a brain and you must set it by that compass.

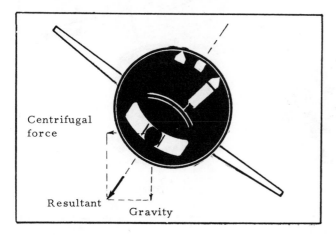

Fig. 2-17. The ball in the turn and slip is kept centered in a balanced turn by the forces acting upon it.

Turn and Slip Indicator

The turn and slip indicator is actually two instruments. The slip indicator is merely a liquid-filled, curved glass tube containing an agate or steel ball. The liquid acts as a shock dampener. In a balanced turn, the ball will remain in the center as centrifugal force offsets the pull of gravity.

In a slip, there is not enough rate of turn for the amount of bank. The centrifugal force will be weak, and this imbalance will be shown by the ball's falling down toward the inside of the turn.

The skid is a condition in which there is too high a rate of turn for the amount of bank. The centrifugal force is too strong, and this is indicated by the ball's sliding toward the outside of the turn. Usually, a turn in an airplane is considered to be balanced if

more than one-half the ball is within the indicator marks.

The turn part of the turn and slip indicator, or "needle" as it is called, uses precession to indicate the direction and approximate rate of turn of the airplane.

Shown in Figure 2-18 is the reaction of the turn and slip indicator to a right turn. The case is rigidly attached to the instrument panel and turns as the airplane turns (1). The gyro wheel (2) reacts by trying to move in the direction shown by (3), moving the needle in proportion to the rate of turn (which controls the amount of precession). As soon as the turn stops, the precession is no longer in effect, and the spring (4) returns the needle to the center. The spring resists the precession and acts as a damper, so the nose must actually be moving for the needle to move.

Older turn and slip indicators are calibrated so that a "standard-rate turn" of 3° per second will be indicated by the needle's being off center by one needle width. This means that by setting up a standard-rate turn, it is possible to roll out on a predetermined heading by the use of a watch or clock. It requires 120 seconds or 2 minutes to complete a 360° turn. The latest types of turn and slip indicators are calibrated so that a double needle-width indication indicates a standard-rate turn. These are usually noted as such on the instrument (4-minute turn) or have "doghouses" on each side (Figs. 2-17 and 2-19).

If your heading is 030° and you want to roll out on a heading of 180°, first decide which way you should turn (to the right in this case). The amount to be turned is 180° - 030° = 150°. The number of seconds required at standard rate is $\frac{150}{3}$ = 50. If you set up a standard-rate turn and hold it for 50 seconds and roll out until the needle and ball are centered, the heading should be very close to 180°.

One thing often brought up in the written test for the instrument rating — and missed — is that the needle deflection tells whether the turn is standard rate or not and the ball has nothing to do with it. If

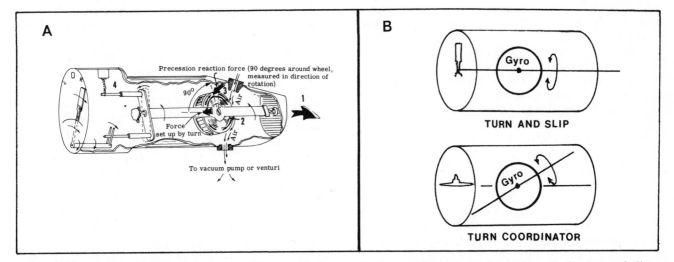

Fig. 2-18. A. A vacuum-driven turn and slip indicator. Most turn and slips nowadays are electric-driven. B. The turn and slip and turn coordinator reactions to precession. The turn coordinator reacts to roll as well as yaw.

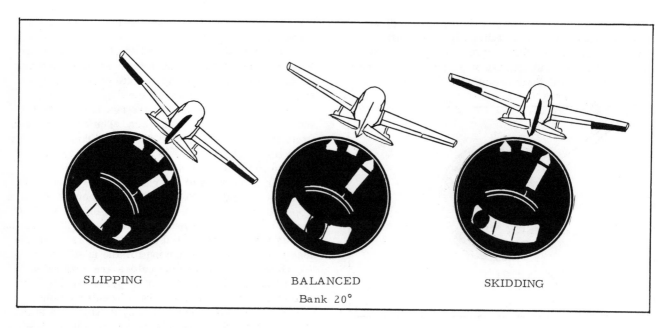

SLIPPING   BALANCED   SKIDDING

Bank 20°

Fig. 2-19. The needle indicates the rate at which the nose is moving; the ball indicates the quality of the turn.

the needle is deflected the proper amount for a standard-rate turn, the nose of the airplane is moving around at a rate of 3° per second. The turn may be slipping, skidding, or balanced. The ball indicates the *quality* of the turn. Figure 2-19 shows three variations of an airplane making a standard-rate turn to the right. The airspeeds are the same (130 knots) in each case, requiring 20° of banks for a *balanced* standard-rate turn.

The advantage of the turn and slip over other gyro instruments is that it does not "tumble" or become erratic as certain bank and pitch limits are exceeded.

A disadvantage of the turn and slip is that it is a rate instrument, and a certain amount of training is required before the pilot is able to quickly transfer the indications of the instrument into a visual picture of the airplane's attitudes and actions.

The gyro of the turn and slip, like the other gyro instruments, may be driven electrically, or by air, using an engine-driven vacuum pump or an outside-mounted venturi.

An interesting note is that the turn and slip becomes less accurate as the bank increases. For instance, in a level turn at a 90° bank (if you could hold it) the needle should *theoretically* come back to the center after the turn was established, indicating that the airplane is not turning at all. (You are doing a loop in a horizontal plane.)

Turn Coordinator

The turn coordinator is being used in some later airplanes to replace the turn and slip. The wheel reacts to precession around an axis that is tilted 30° upward compared to the turn and slip. (Look back to Figure 2-18.) Once the roll is stopped (the bank is established) the yaw rate is indicated (Fig. 2-20).

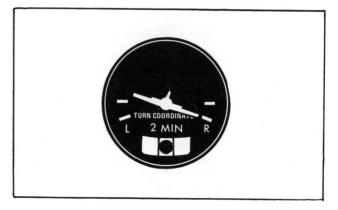

Fig. 2-20. A turn coordinator. A balanced, standard rate turn to the right.

The Outside Air Temperature Gage

This instrument falls into a category of its own, but is very important for instrument flying and should be covered as such.

The outside air temperature gage (or free air thermometer) will assume much greater importance now that you'll be flying IFR. Whereas before the O.A.T. has been mostly a matter of curiosity, it will now be a matter of vital interest in the temperature range where icing may occur. The usual type of thermometer used is that of the bimetal direct reading type. The fact that two dissimilar metals have different expansion (or contraction) rates with temperature change makes possible a comparatively simple method of registering this change.

The two strips of metal are welded together in the form of a coil spring. One end is anchored, and the other is attached to an indicating hand. The thermometer may read in Centigrade or Fahrenheit and is marked in both scales for most instruments. The probe or pickup is in the free airstream, and

20

the dial faces into the cockpit for ready reference. (The instrument is normally at a corner or at the top of the windshield.)

Because of errors in the individual instrument and effects in location (it *should* register the exact ambient or true temperature of the air), you should look for the possibilities of structural icing when in visible moisture and the temperature is down to with within a few degrees of freezing.

ENGINE INSTRUMENTS

Tachometer

The centrifugal tachometer operates on the same principle as a car speedometer. One end of a flexible shaft is connected to the engine crankshaft and the other connected to a shaft with counterweights within the instrument. The rate of turning of the crankshaft (and cable) causes expansion of the counterweight system. The instrument hand is mechanically linked to the counterweight assembly so that the engine speed is indicated in revolutions per minute.

For direct-drive engines, the engine and propeller rpm are the same (Lycoming O-320, O-540, O-360). The geared engine (Lycoming GO-480, etc.) has different engine and propeller speeds, and this is noted in the Airplane Flight Manual (the propeller rpm is less than the engine rpm). The tachometer always measures *engine* rpm, and this is the basis for your power setting.

Another type of tachometer is the magnetic, which utilizes a flexible shaft which turns a magnet within a special collar in the instrument. The balance between the magnetic force and a hair-spring is indicated as rpm by a hand on the instrument face. This type of tachometer does not oscillate as sometimes happens with the less expensive centrifugal type.

A third type is the electric tachometer which depends on a generator unit driven by a tachometer drive shaft. The generator is wired to an electric motor unit of the indicator, which rotates at the same rpm and transmits this through a magnetic tachometer unit which registers the speed in rpm. This type of tachometer is also smoother than the centrifugal type. (It doesn't depend on the electrical system of the airplane.)

Manifold Pressure Gage

For airplanes with controllable pitch propellers (which includes constant speed propellers), this instrument is used in combination with the tachometer to set up desired power from the engine. The manifold pressure gage measures absolute pressure of the fuel-air mixture going to the cylinders and indicates this in inches of mercury.

The manifold pressure gage is an aneroid barometer like the altimeter; but instead of measuring the outside air pressure, it measures the pressure in the intake manifold. When the engine is not run-

ning, the outside air pressure and the pressure in the intake manifold are the same, so that the manifold pressure gage will indicate the outside air pressure as a barometer would. At sea level on a standard day, this would be 29.92 inches of mercury; but you can't read the manifold gage this closely and it would appear as approximately 30 inches.

You start the engine with the throttle cracked or closed. This means that the throttle valve or butterfly valve is nearly shut. The engine is a strong air pump in that it takes in fuel and air and discharges residual gases and air. At closed or cracked throttle setting, the engine is pulling air (and fuel) at such a rate past the nearly closed throttle valve that a decided drop in pressure is found in the intake manifold and is duly registered by the manifold pressure gage. As the engine starts, the indication of 30 inches drops rapidly to 10 inches or less at idle. It will never reach an actual zero, as this would mean a complete vacuum in the manifold (most manifold pressure gages don't even have indications of less than 10 inches of mercury). Besides, if you tried to shut off all air (and fuel) completely, the engine would quit running.

As you open the throttle, you are allowing more and more fuel and air to enter the engine, and the manifold pressure increases accordingly.

The unsupercharged engine will never indicate the full outside pressure on the manifold gage in the static condition. The usual difference is 1 to 2 inches of mercury. The maximum indication on the manifold pressure gage you could expect to get would be 28 to 29 inches. Ram effect may raise the manifold pressure because of "packing" of the air in the intake at higher speeds. In some such instances, it may be possible for the manifold pressure to exceed the atmospheric pressure at full throttle.

The supercharged engine has compressors that bring the air-fuel mixture to a higher pressure than the outside air before it goes into the manifold. This makes it possible to register more than the outside barometric pressure and results in more horsepower being developed for a given rpm, as horsepower is dependent on rpm and the amount of fuel-air (manifold pressure) going into the engine. This type of supercharging means that the compressors are an integral part of the engine. The compressing is done *after* the mixture goes through the carburetor.

The turbo-supercharger is an exhaust-driven blower that compresses the air before it is mixed with the fuel and can be furnished with the new engine or added to an airplane with an unsupercharged engine as an afterthought. (It's not as casual a procedure as that last sentence might indicate; there are a lot of testing and certification procedures done before it is approved for a particular model of airplane.)

Oil Pressure Gage

The oil pressure gage consists of a curved

Bourdon tube with a mechanical linkage to the indicating hand which registers the pressure in pounds per square inch (Fig. 2-21). As shown, oil pressure tends to straighten the tube, and the appropriate oil pressure indication is registered. This is the direct pressure type gage.

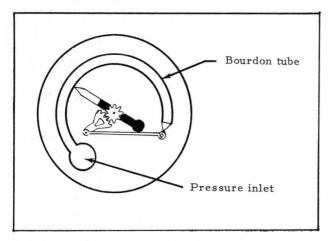

Fig. 2-21. The oil pressure gage.

Another type of oil pressure gage uses a unit containing a flexible diaphragm which separates the engine oil from a nonflammable fluid which fills the line from the unit into the Bourdon tube. The oil pressure is transmitted through the diaphragm and to the Bourdon tube by this liquid because liquids are incompressible.

Oil Temperature Gage

The vapor type is the most common type of oil temperature gage in use. This instrument, like the oil pressure gage, contains a Bourdon tube which is connected by a fine tube to a metal bulb containing a volatile liquid. Vapor expansion due to increased temperature exerts pressure — which is indicated as temperature on the instrument face.

Other types of oil temperature gages may use a thermocouple rather than a Bourdon tube.

Cylinder Head Temperature Gage

The cylinder head temperature gage is an important instrument for engines of higher compression and/or higher power. Engine cooling is a major problem in the design of a new airplane. Much flight testing and cowl modification may be required before satisfactory cooling is found for all airspeeds and power settings. The engineers are faced with the problem of keeping the engine within efficient operating limits for all air temperatures.

The cylinder head temperature gage usually warns of any possible damage to the engine long before the oil temperature gage gives any such indication.

The "hottest" cylinder, which is usually (though not always) one of the rear ones in the horizontally opposed engine, is chosen during the flight testing of the airplane. A thermocouple lead replaces one of the spark plug washers on this cylinder.

The cylinder head temperature gage uses the principle of the galvanometer. Two metals of different electrical potentials are in contact at the lead. Since the electric currents of these two metals vary with temperature, a means is established for indicating the temperature at the cylinder through electric cables to a galvanometer (cylinder head temperature gage), which indicates temperature rather than electrical units.

Some pilots use cylinder head temperature as an aid in proper leaning of the mixture. Generally, richer mixtures mean lower head temperatures, all other things (airspeed, power settings, etc.) being equal. But the engine may not be developing best power at the extremes. Too rich a mixture means power loss plus excessive fuel consumption. Too lean a mixture means power loss plus the possibility of engine damage. Special instruments (Exhaust Gas Temperature gages) have been designed to monitor engine performance and fuel-air ratio (see references at end of this book). The usual procedure for the operation of these instruments is to use a probe in the exhaust to measure the temperature of the exhaust gases. When the mixture is leaned from full rich, the exhaust temperature will increase, peak, and then decrease with further leaning. The idea is to get the mixture to the fuel-air ratio for continuous operation and also have an indication for best mixture for takeoff and climb under different situations, such as taking off at a high density altitude, climb, etc.

Read the EGT Manual or the *Pilot's Operating Handbook* for your airplane because procedures may vary between engine and/or airframe manufacturers.

Fuel Gage

The electric transmitter type fuel gage may be considered to have the following components: (1) the float and arm, (2) the rheostat type control, and (3) the indicator, a voltmeter indicating fuel either in fractions or in gallons. The float and arm are attached to the rheostat, which is connected by wires to the fuel gage. As the float level in the tank (or tanks) varies, the rheostat is rotated, changing the electrical resistance in the circuit, which changes the fuel gage indication accordingly. This is the most popular type of fuel measuring system for airplanes with electrical systems.

Fuel gages of any type are not always accurate, and it is best not to depend on them completely (if at all). A good visual check before the flight and keeping up with the time on each tank (knowing your fuel consumption) are the best policies. Making frequent checks on the fuel gage as a cross reference is a good idea; the sudden dropping of the fuel-level indication may be caused by a serious fuel leak, and you'd like to know about this (particularly when IFR).

Clock

This is a required instrument for IFR work and will be used on every flight. As simple as it seems, know whether it's electrical or wind up, for instance. The clock will gain great significance for you during your instrument flight training.

A LOOK AT SOME AIRPLANE SYSTEMS

This section is intended to be a check list to bring to your mind some items of interest not only for the instrument rating flight test but also in regard to your actual instrument flying later. To cover the theory and operation of each system in detail is impossible — and the *Pilot's Operating Handbook* or operating instructions on the particular type or make of equipment will cover the operations procedures in detail.

Electrical System

The system discussed here is the battery-generator (older airplanes), or battery-alternator combination which, even in VFR conditions, is important (for instance, loss of cockpit and instrument lighting at night can be extremely serious), but when you are flying in actual instrument conditions, electrical failure could result in a fatal accident. It's your job to know just what equipment depends on this system and what your actions should be in case of trouble.

Figure 2-22 shows a simplified idea of a 12-volt airplane electrical system for a single-engine airplane with an engine-driven generator or alternator of 50-ampere capacity and a 35-ampere-hour battery. The dashed lines show the additional equipment needed for a light twin.

The battery stores electrical energy, and the alternator (generator) creates current and replenishes the battery as necessary, directed by the voltage regulator, which is the "automatic valve" to assure proper current flow to the battery. A master switch is provided to close the circuit or "energize" the electrical system, and a reverse current relay is necessary to insure that the current doesn't "back up" and reverse its flow as far as the battery-generator combination is concerned. For older light twins, a paralleling relay is installed so that the two generators are carrying an equal share of the load. The generators may each have a switch to take them out of the system and for checking purposes before take-off. Light twins using alternators (the vast majority) do not require a paralleling relay as can be seen in Figure 2-22.

Circuit breakers or fuses are installed to insure that the various circuits are not overloaded, with a resulting overheating and possible electrical fire. The circuit breakers "pop" out when an overload occurs and break the connection between the battery-alternator/generator system and the item causing the problem. These may be pushed back in the panel to reestablish current flow. It's best to allow a couple of minutes for cooling before doing this. Continual popping of a circuit breaker means a problem, and corrective action (electrical equipment check by an expert) should be taken in this case. Some pilots figure that the third time in a row that a circuit breaker pops out is just cause for leaving it out. To *hold* a circuit breaker in is to ask for strange smells, a smoky cabin, and increased adrenalin flow in the occupants of the airplane.

It would be well for you to know what equipment is protected by circuit breakers and where all of the circuit breakers are located. (Generally, they'll be in one area on the instrument panel or side panel, but there may be one or two scattered at random spots in the cabin.) Some pilots memorize the exact location of each circuit breaker for quick reference, but the main thing to know is that such and such an item has a circuit breaker and check — by

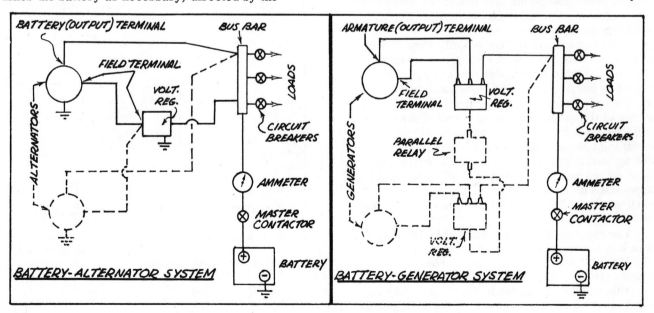

Fig. 2-22. Schematic drawing of a battery-alternator and an older battery-generator system.

looking at the C/B panel — for a popped breaker should this equipment suddenly fall down on the job. Some circuit breakers can be pulled out to shut off a circuit if an overload is suspected or the pilot wants that item out of operation for some reason.

The 35-ampere-hour battery and 50-ampere alternator assure that the alternator has a capacity to keep the battery up to its peak strength. Very few VFR pilots stop to consider that they could overload some systems by turning everything electrical on at the same time.

Figure 2-23 gives some comparative expected, continuous, intermittent, and radio loads of a high-performance, single-engine, four-place airplane. Notice that operating a number of the larger power stealers at once could well exceed the capacity of the electrical system. For instance, operating the landing lights, rotating beacon, pitot heat, and DME could put a real strain on the system (36.1 amps) at a particular time. The main idea of Figure 2-23 is not to give actual figures but to show comparative power requirements. For instance, if you needed pitot heat and were at the limit of available electrical power, it might be well to shut down the DME during this period. Or you might turn off several items totaling the same load as the DME. Notice that the autopilot requires very little current. The items covered in Figure 2-23 are based on airborne requirements, or when the 14.3-volt alternator-battery system is in operation. (The alternator is in action and at its peak output.)

The big current users are the starter and starter solenoid which, of course, are not of particular interest in flight (unless the engine quits) but may require from 50-200 amps and 10 amps respectively. (The starting vibrator for this composite airplane would require 2.5 amps.)

The Omni/Loc head noted in the listing is the navigation indicator (azimuth, TO-FROM indicator, and the left-right needle) as you see it on the panel. It needs power to operate also but not as much as the VOR receiver or communications transmitter and receiver, as shown.

Take a good look at Figure 2-23. Note that this sample autopilot has a very low current requirement. The average pilot often has an exaggerated opinion of how much current an autopilot uses and might shut it off, leaving greater current stealers on — and making his job of flying tougher. More complex autopilots *could* be big current users, however. At night on actual instruments, the rotating beacon should be off anyway, and if you want to conserve electrical power, the wing and tail navigation (position) lights could be dispensed with under that condition. Of course, when you break out VFR, you'd better have the lights back on.

The cigar lighter is no slouch in taking current, even if it is only a temporary drain. If that guy in the right seat needs to smoke, he can use matches or a lighter or forget it if you need the juice.

Talk to some of the local electronics and electrical systems pros to get some pointers on your

A. MAXIMUM PROBABLE CONTINUOUS LOAD (LESS RADIO EQUIPMENT)		
Item	Number Used	Total Current in Amps at 14.3 Volts
Heated pitot	1	7.3
Rotating beacon	1	5.0
Navigation lights (wing)	2	3.3
Navigation lights (tail)	2	2.3
Turn & bank	1	1.1
Instrument spotlight	1	1.0
Master contactor	1	0.5
Autopilot	1	0.3
Flap indicator	1	0.1
Fuel gauge	1	0.1
Gear-up light	1	0.1
Cyl. head & oil temp gages	1	Negligible

B. INTERMITTENT LOADS		
	Number Used	Total Current in Amps at 14.3 Volts
Landing lights	2	17.0
Cigar lighter	1	8.0
Flap motor	1	5-15*
Gear motor	1	5-15*
Gear solenoid	1	2.5
Flap solenoid	1	2.5
Dome lights	2	1.3
Landing gear horn	1	1.0
Fuel pumps (carb.)	2	0.5
Stall warning	1	0.2
*Approximate range of values		

FUEL PUMP (Fuel injection)	
Condition	Total Current in Amps at 14.3 Volts
Start	4.6
Taxi	8.2
Take-off	7.0
Cruise	6.9
Land	8.7

C. MAXIMUM PROBABLE RADIO LOAD			
	Number Used	Total Current in Amps at 14.3 Volts	
Comm/Nav	2	Rcv 9.4	Xmt 14.5
DME	1		6.8
Glide slope receiver	1		2.6
Omni/loc head	2		0.7
ADF	1		0.7
Omni converter	1		0.4
Marker beacon	1		0.2
Junction box	1		0.1

Fig. 2-23. Probable continuous, intermittent, and radio loads in order of current. The figure of 14.5 amps in Part C is the load for both receivers operating and both transmitters "warmed up" (on) at 4.7 amps each, *plus* one transmitter (5.1 amps) transmitting. The current requirements will naturally vary between makes and models of radio equipment.

particular equipment. It's better to know the system *now,* and know where a circuit breaker is located, than to have to learn the hard facts when smoke starts easing out from under the instrument panel — and you're on solid IFR.

De-icing and Anti-icing Equipment

You should be familiar with the operation of the de-icing and anti-icing equipment of any airplane you plan to use for actual instrument flight.

Electric Prop De-icers

If the airplane has electric prop de-icers, you should know the current drain while they are in use. (Multiengine airplanes are the only ones likely to have this type of equipment.)

The Airplane Flight Manual or *Pilot's Operating Handbook* will have a Supplement attached to it explaining the operations and limitations of the equipment and will probably include such items as:

Description and Operating Principles — How the system operates, order of cycle of heat (it is not continuous for any one section of a particular blade but alternates).

Operating Procedures — Normal procedures, such as how to turn the system on, what to watch for on the de-icing system ammeter, expected current requirements (for one light twin, both prop de-icers operating requires 22–26 amps, as noted in the Supplement). Emergency procedures can include the procedures to follow in the event of abnormal de-icer ammeter indications and precautions such as turning off noncritical equipment in the event of excessive power requirements as might occur after the loss of one generator (or engine).

Pneumatic De-icing Systems

The light twin normally uses pneumatic "boots" for wing and tail leading edges (larger airplanes sometimes pipe hot air inside the leading edge of the wing). There is a great deal of discussion going on about the de-icing of single-engine airplanes, but not too much has been done in this regard so the light twin is the airplane being discussed here.

This equipment will also have a Supplement to the Airplane Flight Manual which must be kept in the airplane at all times. The Supplement will list such things as:

Preflight Check — Operation of the boots and the check for normal operation. There may be such notes as "Limit the preflight check to two cycles to reduce wear and premature failure of the vacuum pumps."

Normal Operations — Included here will be such things as suggestions for operation in the various icing conditions and limits of operation.

Placards — A list of the placards or control panel markings for the equipment. Your airplane may have a placard such as "De-icers to be off

during take-off or landing." Others have no such limitation but note in the Supplement that an increase in stall speed may be expected with the de-icers in operation. This item will be discussed further in this book (see Chapter 7).

Emergency Operations — The procedures to take if the timer (which controls the timing of the cycles or inflating of the various portions of the tubes) should fail. There probably will be tips for operation. (The newer lightweight systems, which operate off the engine-driven pressure pumps, are able to operate with only one pump; that is, the capacity of *each* pressure pump is such that it can carry the pressure-driven instruments and the pneumatic de-icer system, if necessary.)

Fluid Anti-icing Systems

Some airplanes use fluid as a means of combating propeller ice. The fluid usually has an alcohol base with an additive to thicken it to prevent quick evaporation and excessive runoff.

The fluid is piped from a reservoir to the prop where it is distributed along the blades by centrifugal force. The pilot is able to regulate the rate of flow by a control in the cabin. If your airplane has such a system, your job will be to have a good idea of the capacity (and flow rate) of the fluid. You should also know how to replenish it and what type of fluid is to be used. (Part of your preflight check for actual IFR work will be to check the fluid level.)

Note that this is an *anti*-icing system, meaning that it's to be used more to prevent ice than to get rid of it after it has become well formed.

Read and know the instructions concerning the use of your particular equipment.

Later in the book the anti-icing and de-icing systems will be covered as they pertain to actual icing situations.

Summary

To cover all the different makes and models of airplane equipment and systems is impossible. It is recommended that you write the various airframe, engine, or equipment manufacturers listed at the end of this book for information. You should know the operation of the particular type of electrical system, radio, de-icing, and other so-called auxiliary equipment in your airplane. For instance, in addition to the systems just covered, you probably will have to be able to operate an autopilot or oxygen system as you progress in your instrument flying.

To repeat, the time to learn the systems is while you are on the ground and before something happens in flight. Read the Supplements to the Airplane Flight Manual and other material available from the manufacturer.

THE AIRPLANE PAPERS

This might seem a strange place to cover the airplane papers — in a chapter on instruments and systems — but the Airplane Flight Manual is closely tied in with flight limitations, instrument markings, and placards and should be covered with the rest of that type of information. Since that document is being discussed, it would be a good idea to review the other ones as well.

There are three documents that must be in the airplane at all times:

1. The Airplane Flight Manual (or equivalent information).

2. The Certificate of Airworthiness.

3. The Certificate of Registration.

If the airplane has radio transmitting equipment aboard (don't go IFR without it), the Federal Communications Commission requires a Radio Station License.

Airplane Flight Manual

Part 23 of the Federal Aviation Regulations states that certain information must be furnished to the pilot.

Airspeed Limitations — Airspeed limits and color coding of the indicator must be furnished, plus such information as the maneuvering speed (V_A) at the maximum certificated weight and the maximum airspeed (C.A.S. or E.A.S. as applicable) for landing gear operation (V_{LO}). In some cases, an airplane is allowed to fly at a higher speed *after* the gear is fully extended (V_{LE}) but is restricted to the lower speed while the gear is being extended or retracted. Most manufacturers use the lower operating speed as a limit for the conditions of gear full down for any gear operations. The max flap extended speed V_{FE} must be marked on the airspeed indicator.

Power Plant Limitations — The power plant limitations (max allowable rpm, oil pressure, and oil temperature ranges, etc.) are given here in addition to the fact that the engine instruments are marked. Special information that could be given on an airplane with fixed-pitch metal prop will be noted, such as "Do not operate the engine between 1870 and 2050 rpm." It would also be noted by a placard on the instrument panel. This limitation is to be found on certain older type propeller installations and is due to harmonics (and fatigue) possibilities being set up in a certain rpm range.

Weight and Balance — The maximum weight, empty weight and empty weight center of gravity position, the useful load and the composition of the useful load, including the total weight of the fuel with full tanks, are given here.

The established center of gravity limits must be furnished. Certain information, such as sample loading problems, will be furnished unless the loading space of the airplane is adequately placarded or arranged so that no reasonable distribution of the useful load will result in a center of gravity outside the stated limits.

Loading information, which includes the weight and location of each item of equipment installed when the empty weight and empty weight C.G. was determined, is furnished.

In the event of alterations or repair of equipment on the airplane which results in a change in empty weight and/or empty weight C.G. since the plane left the manufacturer, FAA Form 337 must be attached as a part of the weight and balance information (and, naturally, supersedes the original data.) The new empty weight and empty weight C.G. is written in the aircraft log book.

Maneuvers — Normal category airplanes will have such statements as "Deliberate spins prohibited" or other limitations. For utility category airplanes, the authorized maneuvers, such as normal spins, chandelles and lazy eights, and others will be listed. The maximum positive and negative limit load factors will be noted.

Kinds of Operation — If the airplane is limited to VFR day operations, this will be noted. Any installed equipment that affects any operation limitation must be listed.

Operating Procedures — Normal and emergency procedures and other pertinent information must be furnished.

Performance Information — Such information as the conditions under which full amount of usable fuel in each tank can be used is noted in the AFM.

As covered earlier, certain equipment requires Supplements to the Airplane Flight Manual. Information concerning autopilots, de-icers, and other items that could affect safety of flight must be carried with the AFM and be available to the pilot *in flight* should the need arise.

Pilot's Operating Handbook

Starting with 1976 models, manufacturers are publishing a *Pilot's Operating Handbook* for their airplanes which could be considered combination *Owner's Handbooks* and Airplane Flight Manuals. One manufacturer is printing *Pilot's Operating Handbooks* for its older models as well.

The idea is to "put it all together" so that day-to-day, normal operating procedures are arranged so that the material required in an Airplane Flight Manual makes for easier pilot use. Also, the POH's for all airplanes (12,500 pounds max certificated weight is the top limit for now) are arranged in the same order for quick reference as needed, for instance:

Section 1 — General Information (weights, fuel and oil capacity, dimensions, etc.)

Section 2 — Limitations (airspeed, powerplant, weight and C.G. limits, maneuvering and flight load and other limits, and a listing of placards)

Section 3 — Emergency Procedures, with amplified procedures at the end of the Section (engine failures, fire, icing, electrical problems and landing with a flat main tire)

Section 4 — Normal Procedures (checklists for preflight, start, taxiing and all other aspects of normal flight, plus amplified procedures at the end of the Section)

Section 5 — *Performance*, (take-off, landing, cruise, plus stall speeds and other information.)

Section 6 — *Weight and Balance and Equipment List*

Section 7 — *Airplane and Systems Descriptions*

Section 8 — *Handling, Service and Maintenance*

Section 9 — *Supplements* (optional systems description and operating procedures, such as for oxygen, radio and autopilot systems).

Some of the models do not require that the POH be carried in the airplane at all times, but furnish separate information as required in an Airplane Flight Manual (which information must be in the airplane at all times). Other models require that the POH be carried in the airplane because it contains the information required for an AFM by the Regulations the airplane was certificated under.

Certificate of Airworthiness

This document must be *displayed* so that it can be seen readily by pilot or passengers. The Airworthiness Certificate will be valid indefinitely as long as the airplane is maintained in accordance with the Federal Aviation Regulations. This means that each aircraft must have had an annual inspection within the preceding 12 calendar months. If an aircraft is to be operated for hire, it must also have had an inspection within the last 100 hours of flight time — the inspection being in accordance with the Regulations (and either done by or supervised by a certificated mechanic). One of the 100-hour inspections may be used as an annual inspection by following certain procedures and noting the fact in the aircraft and engine logs.

Some airplanes may use the *progressive inspection* in which the airplane is continuously inspected after the owner shows that he can provide proper personnel, procedures, and facilities for it. The purpose is to permit greater utilization of the aircraft. This type of inspection eliminates the need for annual and 100-hour inspections during the period this procedure is followed.

Certificate of Registration

This document must be in the airplane and has such information as the owner's name and address, manufacturer, model, registration number, and manufacturer's serial number. When the airplane changes hands or the registration number is changed, a new certificate of registration must be obtained.

Aircraft Radio Station License

This is required by the Federal Communications Commission for any transmitting equipment on board. Such equipment is listed on the form. (DME — Distance Measuring Equipment — is included with the communication transmitters since it transmits a signal to the ground equipment as is also the situation for a transponder.)

Logbooks

There must be a logbook for the airframe and each engine. Entries are made for maintenance, alterations, repair, and required inspections. The logbooks are required to be available for inspection by authorized persons and are usually kept in the airplane as the most logical place. Make sure the logs for your airplane are kept up to date.

Chapter 3

A REVIEW OF AIRPLANE PERFORMANCE, STABILITY AND CONTROL

THE FOUR FORCES

THE AIRPLANE has four forces acting on it in flight: weight, thrust, lift, and drag. As a neophyte instrument pilot, your job will be to see that these forces are balanced (or not balanced) to obtain the required performance. This will be done by your reference to the various instrument indications.

Weight

The man who develops a device that can turn off the force of gravity as desired will soon be a multibillionaire and also put a lot of aerodynamicists out of work. The problem of keeping airplane weight down has probably caused more grief in the aircraft industry than any other single factor.

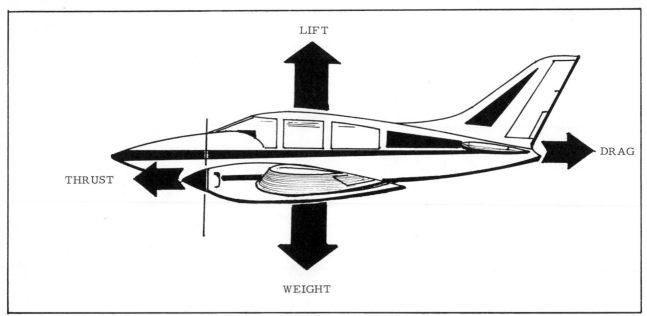

Fig. 3-1. The four forces.

Suffice it to say that weight exists and is considered always to act downward toward the center of the earth. Which means that while the other three forces (lift, thrust, and drag) may operate in various directions as the plane is maneuvered, the direction of the weight vector remains constant.

Weight is considered to work through one point of the airplane — the center of gravity.

Thrust

Thrust is one of the four forces that acts on an airplane in flight and may be produced by a propeller, jet, or rocket. One of the statements often made in introducing the concept of the four forces is that thrust equals drag in straight and level, unaccelerated flight, but this actually depends on the attitude (speed) of the airplane. At slow speed near the stall in level flight and in climbs, thrust is greater than drag, as could be seen by analysis. Only under conditions where the thrust line is parallel to the line of flight is thrust actually equal to drag in straight and level flight.

Torque

The propeller airplane has the problem of "torque" which is a misnomer as far as the majority of the actual forces working on the airplane are concerned, but the term will be used here to cover the several forces or moments that tend to cause the nose to yaw left at high power settings and low speeds.

In instrument flying, during the climb and at low speeds with power, the beginner usually forgets about this factor. He watches the airspeed, his white knuckles, or the wing attitudes and forgets that such a thing as "torque" ever existed. He suddenly wakes up to realize that the airplane has slipped about 60° off heading — just before the check pilot calls his attention to it by a brisk rap with the fire extinguisher.

Torque is the result of several factors. The fact that the propeller is a rotating airfoil means that it is subject to stalls, induced drag (to be covered later), and the other problems associated with airfoils.

Slipstream Effect — For the single-engine airplane, the most important factor is that of the rotating slipstream.

The propeller in producing thrust takes a comparatively large mass of air and accelerates it rearward, which results in the equal and opposite reaction of the airplane moving forward. This law was discovered by Isaac Newton (1642-1727) a couple of hundred years before the first airplane flew successfully. Because of the rotation of the propeller and because of its drag forces, a rotating motion is imparted to the air mass as it moves rearward from the prop. The rotating airstream exerts a force on the left side of the fin and rudder, which results in a left-yawing tendency. (If the airplane had a fin and

rudder of equal size and position on the bottom, the *yawing* forces would tend to be balanced (Fig. 3-2).

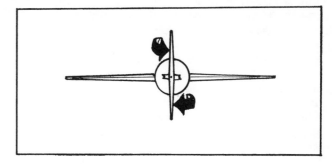

Fig. 3-2.

Of course, it's not all that simple; the varying shape of the fuselage and interference of the wings can affect the rotational path of the slipstream.

For the multiengine airplane with a single fin and rudder, the slipstream effect is not so critical a factor as found for the single-engine type. (You've no doubt flown some of the tricycle gear, light twins and noted that "torque" or left-turning tendency was not as strong on take-off as for some single-engine planes.)

The manufacturer may correct for this slipstream effect on the single-engine plane by one of two ways so that at cruise (the regime in which the airplane operates the majority of its flight time) the airplane does not tend to yaw to the left.

One method is to offset the fin, so that at cruise it has a zero angle of attack in reference to the combination of slipstream and free-stream velocity; therefore, no yawing tendency will be present.

Some manufacturers "cant" the engine or offset the thrust line a few degrees, which results in the same effect of no yawing tendency at cruise. The airplanes you'll be using for instrument training will most likely have rudder trim, and it will assume added importance with speed changes while flying on the gages.

Precession — Back in the discussion of the gyro instruments, precession was mentioned as the factor in the operation of the turn and slip instrument. Precession will affect the airplane only during a change of attitude and is not a factor in steady-state flight. Part of your training may include an ITO (instrument take-off), and precession could give you a little trouble in the tailwheel airplane if you try to raise the tail too quickly. The propeller arc acts like a gyro wheel and resists any tendency to change its plane of rotation.

As seen from the cockpit, the propeller is rotating clockwise. When the tail is raised, it is as if a force were exerted on the top of the propeller arc from behind. Because precession acts at a point 90° around the wheel (or propeller), the airplane acts as if a strong force were acting from behind on the right side of the prop arc (check Figure 2-13 again). The result is a pronounced left-turning tendency and the more abruptly the tail is raised, the worse the effect. Precession in this case is additive to the other left-turning factors of torque, and control could be marginal for a few seconds. Because of its attitude, the tricycle gear airplane doesn't normally have precession problems on take-off.

Asymmetric Disk Loading or P-Factor — This is a situation usually encountered in the climb or during slow flight and results from the fact that the relative wind is not striking the propeller disk at exactly a 90° angle. This results in a difference in angle of attack between the two (or three) blades. The down-moving blade, which is on the right side as seen from the cockpit, has a higher angle of attack than "normal" and consequently higher thrust. Whereas the opposite is the case for the up-moving (left side) blade. The result is a left-turning moment (Fig. 3-3).

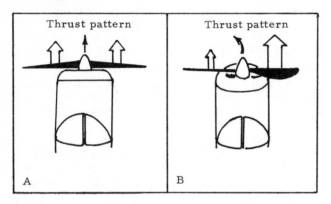

Fig. 3-3. Asymmetric disk loading (P-Factor) effects.

This effect can also be encountered in yaws (a left yaw would give a nose-down tendency, a right yaw, the opposite). P-Factor also is credited for the fact that the left engine of light twins is the worst engine to lose (or it is the *critical* engine). Perhaps it should be said, rather, that the left engine is the critical one for twins with both propellers turning clockwise (as seen from the cockpit) as Figure 3-4

shows. Many engineers and pilots think the role of P-Factor is over-rated in the establishing of the critical engine and believe that slipstream effects are a major factor here also.

Equal and Opposite Reaction — This is the effect that comes closest to the term "torque" and is a factor in Newton's law of "for every action there is an equal and opposite reaction." However, the strength of this effect is overrated for airplanes of higher power loadings (or lower horsepower per pound of airplane weight). The airplane tends to rotate opposite to the propeller. This can be corrected by "wash-in," or a higher angle of incidence, and higher drag on the left wing. For light, fabric-covered airplanes, this is normally a part of the rigging procedures before the airplane leaves the manufacturer. Metal airplanes have bendable tabs on the aileron(s) or aileron trim tabs to do the job. This rigging also can contribute slightly to a left-turning tendency.

These factors make up what pilots call torque. As an instrument trainee, you'll be surprised how smoothly it can sneak into the picture when everything else is going so well.

Lift

Lift is made up of the following factors and has the equation: Lift = $C_L \frac{\rho}{2} V^2 S$ or (L = $C_L x \frac{\rho}{2} x V^2 x S$):

C_L — *Coefficient of Lift* — A dimensionless factor (not measured as pounds, feet, etc.) which increases in direct proportion to the angle of attack (which you will remember is the angle between the chord line of the airfoil and the relative wind) until the stall angle is reached. Check Figure 3-5.

As lift is a combination of C_L and airspeed — plus the other factor of S, or wing area — the airplane with the airfoil and flap combination which has the greatest possible coefficient of lift (or C_{Lmax}) will be able to fly (or land) at a slower airspeed than

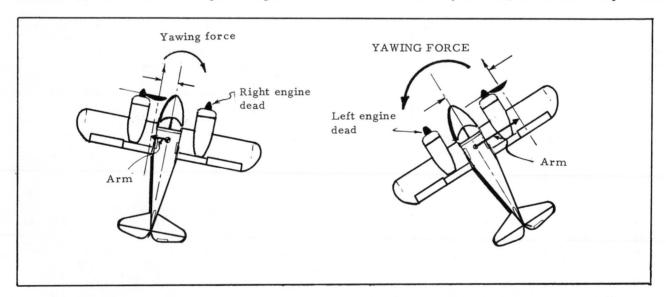

Fig. 3-4. The critical engine. Slipstream effect, as well as P-Factor, has an effect in establishing the critical engine.

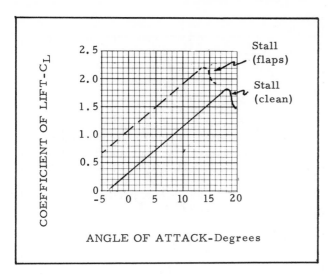

Fig. 3-5. Coefficient of lift versus angle of attack.

another airplane of equal weight and wing area. The dashed line in Figure 3-5 shows the effect of flaps on coefficient of lift for a particular airfoil-flap combination.

An expression often used concerning flaps is that they are designed to increase lift. The purpose and normal use of flaps is to allow you to maintain the required lift at a lower airspeed, or put more technically, it allows a greater C_{Lmax}, which means that the airplane can fly at a lower minimum speed. On the approach and landing, for instance, the forces acting perpendicular to the flight path are balanced as in the climb. With the flaps down on approach and landing, the "up" and "down" forces remain in equilibrium — which means a steeper approach path and slower landing speed. Only if positive "g's" are being exerted would lift be expected to be greater than weight.

$\frac{\rho}{2}V^2$ — The equation for *dynamic pressure*, called "q" by the engineers and discussed in the last chapter in the section on the airspeed indicator. Dynamic pressure is measured by the airspeed indicator, but instead of being expressed in pounds per square foot, it is indicated as mph or knots. The symbol "ρ" is the air density in slugs per cubic foot. The slug is a unit of mass and is found by dividing the weight of an object (in pounds) by the acceleration of gravity, 32.2 feet per second per second. At sea level, the standard air density is 0.002378 slugs per cubic foot.

The V^2 in the equation, you remember, is the *true* velocity of the air particles (squared), so that the dynamic pressure is a combination of one-half the air density times the true air velocity in feet per second (squared).

S — The *wing area* of the airplane in square feet.

Lift is the least understood of the four forces acting on the airplane. Contrary to popular belief, the pilot in normal unaccelerated flight (not pulling any "g's") has little control over lift, because when the power and airspeed are set to obtain the required performance, lift automatically assumes the correct value.

The biggest fallacy is the belief that lift is greater than weight in a steady-state (normal) climb, and "excess lift is what makes the airplane go up." This is not the case at all. On the contrary, if we assume that lift is equal to weight in straight and level, unaccelerated flight, then lift must be *less* than weight in the climb.

Figure 3-6 shows the forces acting on an airplane in a steady-state climb; that is, the airplane is moving along a constant climb path at a constant speed. This means that the airplane and pilot are subjected to the normal 1 "g." For such a condition to exist, the forces as measured perpendicular, or 90°, to the flight path must be in equilibrium. As true weight, or gravity, always acts "downward," or toward the center of the earth, its direction never changes with changes in the airplane's attitude. Lift always acts perpendicular, or 90°, to the relative wind or flight path, and Figure 3-6 shows what must be happening to maintain the steady-state climb. The angle of attack required to provide the flight path shown is omitted here to avoid complication.

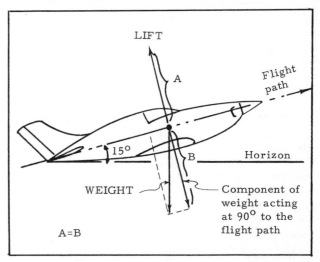

Fig. 3-6. A simplified look at the lift and weight forces acting on the airplane in a steady-state climb.

As you can see, the lift force must *equal* the component of weight acting at 90° to the flight path and, therefore, must be less than weight. The weight can be broken down into two vectors, one acting perpendicular, the other acting parallel and backwards along the flight path. The thrust required to maintain the steady-state condition is equal to the "rearward" component of weight *plus* the aerodynamic drag developed at the climb speed. While thrust is not of prime importance in this discussion, it should be noted that for equilibrium to exist, the forces acting *parallel* to the flight path (which is the reference axis) must be in equilibrium as must be the case for those acting perpendicular to it (lift and weight component).

If you are interested in the mathematics of the problem: Lift = Weight x Cosine of the Climb Angle, or $L = W\cos \gamma$. Assuming (again) that lift equals weight in straight and level flight, an airplane weighing 3000 pounds would require 2898 pounds of lift to maintain a steady-state climb at a 15° angle to the horizon. (The cosine of 15° is 0.966.) Lift is 102 pounds less than weight in such a condition. If lift isn't what makes the airplane climb, what does do it? *Power* — and this is one of the things that must be remembered in instrument flying — *power plus attitude equals performance.* In wings-level flight, it can also be said that power plus airspeed equals performance, but the first statement covers all possibilities better.

Suppose you are flying straight and level at cruise and decide to climb. At the point you *ease* the nose up, the angle of attack is increased without an instantaneous decrease in airspeed, and temporarily, lift *is* greater than weight, but only a very sensitive accelerometer, or g meter, would show that more than the normal 1 g is being pulled. If you make a very abrupt transition to the climb attitude, this is quite evident. Your normal transition to the climb is slow enough so that the very small amount of added g is not noticeable.

As you increased the angle of attack and obtained this "excess" lift, drag increased also and the airplane starts slowing immediately, which has the effect of "decreasing" lift again. As you are maintaining a steady climb speed and attitude, lift must settle down to the proper value to balance the component of weight as shown in Figure 3-6. You are then back in 1 g flight.

A measure of g forces is the ratio of lift to weight. If the forces acting in the direction of lift are greater than weight, then positive g's are being pulled; if the forces acting in the direction of lift are less than weight, negative g's result. The reason the term "forces acting in the direction of lift" was used is that other factors may be introduced (such as down forces on the tail, etc.).

Drag

The total drag of an airplane is made up of a combination of two main types of drag — parasite and induced. Drag acts in an opposite direction to the direction of flight (Fig. 3-7).

Parasite Drag

Parasite drag is not caused by just one factor but three:

Form Drag — This is a result of the fact that a form (the airplane) is being moved through a fluid (air). A blunt object will naturally have more form drag than a streamlined one. Examples of added form drag are extended landing gear, antennas, etc. These objects will also have interference drag at their junctions with the airplane and skin friction drag.

Skin Friction Drag — This is a result of the air moving over the aircraft skin and is one argument for a waxed and clean airplane.

Interference Drag — This is caused by aerodynamic interference and burbling between components and is found, for instance, at the junction of the wing and fuselage, stabilizer and fuselage, etc.

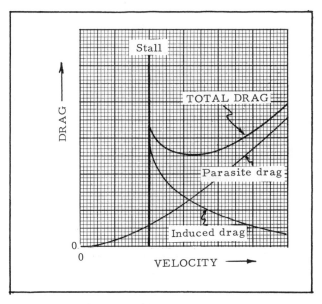

Fig. 3-7. A drag versus velocity curve.

Parasite drag increases with the square of the airspeed; double the airspeed, and parasite drag increases 4 times. Triple the airspeed, and parasite drag increases 9 times. Naturally, parasite drag is greater for the gear-down configuration, and a lot of antennas sticking out can cost a few knots at cruise. (You have a choice between no radios or a few less knots, which isn't a choice at all for instrument flying.)

Induced Drag

This drag is caused by the fact that the wing is creating lift. In creating lift, the relative air is deflected downward, and wing-tip vortices are formed which result in a drag force. As you can see in Figure 3-7, induced drag *increases* as the airplane flies slower and is greatest just at, or above, the stall. It's directly proportional to the square of the coefficient of lift, so that flying at lower airspeeds, induced drag may increase several times its original value. By decreasing the airspeed, the C_L required to maintain a constant value of lift is increased, hence induced drag has increased. Slow flight with a high coefficient of lift with no flaps being used creates the worst wake turbulence, so keep this in mind when following a heavy airplane on an approach. The vortex strength also depends on the "span loading", or weight per foot of wingspan (Fig. 3-8).

THE POWER CURVE

Force, Work, Power, Horsepower

Maybe it's been a few years since you had physics, so a quick and dirty review of the above terms might be in order.

A *force* is considered to be a pressure, tension, or weight. The fact that a force is being exerted

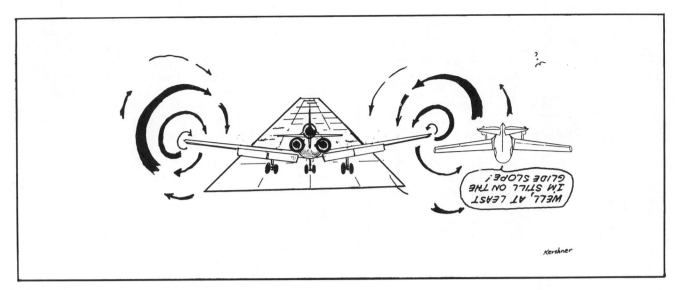

Fig. 3-8. Wing-tip vortices are most critical at low altitudes.

doesn't mean that *work* is being done. You can press against a brick wall with great force all day and from the viewpoint of physics, haven't done any work at all. (Tell this to those aching muscles.) *Work* is done when something moves. If you lift a 1100-pound weight to a height of 10 feet, you've done 11,000 foot-pounds (ft-lbs) of *work*. Or you can raise an 11-pound weight 1000 feet and will also have done 11,000 ft-lbs worth of *work*. Notice that nothing is said about time. You can do the job in 1 second or 24 hours; the work done is the same. *Power* is a different matter; that's where *work per unit of time* comes in. If the 11,000 ft-lbs of *work* is done in one second a great deal more *power* is used than if a full day were taken. The most familiar measurement of power is *horse-power*, and this is established as 550 ft-lbs of work being done in one second, or 33,000 ft-lbs of work per minute. Then, to do 11,000 ft-lbs of work in *one second* requires the developing of 20 horsepower for that period.

The type of horsepower most familiar to the pilot is that of *shaft* or *brake horsepower* (BHP) — that horsepower being developed as measured at the crankshaft by means of various devices such as a torque meter, dynamometer, or prony brake. (Wags have suggested that it should be called a "pony brake" as, after all, it's measuring *horse*power.) Brake horsepower is used to set up power on the power chart because BHP is considered to be constant for all speed ranges. It is comparatively easy to measure as a combination of manifold pressure and rpm. *Thrust horsepower* is a term of more interest to aeronautical engineers. Thrust horsepower (THP) is the horsepower being developed by the fact that a force (thrust) is moving an object through the air at some rate (velocity).

$$\text{THP} = \frac{\text{Thrust (lbs) x Velocity (fps)}}{550}.$$

If the propeller is exerting 1000 pounds of thrust to move an airplane through the air at a constant 275 feet per second, the THP being developed is:

$$\frac{1000 \times 275}{550} = 500 \text{ THP}.$$

If you prefer to think in terms of miles per hour: THP $= \frac{TV}{375}$. (This is because 550 fps = 375 mph.) To convert mph to fps, multiply by 1.467, or 1.467 x 375 mph = 550 fps. Roughly, any value in feet per second is one and a half times its value in miles per hour. For knots, use THP $= \frac{TV}{325}$.

As a prop pilot, you have no direct way of measuring the thrust, so power is used as a measure of what the engine is contributing to the process of flight. Figure 3-9 is a graph of thrust horsepower required versus indicated (calibrated) airspeed for a particular airplane at sea level.

Notice that the power-required curve has more than a passing resemblance to the drag curve. Instead of being expressed in terms of parasite drag and induced drag, Figure 3-9 is shown as parasite power required and induced power required, combined to make up the curve of total power required.

Let's take another look at thrust horsepower, or the horsepower actually being developed by the propeller in moving the airplane through the air. The propeller is only up to 85 per cent efficient in utilizing the brake or shaft horsepower. So that for a specific airspeed, the following might apply in finding how much thrust horsepower is being developed:

$$0.85 \times \text{BHP} = \frac{TV(\text{mph})}{375}, \text{ or } 0.85 \times \text{BHP} = \frac{T \times V}{375}.$$

Assuming that thrust equals drag in unaccelerated straight and level flight, the thrust horsepower equation could be written $\text{THP}_{required} = \frac{DV\text{mph}}{375}$. As drag in the cruise area is roughly proportional

33

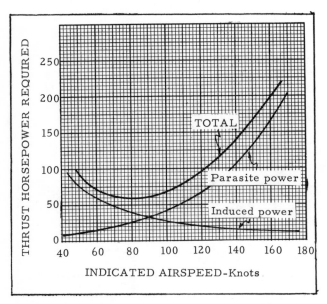

Fig. 3-9. Thrust horsepower required versus velocity.

to the square of the velocity, the horsepower required is multiplied by another V, so that horsepower required is a function of V^3, or the cube of the velocity. Boiled down, this means that to double the speed in cruise or top speed area, *eight* times the horsepower is required for a particular airplane. This goes for *brake or thrust* horsepower, but again, as you are only really interested in brake horsepower, this would be the item of interest.

Figure 3-10 is a brake horsepower required and available versus I.A.S. for a fictitious airplane at sea level and at gross weight. Brake horsepower and indicated (calibrated) airspeed are used here because these are the two items that are used by the pilot to obtain the desired performance in climbs, cruise, and descents.

Notice that the airplane can fly at two speeds for most power setting percentages. For instance, at 65

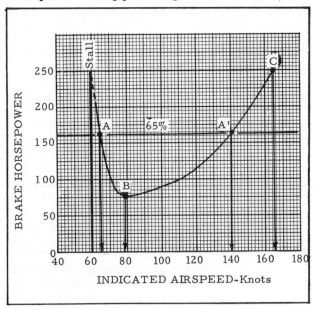

Fig. 3-10. Brake horsepower versus velocity.

per cent power, the airplane can fly straight and level at 65 knots and also at 140 knots as shown by A and A'. This works for all power settings down to that of minimum power required, which will run at about 35-40 per cent normal rated power (brake horsepower) at gross weight for most airplanes of the type you'll be flying on instruments. It would take the least horsepower to maintain altitude at the airspeed under Point B.

The airspeed at Point B, in theory, would be the one to use at gross weight in the instrument holding pattern. The term "in theory" was inserted because other factors may enter. Your airplane may have poor handling characteristics at fairly low speeds. The propeller low-rpm characteristics may not be of the best plus the fact that turbulence could cause additional handling problems, so that a slight increase in airspeed to slightly above that given for Point B might be better. Another factor is that the brake-specific fuel consumption, or pounds of fuel burned per brake horsepower per hour, normally increases at both ends of the power setting range. By adding some power you may decrease the fuel consumption per horsepower to such an extent that the efficiency of the engine is increased.

Looking back to Figure 3-10, Point C shows the top speed (level flight) of the airplane at the particular altitude. As Figure 3-10 was drawn for sea level, this would be the *absolute maximum level flight speed* for the airplane with an unsupercharged engine (or engines). While Figure 3-10 hints that, in theory, there is also a corresponding low speed for 100 per cent power, it is highly unlikely that the stall characteristics of the airplane would allow it to fly at such a low speed that 100 per cent power would be necessary to maintain level flight; the stall break would occur first.

Setting Power

The Unsupercharged Engine

Along with theory should come some practical application, and Figure 3-11 is a true airspeed versus standard (density) altitude chart for a four-place retractable gear airplane at maximum certificated (gross) weight. The airplane is powered by a 250-horsepower unsupercharged engine.

Point 1 shows that for the unsupercharged engine the maximum level flight airspeed is found at sea level, and the level flight speed at maximum power decreases with altitude. This is because the amount of horsepower available is dropping faster than the gain of true airspeed effects.

It's a different story with power settings of less than maximum; take a look at the line for 75 per cent power. The T.A.S. increases with altitude until at about 7000 feet — Point 2 — the airspeed starts dropping again. This is because 7000 feet is the highest altitude at which 75 per cent power can be maintained at the recommended max cruise rpm of 2400. Point 2 — 7000 feet — would be the full throttle altitude (or critical altitude) for 75 per cent power

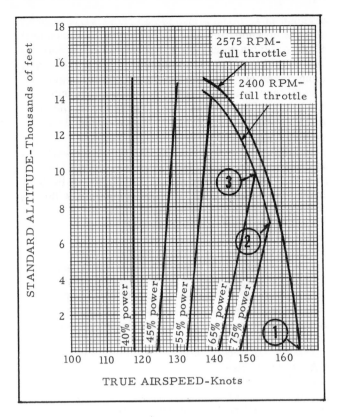

Fig. 3-11. Performance chart. *(Piper Aircraft Corp.)*

means a greater density if the mixture pressure (manifold pressure) remains the same.

2. The exhaust gases have less back pressure (outside pressure) to fight at higher altitudes. The "explosion" in the cylinder is sealed, and some power is required to expel the waste gases. The less back pressure existing, the less power wasted, and more can be used to "drive" the airplane.

Power Setting Table - Lycoming Model IO-540-D, 260 HP Engine

Press. Alt. 1000 Feet	Std. Alt. Temp. °F	143 HP - 55% Rated Approx. Fuel 11.4 GPH RPM AND MAN. PRESS.				169 HP - 65% Rated Approx. Fuel 12.7 GPH RPM AND MAN. PRESS.				195 HP - 75% Rated Approx. Fuel 14.1 GPH RPM AND MAN. PRESS.			
		2100	2200	2300	2400	2100	2200	2300	2400	2200	2300	2400	2500
SL	59	22.3	21.5	20.7	19.8	25.3	24.1	23.2	22.2	26.9	25.8	24.8	24.0
1	55	22.1	21.3	20.5	19.6	25.1	23.9	22.9	22.0	26.6	25.5	24.5	23.7
2	52	21.9	21.0	20.3	19.4	24.8	23.6	22.7	21.8	26.3	25.3	24.3	23.5
3	48	21.7	20.8	20.0	19.2	24.5	23.4	22.5	21.6	26.0	25.0	24.0	23.2
4	45	21.4	20.6	19.8	19.0	24.2	23.1	22.2	21.4	25.7	24.7	23.8	22.9
5	41	21.2	20.3	19.6	18.8	24.0	22.9	22.0	21.1	25.4	24.4	23.5	22.7
6	38	21.0	20.1	19.4	18.6	23.7	22.6	21.7	20.9	-	24.1	23.3	22.4
7	34	20.7	19.9	19.1	18.4	23.5	22.4	21.5	20.7	-	-	23.0	22.2
8	31	20.5	19.6	18.9	18.2	-	22.1	21.2	20.5				21.9
9	27	20.3	19.4	18.7	18.0	-	21.9	21.0	20.3				
10	23	20.0	19.2	18.5	17.7	-	-	20.7	20.0				
11	19	19.8	18.9	18.2	17.5				19.8				
12	16	19.6	18.7	18.0	17.3								
13	12	-	18.5	17.8	17.1								
14	9	-	-	17.5	16.9								
15	5	-	-	17.3	16.7								

To maintain constant power, correct manifold pressure approximately 0.17" Hg. for each 10° F variation in induction air temperature from standard altitude temperature. Add manifold pressure for air temperature above standard; subtract for temperatures below standard.

Fig. 3-12. Power setting table.
(Lycoming Division of AVCO.)

at 2400 rpm. If, for instance, you prefer 65 per cent power for cruise on an instrument flight (and this is certainly more economical and easier on the engine), then 10,000 feet — Point 3 — would be the full throttle altitude and would be the altitude to fly for the most airspeed and range for that power setting. (For 75 per cent power, 7000 feet would be the magic altitude to pick.) This, naturally, doesn't take into account such things as wind at that altitude or assigned altitudes on IFR, but it would still be best as far as getting the max *true airspeed* is concerned.

You'll find that full-throttle operation will probably produce a maximum manifold pressure of 28-29 inches of mercury in standard sea level conditions (unsupercharged engine). As the barometric pressure drops about one inch of mercury per thousand feet, you would at some altitude run out of the manifold pressure necessary to maintain the desired percentage of power. You've reached the full-throttle or critical altitude for that power setting.

Figure 3-12 is the power setting table for the Lycoming IO-540-D engine (260 horsepower) using a constant-speed prop. For a given rpm, less manifold pressure (mp) is required for a specific percentage of power, as altitude increases. Note that at 65 per cent power, using 2300 rpm, 23.2 inches mp is required at sea level. At 7000 feet only 21.5 inches mp is required to maintain 65 per cent power. There are mainly two reasons for this:

1. The air is cooler at higher altitudes, and if you used the same mp as you carried at sea level, the mixture density and the horsepower developed would be greater, because a lower temperature

Most pilots prefer a particular rpm for cruise at either 65 or 75 per cent, and looking at Figure 3-12, an interesting fact comes to light concerning the manifold pressure drop. For 2300 rpm at sea level (again), 23.2 inches is needed for 65 per cent power, and 25.8 inches is necessary to obtain 75 per cent power. Another look shows that the required manifold pressure to maintain the chosen power at 2300 rpm drops about $\frac{1}{4}$ inch per thousand feet. You can subtract $\frac{1}{4}$ inch per thousand from the sea level manifold pressure and get the power setting table figure. If you want the mp required for 65 per cent at 5000 feet, you would subtract (5 x $\frac{1}{4}$) or $1\frac{1}{4}$ (1.25) inches from the sea level figure of 23.2, for an answer of 21.95. (The table says 22.0, but you can't read the mp gage that closely anyway.) For 5000 feet at 75 per cent, it would be 25.8 - (5 x 0.25) or 25.8 - 1.25 = 24.55 inches (table figure — 24.4 inches). For cruise information, if you use 2300 rpm, you could remember 23.2, 25.8 and $\frac{1}{4}$ inch drop per thousand feet and not have to refer to the power setting table every time when using 65 or 75 per cent power respectively.

Another tip for setting power for cruise after level-off (VFR *or* IFR) is to leave the power at the climb value until the expected cruise I.A.S. is approached. This expedites the transition. In addition, if the cruise power were set just as the plane leveled off at a comparatively low speed right out of the climb, it would mean resetting the mp as cruise speed is approached (you'll have to throttle back slightly). Why? Because ram effect will be packing in more air at the higher speed, the mp will increase, and you've just given yourself another little chore.

Figure 3-12 shows that corrections for

variations from standard temperatures must be taken into account. This follows the same principle mentioned concerning the lower mp required to maintain the same power at higher altitudes. This is because of the fact that one of the real measures of horsepower developed is the mixture *density*. The temperature being higher or lower would vary the density if the manifold pressure were held the same. As there is no simple way to measure mixture density, the manifold pressure must be varied to take care of temperature variations. In general, the *equation of state* (gases) puts the relationship of pressure, temperature, and density this way:

Pressure Constant — Temperature increase means a density *decrease* and vice versa. (At a constant pressure, density is inversely proportional to temperature.)

Temperature Constant — Pressure increase means a density increase, or density is directly proportional to pressure.

Density Constant — Temperature increase means a pressure increase, or pressure is directly proportional to temperature.

The "density constant" factor is the one that could cause you to fly into a mountain when on instruments, as covered in Chapter 2 in the section on quirks of the altimeter.

AIRPLANE STABILITY AND WEIGHT AND BALANCE

It's very important that the airplane be in a stable condition on any flight but even more so when flying IFR. The unstable airplane can require constant pilot attention, which means that a lapse of attention, such as chart checking or clearance copying, can spell trouble for the single-piloted aircraft without an autopilot. Even if you have an autopilot, it will work harder and, in correcting, may cause the flight to be rougher than it would be if the airplane were stable. Your knowledge of weight and balance procedures will become even more important.

Figure 3-13 shows the three axes around which the airplane maneuvers. The Federal Aviation Regulations (Parts 23 and 25) require that the airplane certificated in the normal, utility, and transport categories must meet certain minimum requirements of stability about each axis.

But, before getting into the specifics of the airplane, let's take a look at the basic idea of stability.

Static Stability

The term "static" might well express the idea of "at rest." When you are flaked out on the couch for a nap, you could be said to be "static." An object or system that has an initial tendency to return to its original position and resists being offset in the first place may be termed *statically stable*. Figure 3-14 shows three possible static stability situations involving a "system" — a perfectly smooth bowl or hubcap and a steel ball.

Figure 3-14A shows a *positive static stability* or a system that is *statically* stable. If you tried to displace the ball from its center position, it would resist displacement and tend to return to its original position.

System B is *neutrally stable,* or has neutral static stability; it's a steel ball on a flat plate. If a force acts on it, it moves and stops at some new position when the force or the effects of that force are gone. The ball has no tendency to return to the original position.

System C is *negatively stable.* If it is displaced from its balanced position, as shown, it will get

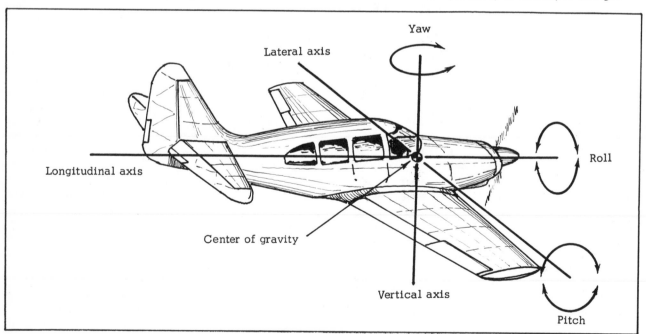

Fig. 3-13. The three axes.

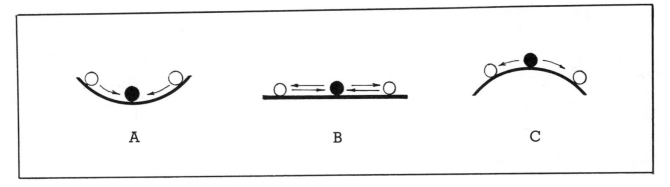

Fig. 3-14. Positive, neutral, and negative static stability.

farther and farther, at an increasingly faster rate, from the original position.

Notice that only in the statically stable system does the ball have any tendency to return to its original position, and this brings up another point:

Dynamic Stability

The actions a body takes in response to its static stability properties show its dynamic stability (dynamic = active). Dynamic stability is considered to be the time history of a body's responses to its static stability condition. In System A in Figure 3-14, the ball would resist any tendency to be moved from its center position and is *statically stable.* When released, it would return toward the center, overshooting and returning in decreasing oscillations until it would again come to rest in its original position. In such a case, it would also be *dynamically stable,* or it would have *positive dynamic stability* (because the actions of the ball returned it to the original position). If you had added an outside force by rocking the hubcap, you might cause the oscillations to continue with the same magnitude *(neutral dynamic stability),* or you could rock it enough so that the oscillations get more violent until the ball shoots over the side *(negative dynamic stability).*

The experiment just accomplished will work only if the system is statically stable to begin with — the ball would hardly oscillate on the flat plate and certainly not if it were on the outside of the hubcap as shown in Figure 3-14C. This leads to the brilliant conclusion that in order for a system to have any oscillatory properties at all, it must be *statically stable* or have *positive static stability.*

LONGITUDINAL (PITCH) STABILITY OF THE AIRPLANE

The rather complicated title of this section basically means taking a look at how the airplane wants to hold its trim airspeed — or its actions in returning to that airspeed.

Longitudinal stability, or stability around the lateral axis (check Figure 3-13 again), is the most important of the three because the pilot can affect it more with placement of weight. Sure, burning more fuel out of one wing tank can affect lateral stability, and moving weight rearward *can* affect directional (yaw) stability (it's doubtful that you could notice it, though), *but* longitudinal stability problems have pranged more airplanes by far than the other two combined.

As a review, an airplane in steady-state flight (such as straight and level, steady climb, or steady glide, etc.) must be in "equilibrium." This means that the summing up of forces and moments must equal zero; the idea of equilibrium of forces was touched on briefly in the last chapter.

You recall that a force can be considered as a pressure, tension, or weight. A "moment" usually is the result of a force or weight acting at some distance from a fulcrum, or pivot point, at a 90° angle to its "arm." A seesaw is a good example of a system of moments, as shown in Figure 3-15.

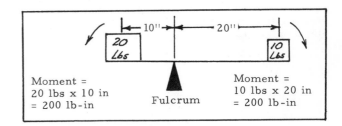

Fig. 3-15. A system of moments in equilibrium.

The two moments are equal, but the distances (arms) and weights are different. Moments may be expressed as foot-pounds (ft-lbs), inch-pounds (in-lbs), pound-feet, or pound-inches. *Either method of expression such as force-distance or distance-force is correct.* But often the procedure for moments is to have the force or weight expressed first (pound-inches) to keep from confusing a "moment" with "work" which, you remember, was expressed as "foot-pounds," using the distance factor first.

You are trying to balance a 200-pound rock with a mechanical lever system as shown in Figure 3-16.

If you are holding the lever at 40 inches from the fulcrum, how much force do you have to exert to "assure equilibrium of the system" (balance the rock)?

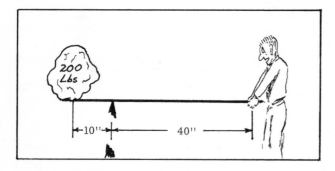

Fig. 3-16. A simple lever system.

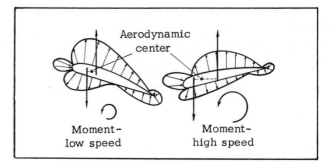

Fig. 3-18. Moments created by a unsymmetrical airfoil at two different airspeeds. The angles of attack and pressure patterns have been exaggerated.

It's assumed that the bar is rigid, and its weight will be neglected.

Since the system is in equilibrium (you're balancing the rock), the two moments are equal. The moment on the rock side of the fulcrum can be found as 200 lbs × 10 in = 2000 lb-in. The moment on your side must also be 2000 lb-in, so you will have to exert *50 pounds* at a 90° angle to the lever at the distance of 40 inches to balance the rock (50 lbs × 40 in = 2000 lb-in).

Take a look at Figure 3-17 to see some of the forces and moments acting on a typical high-performance, four-place, general aviation airplane flying in straight and level cruising flight.

In Figure 3-17, rather than establish the vertical acting forces (lift, weight, and the tail force) with respect to the center of gravity as is the usual case, they will be measured fore and aft from the center of lift. Assume at this point that lift is a string holding the airplane "up," and its value will be found later. The airplane in Figure 3-17 weighs 3000 pounds, is flying at 150 knots I.A.S. (C.A.S.), and at this particular loading, the C.G. (center of gravity) is 5 inches ahead of the "lift line."

Summing up the major moments acting on the airplane (check Figure 3-17 for each):

1. *The Lift-Weight Moment* — The weight (3000 pounds) is acting 5 inches ahead of the center of lift, and this results in a 15,000 pound-inches *nose-down* moment (5 in × 3000 lbs = 15,000 lb-in).

2. *The Thrust Moment* — Thrust is acting 15 inches above the center of gravity and has a value of 400 pounds. The *nose-down* moment resulting is 15 in × 400 lbs = *6000 lb-in.* (The moment created by thrust will be measured with respect to the C.G.) For simplicity, it will be assumed that the drag is operating back through the C.G. Although this is not usually the case, it saves working with another moment.

3. *The Wing Moment* — The wing, in producing lift, creates a nose-down moment which is the result of the forces working on the wing itself. Figure 3-18 shows force patterns acting on a wing at two airspeeds (angles of attack). These moments are acting with respect to the aerodynamic center, a point considered to be located about 25 per cent of the chord for all airfoils.

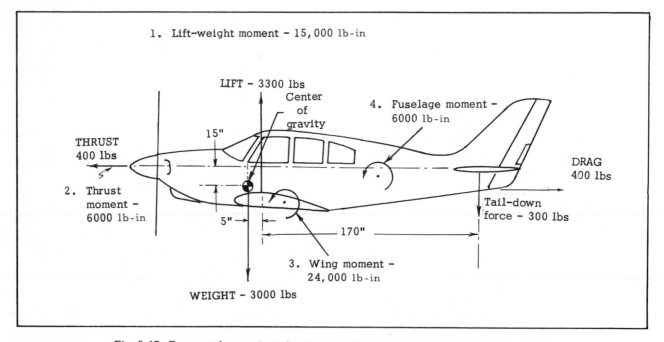

Fig. 3-17. Forces and moments acting on an airplane in straight and level cruising flight.

38

Notice that as the speed increases (the angle of attack decreases), the moment becomes greater as the force pattern varies. The nose-down moment created by the wing increases as the *square* of the airspeed if the airfoil is not a symmetrical type. (There is no wing moment if the airfoil is symmetrical, because all of the forces are acting through the aerodynamic center of the airfoil.)

For an airplane of the type, airspeed, and weight used here, a nose-down moment created by the wing itself of 24,000 pound-inches would be a good round figure. Remember that this would vary with indicated airspeed. *Nose-down moment created by wing = 24,000 pound-inches.*

4. *The Fuselage* — This may also be expected to have a moment about its C.G. because it too has a flow pattern and, for this example, airplane type and airspeed would be about *6000 pound-inches nose-down.* (This is not always the case.)

Summing up the nose-down moments:

Lift-Weight Moment	=	15,000 pound-inches
Thrust Moment	=	6,000 pound-inches
Wing Moment (at 150 knots)	=	24,000 pound-inches
Fuselage Moment (at 150 knots)	=	6,000 pound-inches
Total Nose-down Moment	=	51,000 pound-inches

For equilibrium to exist, there must be a *tail-down* moment of 51,000 pound-inches, and this is furnished by the tail-down force. Figure 3-17 shows that the "arm," or the distance from the lift line to the center of the tail-down force, is 170 inches. So, the moment (51,000 lb-in) and the arm (170 in) are known, and the force acting at the end of that arm (the tail-down force) can be found as $\frac{51,000 \text{ lb-in}}{170 \text{ in}}$ = 300 lbs. The airplane nose does not tend to pitch either way.

Since the forces acting perpendicular and parallel to the flight path must also be in equilibrium, a couple of other steps could be taken to complete the problem. Figure 3-17 shows that the forces acting parallel to the flight path are equal (thrust and drag are each 400 pounds), and a summation of the vertical forces, or forces acting perpendicular to the flight path, is next checked. The "downward" acting forces are weight (3000 lbs) and the tail-down force (300 lbs), for a total of 3300 pounds. The only opposing force is lift, and for equilibrium to exist, this must be 3300 pounds.

Looking back at the idea of the wing moment, you'll notice that it tends to keep the nose down and the airspeed up, and for the unsymmetrical airfoil, this tendency increases with the square of the airspeed. Suppose, as an example, that you moved weight back in the airplane at cruise so that the lift-weight moment is a minus factor — you've moved the C.G. so far back that lift is acting *ahead* of weight. Say, for example, that lift is acting 10 inches ahead of weight at the cruise airspeed of 150 knots; the lift-weight moment is now a *minus* factor (minus means a nose-up moment).

Lift-Weight Moment	=	-30,000 pound-inches
Thrust Moment	=	6,000 pound-inches
Wing Moment (at 150 knots)	=	24,000 pound-inches
Fuselage Moment (at 150 knots)	=	6,000 pound-inches
Total Nose-down Moment	=	6,000 pound-inches

Summing the nose-up and nose-down moments, the result is a 6000 pound-inches nose-down moment which is to be balanced by the tail-down force. The arm is 170 inches so that the required tail-down force is $\frac{6000 \text{ lb-in}}{170 \text{ in}}$ = 35 lbs. You can see that the tail-down force is rapidly disappearing and the airplane is becoming less statically stable.

The wing moment is a function of the *square* of the airspeed, so that if the plane is slowed up to one-half its cruise speed, the wing and fuselage moments would be one-fourth of their values at cruise. As the airplane was slowed to holding speed, you just might find yourself running out of the forward wheel (down elevator) necessary to furnish the nose-down moment. The nose could rise abruptly and a stall could occur, followed by a loss of control (on instruments!).

The treatment given here is that the weight (C.G.) was moved aft *in flight,* and the pilot was able to control it until he slowed down and lost his "helpful" wing and fuselage moments. The realistic view would be that the airplane would have problems on take-off and probably never get to the cruise condition. Your job as a pilot is to make sure that the airplane is loaded properly so that problems don't arise. Remember that even though the airplane is controllable at cruise, and at holding speeds, *in turbulence* control might be marginal when your attention is directed to other things such as taking clearances, checking engine instruments, etc. Sometimes, it seems that ATC has a TV camera in the cabin and knows the exact time that you least want a clearance. ("Okay, boys, he's in turbulence, is picking up ice, and his pencil just rolled all the way back under the back seat; let's give him Clearance 332, that always leaves them climbing the walls.") That's when prior checking of weight and balance could have helped.

WEIGHT AND BALANCE

The *basic* empty weight of the airplane is initially established by the manufacturer and, for a particular airplane, *includes unusable fuel, full oil, hydraulic fluid, and all equipment necessary for flight (and optional equipment).* In other words, the airplane is mechanically ready to fly but lacks usable fuel, occupants, and baggage.

To obtain the empty weight, the airplane is weighed as shown by Figure 3-19.

One thing to consider is whether the airplane was weighed at the factory *before* or *after* being painted. The weight of the paint of a four-place, single-engine, general aviation airplane can be from 10 to 30 pounds

39

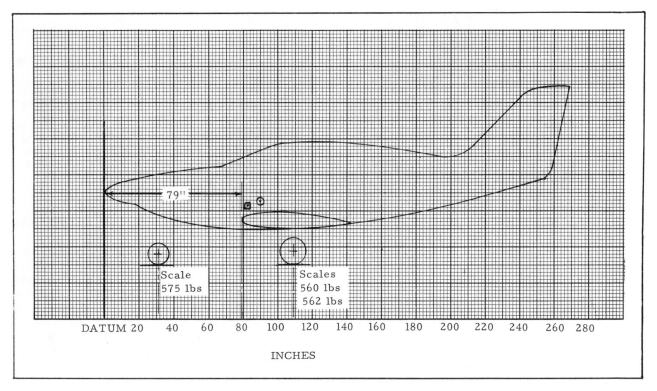

Fig. 3-19. Determining the empty weight and empty weight center of gravity.

(or more) depending on the area covered. If the airplane is unpainted at the time of weighing, the paint weight is added to it on the Weight and Balance Form.

The newly manufactured airplane is placed on scales as shown in Figure 3-19. The total empty weight for this airplane is the sum of the three scale figures and is 1697 pounds. The resulting figure is called the "empty weight as weighed," and the unusable fuel and full oil (and the weight of the paint, if necessary) are added to this to get the "basic empty weight." The airplane in this example is painted, and all radios and optional equipment are installed.

To find the empty weight C.G., the "datum" is used. This is an imaginary point located at, or some distance ahead of, a well-defined spot on the airplane, such as the front side of the firewall, wing leading edge, etc. In this case, the datum is located 79 inches ahead of the straight leading edge of the wing. On airplanes with tapered wings, the datum may be established a certain distance ahead of the junction of the wing leading edge and the fuselage. When the datum is ahead of the airplane as is the case here, all of the arms are positive; when the datum is located at the leading edge of the wing or front side of the firewall, the arm is in a positive direction when aft of the datum and negative when forward.

The empty weight (or "as weighed") C.G. is located by using the principle of moments and using the datum point as the fulcrum or pivot point.

The weight concentrated on the nosewheel is 575 pounds, and it is 31 inches from the datum; hence, its moment would be 575 lbs × 31 in = 17,825 lb-in.

The moment created by the weights on the main

gear would be (560 + 562) lbs × 109 in = 1122 lbs × 109 in = 122,298 lb-in. The two weights on the main wheels are combined, as the two wheels are the same distance from the datum (or should be, if the airplane hasn't been taxied into something).

Rearranged, the problem would look like this:

Weight		Arm (in)		Moment
575	×	31	=	17,825
1122	×	109	=	122,298
Total 1697 lbs				140,123 lb-in

Dividing the total moment by the weight, the empty center of gravity is found to be located at $\frac{140,123}{1697} = 82.6$ inches aft of the datum.

This particular airplane is considered to have 4 gallons of unusable fuel and was weighed with no oil, so the step to get the basic empty weight is:

Item	Weight	Arm (in)	Moment
Empty Weight as Weighed	1697	82.6	140,123
Oil (3 gal)	23	28.0	644
Unusable Fuel (4 gal)	24	90.0*	2,160
Total	1744 lbs		142,927 lb-in

The *basic* empty weight for this particular airplane is 1744 pounds, and the empty C.G. is found by dividing the total moment (142,927 lb-in) by the total weight (1744 lbs). The answer is 81.9 inches, as shown by the square in Figure 3-19. The answers

40

were rounded off to the nearest one-tenth of an inch. What about that mysterious 90* that showed up in the last calculation? That's the distance of the center of gravity of the fuel load from the datum — 90 inches. This information is given in the Weight and Balance Form for the airplane. Incidentally, fuel (gasoline) is considered to weigh 6 pounds per gallon, oil weighs 7.5 pounds per gallon. It's best to use the actual passenger weights.

You could then find out the effects of adding usable fuel, passengers, and baggage as shown below (the maximum certificated (gross) weight is 2900 pounds for this airplane):

Item	Weight	Arm	Moment
Basic Empty Weight	1744		142,927
Fuel (56 gals)	336	90.0*	30,240
Pilot	170	84.8*	14,416
Passenger (front)	170	84.8*	14,416
Passenger (rear)(1)	170	118.5*	20,145
Baggage	200	142.0*	28,400
Total	2,790 lbs		250,544 lb-in

The total moment is divided by the weight to find the center of gravity with that loading to get an answer of 89.8 inches aft of the datum. This is shown by the circle in Figure 3-19. Those arms marked with an asterisk are the same for each airplane of this model and are given on the Weight and Balance Form.

Pilot Operating Handbooks for 1976 models (and later) use the term "basic empty weight" rather than "licensed empty weight," the older term, in case this is a review. Again, the airplane basic empty weight includes unusable fuel *and full oil,* the latter not being considered as part of the empty weight on the earlier models. The *Pilot's Operating Handbook* will have a Weight and Balance form with blanks for the specific airplane.

The Weight and Balance Envelope

You know that the airplane is legal from a weight standpoint; but is the C.G. in the proper range?

Figure 3-20 is an envelope for the four-place, high-performance airplane used in the example.

The range is from 80.5 inches to 93 inches aft of the datum, and it can be found that the loading checked earlier is within safe limits. In the problem, no fourth passenger was taken, but the full allowable 200 pounds of baggage was included. You could work out various combinations of items and check for safe operation by referring to the weight and balance envelope.

The forward C.G. limit is set by control in ground effect, rather than stalls at altitude.

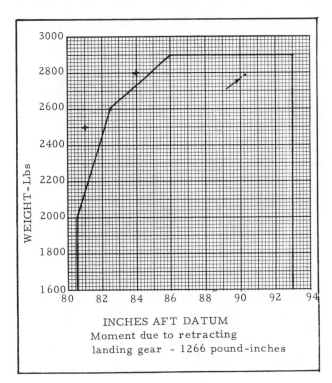

INCHES AFT DATUM
Moment due to retracting landing gear - 1266 pound-inches

Fig. 3-20. Weight and balance envelope.

Notice Figure 3-20 has a remark that the moment due to retracting the landing gear is 1266 pound-inches. The empty weight C.G. was checked with the landing gear down, so that when it is retracted (the nosewheel swings back, and maybe, the main gear also moves back slightly with respect to the datum), the value cited must be added to the total moment of airplane, people, fuel, etc., to get the true C.G. position *in flight*.

The final C.G. would be found by adding the moment of 1266 pound-inches to the total moment to find the C.G. with the gear retracted (which is what you are really interested in). Notice that the total weight would be the same; it's just been moved back, increasing the moment. The new C.G. is

$$\frac{(250,544 + 1266)}{2790} = \frac{251,810}{2790} = 90.3 \text{ inches aft of}$$

the datum (rounded off). While it's not a critical factor for this problem, it could nudge you over the line if the C.G. was right at the rearward limits to begin with. The dot in Figure 3-20 shows that the final C.G. is within safe limits. This practice is being eliminated by some manufacturers, but is mentioned to cover a factor you might not have considered.

The baggage placard is to be respected; not only could the C.G. be moved too far aft but also the baggage compartment floor could be overstressed. The baggage compartment floor area is designed to withstand a certain number of g's with 200 pounds (or whatever the placard limit indicates), and if you pull that same number of g's with, say, 400 pounds in it, what used to be the baggage compartment may be just a memory. The allowable positive and negative load factors and airplane categories are given in the *Pilot's Operating Handbook.*

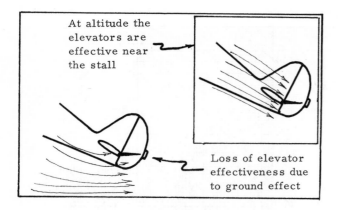

At altitude the elevators are effective near the stall

Loss of elevator effectiveness due to ground effect

Fig. 3-21. Ground effect and elevator effectiveness.

Forward C.G. Considerations

The envelope in Figure 3-20 is not square but has cutoff areas in the upper left-hand corner. For instance, you would not be legal flying at 2800 pounds with the C.G. located at 84 inches, or at 2500 pounds with a C.G. at 81 inches aft of the datum (as shown by the crosses).

Everybody worries so much about the rear limit of the envelope that the idea of a forward limit is forgotten. For simplification, the rear limit is basically established to insure stability, and the forward limit is to insure proper controllability.

Suppose you kept moving weight forward while in flight; it would require more and more up-elevator (or trim) to maintain "longitudinal equilibrium" (keep the nose up). You could reach a situation where *full* up-elevator would be necessary in the cruise regime.

When power was chopped, the nose would drop again, and the airspeed would pick up. You'd be in trouble trying to land the airplane in this exaggerated situation. Actually, the allowed forward limit of the airplane would be even farther forward if it were not for ground effect.

At altitude, the elevators may be still effective near the stall with a well-forward center of gravity but lose effectiveness because of ground effect as shown by Figure 3-21.

Figures 3-22 and 3-23 show an approach to the weight and balance computations for another airplane. Instead of thinking in terms of pounds and inches aft of the datum as in the case for the envelope of Figure 3-20, this one uses the graph form of pounds versus moment (pound-inches) as the criterion. If a certain airplane weight comes up with a moment (which is distance times pounds) that falls within the envelope, it really doesn't matter whether you know the exact position of the C.G. — in inches — or not; it is in a safe range.

Figure 3-22 shows a simplified loading graph. The slope of the lines represents the effects of distance on the moment. Notice that a pilot and passenger weighing a total of 340 pounds in the front seat created a moment of 12,200 pound-inches as compared to about 24,000 pound-inches created by rear passengers totaling 340 pounds. (Notice that 120 pounds of baggage, because of its location, makes a moment nearly as great as the 340 pounds worth of people in the front seat.)

This sample airplane is a four-place single-engine type and has a gross weight of 2800 pounds. It's given that the airplane has a basic empty

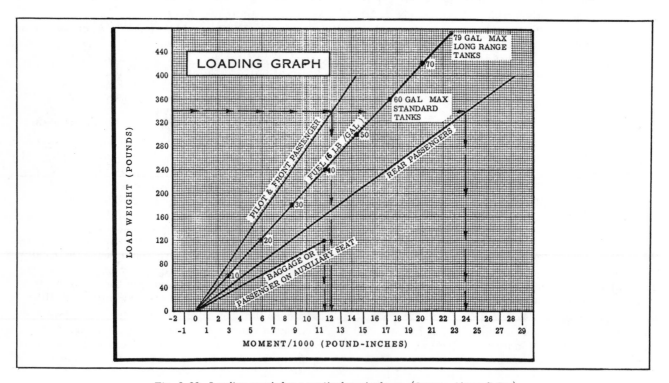

Fig. 3-22. Loading graph for a particular airplane. (*Cessna Aircraft Co.*)

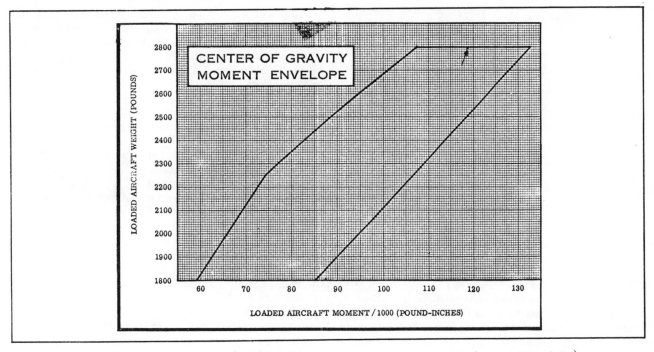

Fig. 3-23. Center of gravity moment envelope for the airplane of Figure 3-22. (*Cessna Aircraft Co.*)

weight of 1682 pounds with an empty weight moment of 57,600 (or 57.6 × 1000) pound-inches. The datum is located on the front face of the firewall, and you will notice that the oil has a negative moment because it is *ahead* of that point. You could make up a table. Refer to Figure 3-22 for the moments resulting for the various items.

Item	Weight	Moment
Basic Empty Weight (given)	1682	57.6
Pilot and Passenger (front)	340	12.2
Fuel — 60 gals (at 6 lbs/gal)	360	17.3
Rear Passengers	340	24.1
Baggage	78	7.6
Total	2800 lbs	118.8 lb-in

(More properly, this is 118,800 pound-inches.)

Figure 3-23 is the center of gravity moment envelope for this airplane. You can see by the arrow that the airplane is legal both from a weight standpoint and center of gravity, or total moment, consideration. Notice how the envelope "leans" to the right as the weight increases. At 1800 pounds, a large moment (say, 90,000 lb-in) would mean that the C.G. was quite far back from the datum (50 inches to be exact), whereas at a higher weight of 2250 pounds, a moment of 90,000 pound-inches is quite acceptable because the arm is shorter (40 inches aft of the datum), and the C.G. is not in a

rearward critical condition. If, out of curiosity, you wanted to find the exact location of the center of gravity of the airplane just discussed, you could divide the total moment (118,800 lb-in) by the total weight (2800 lbs) and find that the C.G. is located at approximately 42.4 inches aft of the datum. Notice that if you wanted to take the time and trouble, you could convert either type of envelope (Figure 3-20 and 3-23) to the other form.

SUMMARY OF THE CHAPTER

There's plenty more to be said about stability, weight, and balance. This chapter covered only longitudinal stability because it is the one area that the pilot can affect most easily. Directional and lateral stability are pretty much built in at the factory, and the airplane must meet certain minimum requirements in that regard. As far as longitudinal stability is concerned, the airplane will meet the requirements for safe operation as required by the FAA *only* if the weight and center of gravity are kept within the envelope. Be sure that you can work the Weight and Balance for the airplane you are using.

Don't load the airplane so that it's right on the end of the envelope. It could be acceptable in smooth air, but turbulence and icing could cause the handling to become marginal or cause additional fatigue over the duration of the flight — and an approach in conditions of a 200-foot ceiling and 1/2-mile visibility needs all the attention and alertness you can give it.

43

Chapter 4

BASIC INSTRUMENT FLYING

BACKGROUND

IT'S QUITE POSSIBLE that you can fly an airplane both VFR or on instruments without knowing the theory of performance. But if you are going to the trouble of learning a new area of flying, you might as well get the background. A lot of pilots have reacted wrongly to an unusual situation in flying — and paid for it because they didn't know what the airplane could or could not do.. Even experienced pilots have gotten into a bind, for instance, by unconsciously trying to stretch a glide or descent by pulling the nose up farther instead of adding power in a situation where outside distractions have become almost intolerable: You are making an ILS approach and have picked up a lot of ice, the ceiling and visibility are right on minimums, it's turbulent, and there's a lot of communication between you and the tower. You might end up close to a stall by trying to stay up on the glide slope by using the elevators and not enough power.

It's funny, but pilots pay lip service to the statement "Power controls altitude, and the elevators control airspeed" and can (and will) rattle it off as a schoolboy recites the Preamble to the Constitution. Unfortunately, too often neither the pilot nor the schoolboy stops to think about what the words really mean.

Of course, when you are at cruise and pull back on the wheel the airplane climbs, so, "obviously" the elevators make it climb. Not so, you are moving to a different (slower) airspeed using the same power

setting so excess horsepower — and energy altitude — are working to gain altitude. In practicality, at *high* speeds, you move the nose up to climb. Don't do this near the stall.

There are several ideas concerning instrument flying which often are brought up to the pilot when he starts his training for the rating. One is that somehow, as soon as the pilot is unable to fly by outside references, the airplane is subjected to a new set of "laws." For instance, in using the needle and ball, it is sometimes stressed that the pilot doesn't use the instrument to make turns as done in VFR flying. The airplane is "turned" with the rudder, and the ailerons are used to keep the ball centered (which finally results in the same effect but goes at it in a backwards fashion). The flight instruments of today have practically the same presentation as they had twenty years ago, requiring the establishment of a "scan" between the several instruments. The problem is that the arrangement of the flight instruments often varies between airplanes.

It is the job of the instruments to provide a picture so that you can "see" and control the actions of the airplane, and the theory and manner of control are exactly the same as those used with visual references.

Instrument Scan (the Big Picture)

Shown in Figure 4-2 is one established arrangement of flight instruments in which the attitude gyro is the nucleus. This is a good logical arrangement

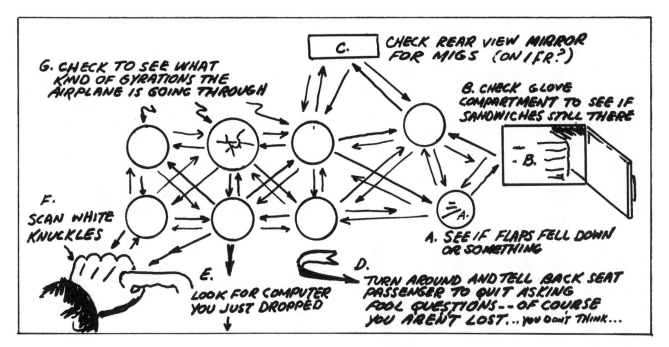

Fig. 4-1. A constant instrument scan is important for safe IFR flight.

and makes for a reasonably simple scan (Fig. 4-2).

Notice that the airspeed is in the upper left-hand corner of the instrument panel. This is a good spot for that instrument because, during take-off or on final approach, the pilot doesn't have to move his scan so far in checking the runway and airspeed. Some older model airplanes had the airspeed indicator at the position of the rate of climb, as shown in Figure 4-2.

Maybe you don't like to use somebody else's idea for the order of looking at the gages, so use your own. You can work this out during your basic instrument flying. The main thing is to cover all of the flight instruments (and later you'll have to monitor the navigation instruments also) and *not* let your

attention stick on one instrument. (This problem of staring at one instrument is quite common in the beginning for instrument trainees.)

The guess is that you'll have less trouble maintaining altitude than with keeping up with the heading. Why? Because you've been flying many hours keeping an eye on the altimeter — which is the only instrument for checking altitude — whereas quite a bit of your directional references have been outside the airplane. As for airspeed in the cruise regime, if the power is set and the altitude kept constant, the airspeed is usually ignored, but the altimeter is a prime instrument. It's funny, but heading can sneak off more insidiously than altitude. You know that you can sense altitude change with your eyes shut. (The change in sound is most often a clue in this regard.) Also, the heading can slip off even when the wings are level, as indicated by the attitude gyro, particularly in a climb where torque is a factor.

In keeping with the idea that maintaining a constant altitude probably will be less of a problem than heading, the diving spiral (and altitude loss) is the final result of neglecting the heading (letting the airplane get into a wing-down attitude with heading change).

Don't think that speed of scan does the trick; this is as bad as staring at one instrument too long. You don't try for a speed record when, say, you start playing the banjo. Sometimes in training, you'll have to slow down because you may be looking but not seeing. If you get excited or scared you may be looking from one instrument to another too fast, so slow down.

The scan can be summed up by stating that the necessary instruments should be checked at the right time. Don't try to set up and memorize the primary bank or pitch instrument for a particular

Fig. 4-2. The "Basic T" arrangement for flight instruments. Note that the turn coordinator (or turn and slip) and vertical speed indicator "fill in the gaps."

maneuver, but fly the airplane by "seeing" its actions through the instruments — and flying the airplane comes ahead of *all* other considerations (voice reports, etc.)

For instance, if you've been flying under the hood at *cruise* power at a constant altitude and suddenly notice that the airspeed is low and decreasing, you don't have to be told that the airspeed is a pitch reference. Without looking (assuming no power reduction), you could say that the altitude is increasing (if things haven't gone too far; but the airspeed will tell you this), and the attitude indicator will show a nose-up attitude. The wing attitude is another matter; they may be level or banked, and you could check this by the attitude indicator or turn *and* slip (T/S) or the turn coordinator (T/C). (If the needle or small airplane is deflected and the ball is in the center, the wings are banked. A needle or small airplane deflection with a skid indication may or may not indicate a *bank* — it would certainly indicate some degree of nose movement.) The heading indicator will also show a change in heading, which could be the result of either a "flat" or a banked turn.

Your job is to get the "big picture," as the advertising people would probably put it, through several sources (the various instruments). The present method of instrumentation is crude and unhandy; the pilot should be able to focus his attention on one area, as if he were viewing the ground through a television screen with data such as altitude, heading, and airspeed superimposed on the attitude — and ground presentation.

Now, you have to cross-check one instrument against the other; the attitude indicator, while the center of attention, only tells attitude (which actually is only what it was designed to do), not performance. Show most pilots who've not had instrument training a picture of an attitude indicator with the small reference airplane (wings level) above the horizon, and they'll say flatly that the airplane is in a straight climb. Not the instrument pilot; he's suspicious. What does the airspeed show? Is it to the point of stall or holding steady at the proper climb speed? Is the altitude increasing, or is the airplane in slow flight condition (nose up and low airspeed in level flight)? How much power is being used? The A/I is very valuable in setting up initial positions for required climb or descent attitudes. You'll find, after practice, that you can approximate the required airspeeds by flying the airplane so that the small reference airplane is a certain number (or fraction) of horizon bar widths above (or below) the horizon. The airspeed is used as a more precise measure, however.

After some time practicing basic instrument flying, you'll figure that you have the situation well in hand. Then comes the introduction to radio navigation. Your "new scan" will have to include copying clearances, changing frequencies, and monitoring the navigation instruments.

One thing you can do is spend some time on the ground, sitting in the cockpit and pinning down the instrument arrangement and position of switches, circuit breakers, and other controls, so that there

is no fumbling in flight. Work out the probable scan for *your* airplane's instrument and radio arrangement, and modify it in flight as necessary.

Probably the first thing your instructor will do before you fly that first flight will be to pick certain working airspeeds for your airplane, such as best speeds for approach, climb, holding and turbulence penetration. He'll then find the general power settings required for performance at these speeds and you will use these numbers for a starting point. (When you start working on the instrument rating in earnest you should get an instructor so that you get started out right.) After picking the working speeds he'll introduce you to the pitch and bank instruments.

PITCH INSTRUMENTS

Figure 4-3 shows the pitch instruments and the instructor will "introduce" them to you. Most pilots who haven't worked on the instrument rating before haven't really looked at each instrument in detail in flight.

Fig. 4-3. The pitch instruments.

The instructor will likely give you the word on the pitch instruments one at a time so that you can check the response of, say, the airspeed, altimeter or VSI to various pitch changes of the attitude indicator. He may discuss the instruments in the following order:

Attitude indicator — you'll probably fly this instrument noting the effects of a one-half or one bar-width pitch change. Most non-instrument trained pilots haven't noticed the effects of very small pitch changes as referenced by the A/I. (Most have used the A/I for bank references and noted that in a climb it might indicate two or more bar-widths up from level flight position — and that's about the extent of it.)

Bank won't be a major factor in your introduction to the pitch use of the A/I, but you should keep the wings reasonably level. As a suggestion, once the airplane is established in straight and level cruising flight, (visually) line up the *top* of the wings of the

46

miniature airplane with the top of the horizon line and use this as the level flight reference. The instructor may cover the other flight instruments, and have you fly a couple of minutes (or longer) using the A/I for pitch and altitude information.

When the altimeter is uncovered you can see how well you handled the pitch problem.

Altimeter — The instructor may have you check the altimeter's response to various rates of pitch change; but because you are more familiar with this one you may not spend as much time with it as the other pitch instruments.

Airspeed indicator — You'll find that the airspeed indicator has more to offer as a pitch indicator than you'd thought. Sure, you've been using the airspeed for climbs and descents, but may have been thinking in terms of keeping the indications ± 5 knots of what you wanted. Now you'll start thinking in terms of one knot variations (or maybe even less) as you see how the airspeed not only acts as a pitch reference for straight and level, but can also help keep the proper pitch for maintaining a constant altitude in a standard rate turn. (Ease the nose up to lose about 3% of your airspeed to maintain altitude in a standard rate turn.)

Vertical Speed indicator — You'll likely be flying this instrument for the first time and will get a chance to fly various rates of descent and climb as well as "flying it" in straight and level hooded flight with the other instruments covered. You can see that controlling the VSI can keep altitude within limits. You'll learn how to correct your rates of descent or climb to get a required rate.

BANK INSTRUMENTS

Figure 4-4 shows the instruments used for bank control.

Fig. 4-4. The bank instruments.

Note that the attitude indicator is the only instrument that gives both pitch and bank information. That's the reason for its location in the "Basic T."

The turn and slip indicator is being replaced in many airplanes by the turn coordinator (Chapter 2). You'll get a chance to practice plenty

of timed turns with whichever instrument your airplane has, during the basic portion of your instrument training.

The heading indicator, like the altimeter for the pitch instruments, tells you the result of your bank control. The instructor will probably have you practice timed turns using either the T/S or T/C with the heading indicator covered. Then you'll look at it to check your accuracy. (The A/I may be used to check the validity of the bank thumb rule for a standard rate turn with the heading indicator covered.)

STRAIGHT AND LEVEL FLYING

On instruments, as in VFR flying, the majority of your time will be spent flying straight and level. If you wander all over the sky, ATC — particularly the enroute radar controllers — will be wondering what is going on up there.

The ability to make transitions from cruise to holding speeds and back is of particular importance. In actual instrument flying, you *may* have to talk and listen on the radio, compute ETA's, and absorb other information *at the same time* that you are making a transition.

Figure 4-5 shows the airplane's position on the power curve and the instrument indications for normal cruise at 65 per cent at a density altitude of 5000 feet.

Remember that the attitude indicator if set properly can be a valuable aid in establishing the nose attitude for straight and level.

After you have pretty well gotten the word on straight and level in the cruise area, check the power setting required to maintain a constant altitude at a speed of 30 per cent above the power-off stall speed (gross weight), given as calibrated airspeed. (This is for single-engine, retractable gear types and light twins.) For single-engine planes with fixed gear, a speed of 20 per cent above the stall is recommended. Find out how much power is required to maintain a constant altitude at this speed. It will be at *approximately* the minimum power point as indicated by A in Figure 4-5.

In making level-flight transitions, lead the power setting by 3 inches of manifold pressure, if possible. Say it takes 22 inches for cruise and 13 inches for the holding speed at a particular weight and altitude. When making the transition from cruise to holding, throttle back to 10 inches. Then as the airspeed approaches the proper value, set up the required 13 inches. For going back to cruise, set up 25 inches (if you can get it at your altitude), and as the airspeed reaches cruise, set the power to 22 inches again.

The manifold pressure and tachometer settings given throughout this book are arbitrary. Your airplane will have its own requirements, and the basic figures will vary with weight and density altitude.

Instrument trainees often have trouble with straight and level flight under the hood (or on actual

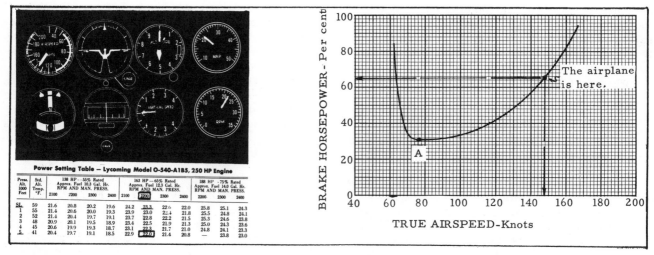

Power Setting Table — Lycoming Model O-540-A1B5, 250 HP Engine

Press. Alt. 1000 Feet	Std. Alt. Temp. °F.	138 HP - 55% Rated Approx. Fuel 10.3 Gal. Hr. RPM AND MAN. PRESS.				163 HP - 65% Rated Approx. Fuel 12.3 Gal. Hr. RPM AND MAN. PRESS.				188 HP - 75% Rated Approx. Fuel 14.0 Gal. Hr. RPM AND MAN. PRESS.		
		2100	2200	2300	2400	2100	2200	2300	2400	2200	2300	2400
SL.	59	21.6	20.8	20.2	19.6	24.2	23.3	22.6	22.0	25.8	25.1	24.3
1	55	21.4	20.6	20.0	19.3	23.9	23.0	22.4	21.8	25.5	24.8	24.1
2	52	21.4	20.4	19.7	19.1	23.7	22.8	22.2	21.5	25.3	24.6	23.8
3	48	20.9	20.1	19.5	18.9	23.4	22.5	21.9	21.3	25.0	24.3	23.6
4	45	20.6	19.9	19.3	18.7	22.3	22.3	21.7	21.0	24.8	24.1	23.3
5.	41	20.4	19.7	19.1	18.5	22.9	22.0	21.4	20.8	—	23.8	23.0

Fig. 4-5. Normal cruise—straight and level at 5000 feet. The instrument indications and the power settings are shown. Note that the I.A.S. is 137 knots for the T.A.S. of 148 knots at 5000 feet (standard) altitude. (The altimeter setting here is 29.92 and the outside air temperature is +5°C.)

instruments) because they don't trim the airplane properly. It's usually a throwback to their VFR flying — they never learned how to trim an airplane properly. Too many students and private pilots, when leveling off for cruise from a climb, take their *hands off* the controls and try to "catch" the proper nose position by juggling the trim control. This takes a lot of time and effort that would be used more wisely in checking the instruments. Needless to say, these people use the same crude technique in making transition from cruise back to slow flight. Instructors have been known to fall asleep waiting for a transition to be completed by the student. The proper method for level-off is to hold the nose where it belongs with reference to straight and level flight, get the cruise power and airspeed established, and *use the trim to take care of the pressure you are exerting on the wheel or stick.* The same applies in slowing up — let the trim take the pressure.

Practice straight and level flight at minimum controllable airspeeds also.

THE CLIMB

The flight instrument for initially establishing the proper straight climb schedule is the attitude indicator. You'll have climb power and will have the wings level in order to get the best prolonged climb.

The climb on instruments is what causes a lot of VFR pilots to get confused again about what makes the airplane go up. There have been cases of pilots who swore under oath that power was the factor that made the airplane climb and that the elevator was only used to control the proper airspeed. *But,* when the instructor told them that the airplane would climb better at a slower airspeed, they *throttled back!* Apparently, no matter how much was said about power controlling altitude, etc., they really didn't believe it.

Going back to the power-required curve, Figure 4-6 shows the power required and power available versus airspeed for a fictitious, single-engine,

retractable gear airplane at sea level and at 10,000 feet.

Thrust horsepower is used because this is the actual horsepower working to fly the airplane. The airplane's climb rate depends on the excess thrust horsepower working to "pull it up." Looking at Figure 4-6, you can see that at one point the excess horsepower is greatest, and this is the airspeed recommended for best climb at the conditions of weight and altitude shown in Figure 4-6. Sea level was used as one altitude for simplicity. The power available is that available at the recommended climb power.

Note that at the extreme ends of the curve the excess horsepower decreases pretty rapidly, which means that the airplane has a zero rate of climb at the top speed and near the stall where all of the horsepower is being used to maintain altitude. This

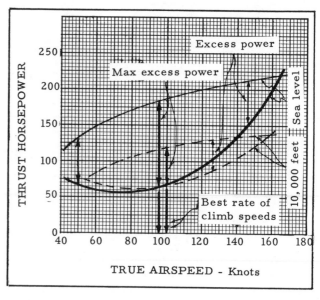

Fig. 4-6. Power-required and power-available (thrust horsepower) curves for a four-place general aviation airplane at sea level and 10,000 feet density altitude.

is not to say the airplane will not climb at all speeds between the two extremes (assuming climb horsepower is being used as shown), but as you approach the limits, the rate of climb will decrease rapidly.

The equation for rate of climb is:

$$\text{Rate of Climb (fpm)} = \frac{\text{Excess Thrust Horsepower} \times 33,000}{\text{Airplane Weight}}$$

This is nothing more than a variation of the basic facts of horsepower discussed in Chapter 3.

For instance, remembering that power is a force or weight moving a certain distance in a certain length of time and that one horsepower is equal to 550 foot-pounds of work per second, or 33,000 foot-pounds per minute, the 33,000 in the equation begins to make sense as we are interested in the rate of climb of the airplane in feet per minute. The excess horsepower is that available to move the airplane's weight upward in a certain period of time. The greater the excess horsepower, the faster that same weight can be moved upward; hence, the greater the rate of climb. The greater the weight to be moved upward for a certain amount of horsepower available, the less the rate of climb. So the equation says that the rate of climb (in fpm) is directly proportional to the excess horsepower available above that required just to maintain altitude (*more* excess hp, more rate of climb), and the rate of climb is inversely proportional to the weight (less weight, more climb, and vice versa).

Suppose you are climbing at the recommended airspeed and power setting but are getting impatient. Believing that the elevators make the airplane climb, you exert back pressure to get a little more climb. The airspeed decreases, and as you can see in Figure 4-6, the excess horsepower (and rate of climb) decrease, and you do the same thing again, so the cycle begins. It ends when the airplane stalls. You've had enough experience by this time to avoid this sort of foolishness.

The dashed lines in Figure 4-6 show the power available and required for the airplane at 10,000 feet standard altitude. The unsupercharged engine loses (roughly) 3 to 4 per cent of its original (sea level) power per thousand feet, and so goes the excess horsepower — and rate of climb. The rate of climb decreases in a straight line with density altitude to the absolute ceiling where the rate of climb is zero. (The service ceiling, you recall, for single-engine airplanes, or multi's with all engines in operation, is the standard (density) altitude at which the rate of climb is 100 feet per minute.)

The maximum rate of climb is found at the airspeed where the maximum amount of excess horsepower is available (about 60 per cent above the flaps-up, power-off stall speed at gross). The maximum angle of climb is found at a lower airspeed where the maximum excess *thrust* is available.

The thrust available to the propeller airplane decreases with increased airspeed for a given power. The best angle of climb is found at the airspeed where the maximum excess *thrust* exists. A plot of

thrust available and drag versus airspeed for the airplane used in Figure 4-6 would show this speed to be lower than that required for max *rate*.

The maximum angle climb is normally found between 10 and 30 per cent above the flaps-up power-off stall speed at gross weight, depending on the airplane. As the name implies, this is the situation where more feet of altitude is gained per foot of forward travel. (The maximum angle climb, however, has a lesser rate of climb than the maximum rate of climb, but you will be more likely to clear obstructions with it.)

When you are cleared to another higher altitude, Air Traffic Control will expect you to climb at the maximum rate until reaching 1000 feet below the new assigned altitude; then, a rate of 500 fpm will be used. So, if you are at 6000 feet and are cleared to 9000, you would climb at your max rate until reaching 8000 and then set up a 500-fpm climb to 9000. Knowing that rate of climb is a function of excess horsepower, should you throttle back at reaching 8000 but still hold the max rate of climb *speed*? (The max rate of climb speed would still give the best rate of climb for any particular power setting or power available, although the absolute maximum rate of climb would only be found at full power.) Figure 4-7 shows two possible procedures in such a case.

In Figure 4-7, you could make a cruise type climb the last 1000 feet and would be making better time along your route. *Or,* you could maintain the max rate of climb speed all the way up and vary the power to get the required 500 fpm the last 1000 feet of climb as shown by the dashed lines.

So, as you can see, the 500-fpm climb can be obtained by increasing speed or decreasing power to get the same rate. (This could lead to the assumption by some pilots that the elevator controlled the change in rate of climb, but *you* know differently; the elevator merely acted as a control to slide you along the power-required and power-available curves until the proper excess horsepower is available to give the required rate of climb.) You don't know how much excess power is needed to obtain 500 fpm, and it wouldn't make any difference if you did. The easiest (but not the most efficient) method is to establish the particular rate of climb and vary the power as needed to maintain that rate — which leads to the following:

Start thinking in terms of power settings and airspeeds for *your* airplane in various maneuvers — a 500-fpm climb is one to keep in mind. Others will be covered as they come up. There are a lot of maneuvers available to get you in the habit of a good scan for *your* airplane.

It would be good to have the power settings required to get climbs of 250, 500, and 1000 fpm (the last climb rate, of course, probably only available at low altitudes). But, practically speaking, 500 fpm is the only one to really have fixed in your mind. The purpose of this book is not to give you data on power settings of various airplanes (it's readily available). This is included to have you understand the principles, so that you could go out to a Curtiss

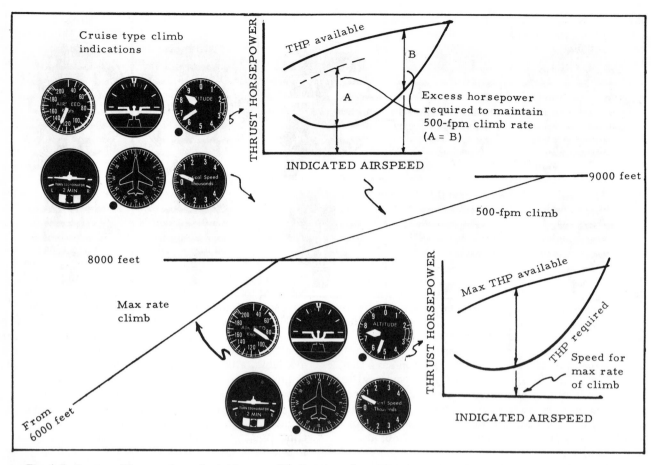

Fig. 4-7. Two possible procedures for setting up a 500-fpm climb (constant power and increase airspeed or constant climb airspeed and decrease power).

Robin or Bleriot and establish the various power settings and airspeeds that would be necessary (after a little experimenting) to obtain the desired performance. You will choose the airspeed that you need for a maneuver and then find out what power is required. For simplicity you might, for instance, want to use the same prop control setting for climbs and descents, so that the only variable would be manifold pressure — if you have a constant-speed prop. That is, if an rpm of 2400 is used for a climb, don't go to the trouble of moving the prop control for descents. If it can be worked without too much loss of performance, it would be well to settle on a *mutual airspeed* for both climbs and descents.

The ordinary vertical speed indicator is a valuable instrument. However, if you had a choice of which flight instrument to lose, this probably would be your choice. It's a fine reference for smooth air and a smooth pilot but is practically useless in turbulence. Under some conditions it would be valuable as an aid to finding the power setting required to get a particular rate of climb or descent, but you'll be introduced to the timed climb later.

This maneuver is good for your scan development and as a method of finding out the amount of power needed to get a specific rate of climb — it will give a rough figure for remembering.

Figure 4-8 shows an example of the first 500

feet of the 500-fpm climb from 8000 to 9000 feet, of Figure 4-7.

The speed for max rate of climb varies with altitude. For airplanes using max rate of climb speeds in the vicinity of 100 knots (I.A.S.), subtract 1 knot per thousand feet for best climb. For max angle climb add 1/2 knot per thousand feet. (It's not quite that much usually, but it's easy to remember.)

Suppose you find that at a certain weight at sea level at 2400 rpm you can get a 500-fpm rate of climb in your airplane by using 20 inches of manifold pressure at the max rate of climb speed of, say, 90 knots. At 5000 feet (same weight), the speed you should hold would be about 85 knots (I.A.S.) for max climb; but check it in the *Pilot's Operating Handbook*. How much power should you use? If you use 20 inches of pressure, that will be too much, because you remember back in Chapter 3 that the manifold pressure must be decreased by roughly 0.25 inches of manifold pressure per thousand feet of *density altitude*. Assuming you're using the same I.A.S., the aerodynamics, or power required, will be the same (in other words, you'll be flying at the same angle of attack, or C_L — and same weight — so the lift and drag and power-required equations will have the same requirements as before). In order to maintain the same margin of excess horsepower and the same 500-fpm rate of climb, at altitude you will vary the

50

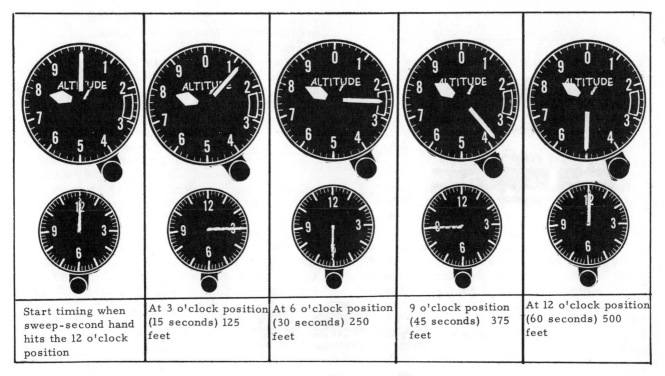

Start timing when sweep-second hand hits the 12 o'clock position	At 3 o'clock position (15 seconds) 125 feet	At 6 o'clock position (30 seconds) 250 feet	9 o'clock position (45 seconds) 375 feet	At 12 o'clock position (60 seconds) 500 feet

Fig. 4-8. The 500-fpm timed climb.

mp as just given. You might check your own airplane's power setting chart to get the amount of mp drop per thousand feet of *density altitude*. You'll find it to be from 0.25 to 0.30 inches of mercury per thousand feet.

As far as weight is concerned, it's already known that more weight requires more power to get the same rate of climb. At 5000 feet, the mp required at the same weight for a 500-fpm climb for the example would be 20 - (5 × 0.25) or 18.75 inches mp. It would be tough to read any manifold pressure gage that closely. Suppose, now, that you are 10 per cent over the weight used to find the original required manifold pressure of 20 inches. This would require an additional 10 per cent (or more) of excess horsepower to have the same rate of climb (referring back to the equation). While you could figure out exactly the power required for every flyable weight at 500 fpm, it's not worth the effort. If the rate of climb is too low, add power. If it is too high, reduce power. The idea is to have some power setting from which to start.

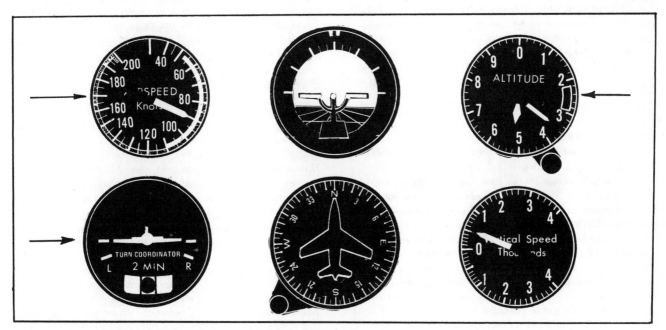

Fig. 4-9. The straight climb (partial-panel). This is also referred to as the "Emergency Panel."

Climbs — T/S or T/C,
Airspeed, and Altimeter

In the Maneuvering by Reference to Instruments part of the flight test, you'll be expected to do turns, climbs, and descents plus straight and level flight using only the T/S or T/C, airspeed, and altimeter for flight instruments. As before, the two climbs of most interest will be the max rate and the 500-fpm climb. It will be most important to have the approximate power setting for the 500-fpm climb in mind.

Figure 4-9 shows a 500-fpm "partial-panel" climb at the best rate of climb speed. The other instruments are included to compare their readings with the three primary instruments (arrows).

The straight climb is a tough maneuver at first. You will have to keep a close eye on the T/S or T/C because this is now your primary direction indicator. The magnetic compass is not much help in the climb, particularly in choppy air. The mag compass can be used to check for large variations in heading, but you remember that on East or West headings acceleration will result in a more northerly reading. This can be used as a check, *but* the needle or small airplane is still the primary heading indicator. If the needle or small airplane is deflected to one side a certain amount, deflect it an equal amount to the other side for what you think is an equal amount of time.

"Torque" will be a particular nuisance in the partial-panel climb. In full-panel climbs where the heading indicator can be used, you'll have a good heading check and can see immediately if torque is giving trouble. You may think that you're correcting very well for torque, but as Figure 4-10 shows, the needle has only to be off a little to cause problems.

The problem is worse for the older type of needle using one width for the standard-rate turn (3° per second). In Figure 4-10A, that's the type that is shown. Figure 4-10B shows that the double-needle-width or doghouse type of needle, being more sensitive, is more apt to give evidence of heading creep.

If you allow the condition as shown in Figure 4-10 to exist for one minute, you'll be 30° off course. If you allow it to go on for six minutes, a 180° turn

would result. You're not likely to allow a constant needle deviation for that long a period, but a little can hurt you.

The 2-minute turn needle (Figure 4-10A) does have one advantage in turbulence in that it doesn't quite have the oscillation of the other type. In rough air, you'll have to average out the needle swings. In very rough air, partial-panel flying can be extremely interesting. (This goes for the turn coordinator also.)

Most instructors do not cage the A/I and H/I but cover them, so that the instruments can be uncovered at any time to check progress or to make a point in training. In such a situation, try climbing partial panel for one minute, starting out on a particular heading. Then uncover the heading indicator to check your heading. Climb another minute and uncover. You can have a good idea from this that careful supervision of the needle can result in a straight path. (It's a good feeling, after several minutes of T/S or T/C and airspeed work, to check the H/I and find that you are still very close to your original heading.)

It's particularly important that you be smooth in partial panel climbs. In general, the same thing applies for leveling off and entering the climb as for VFR work except that you'll have a little more problem with rudder use, altitude and heading during the transitions.

THE DESCENT

In VFR flying, the descent has been only a method of getting down to a new altitude. Sometimes you let down fast (sounds of passengers' eardrums popping) and sometimes slow — and you never really worried about keeping a particular rate of descent. In instrument flying, a controlled rate of descent is one of the most important (if not the *most* important) factors in successful completion of an instrument flight. To make a perfect take-off and climb, to hit every estimate en route, and to hold altitude within 20 feet all the way on an IFR flight are all fine. However, if you don't have the word on a controlled descent, it's all wasted — unless you want to hold until it gets VFR again (this is assuming you have a SAC tanker plane following you around the holding pattern so that endurance is no problem).

The descent is covered right after the climb because there is a tie-in between the two. The climb is the result of excess horsepower, the descent a result of deficit horsepower.

First, look at the descent as a *power-off condition.* Figure 4-11 shows a rate of sink versus velocity curve for our fictitious airplane at maximum certificated weight at sea level and at a higher altitude (dashed line). Basically, the curve is derived by gliding the airplane at various airspeeds at a particular weight and altitude and noting the rate of descent for each speed. Obviously, it would be impractical to glide through sea level in most areas, so other density altitudes are used, and the data extrapolated to sea level. True airspeed is used in Figure 4-11 to give a better look at altitude effects on rate of sink.

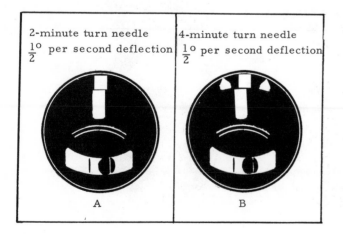

2-minute turn needle	4-minute turn needle
$\frac{1}{2}$° per second deflection	$\frac{1}{2}$° per second deflection
A	B

Fig. 4-10. Two types of turn and slip instruments.

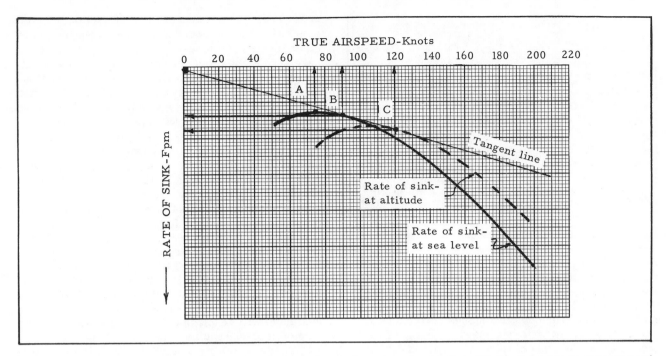

Fig. 4-11. Rate of sink versus velocity curves for a particular airplane at sea level and at altitude. Point A is the true airspeed for minimum sink at sea level. Points B and C represent the true airspeeds for max distance glides at the two altitudes.

This curve looks like the power curve turned upside down. Basically, that's what it is. Point A shows the airspeed for the minimum rate of descent under the conditions stated, or weight and sea level altitude. Incidentally, it might as well be noted at this time that the absolute minimum rate of sink is found at sea level (standard altitude) if all other conditions of weight, airplane configuration, etc., are equal. Then, following this and looking again at Figure 4-11, it can be said that for any *indicated* airspeed the lower the altitude, the lower the rate of sink in the power-off condition (and partial power condition as well, which will be covered later with further explanation). Point B is the maximum distance glide airspeed and is found by extending a line from the origin, tangent to the curve. No other speed will give the shallowest glide angle for the conditions as set up in Figure 4-11. It's interesting to see that the line extended through B also goes through Point C, which leads to the conclusion that the *maximum distance glide ratio* is the same for all altitudes. While at altitude at the speed of Point C, the airplane is sinking faster. It is also moving through the air faster because of a greater true airspeed, and the angle of glide remains the same. The glide ratio is a function of the lift-to-drag ratio (or L/D) of the airplane. L/D varies with angle of attack. At one particular angle, it is the greatest value. If the L/D of an airplane is 9 to 1 at some angle of attack, this is also its glide ratio. Since L/D is a function of angle of attack, or, more properly, C_L (coefficient of lift), it is often expressed as the C_L/C_D ratio because the other factors that affect both lift and drag (dynamic pressure, $\frac{\rho}{2}V^2$, and wing area, S) are the same for any particular situation and cancel out,

leaving only the C_L and C_D as $\dfrac{L}{D} = \dfrac{C_L \, S \frac{\rho}{2} V^2}{C_D \, S \frac{\rho}{2} V^2} = \dfrac{C_L}{C_D}$.

The maximum distance glide is always found at the same C_L.

As far as altitude was concerned, you would see the same I.A.S. for all altitudes to get the C_L/C_D max, or maximum glide distance. But as far as weight is concerned, the I.A.S. for max distance glide must be decreased with decreased weight to maintain the magic coefficient of lift. This is understandable if the lift equation is examined again. Lift $= C_L \, S \frac{\rho}{2} V^2$. If weight is decreased, lift must be decreased in order to maintain the same balance of forces. As only one C_L provides the max C_L to C_D ratio, it will be fixed. The wing area (S) is fixed so that the expression $\frac{\rho}{2} V^2$, or dynamic pressure (or indicated airspeed), must be decreased to maintain the same C_L at the required lower lift.

Following the reasoning given for the climb, the rate of sink might be expressed as: Rate of sink $= \dfrac{\text{Deficit HP} \times 33,000}{\text{Airplane weight}}$. By controlling the amount of deficit horsepower, you can readily control the rate of descent at a chosen airspeed. In making an ILS (Instrument Landing System) approach, the rate of descent must be carefully controlled in order to stay on the glide path.

Figure 4-12 shows part of an ILS Approach Chart for Nashville Metropolitan Airport.

The glide slope is set up at 2.5° to 3° above the horizon, depending on the terrain and obstructions on the final course at a particular airport. The full

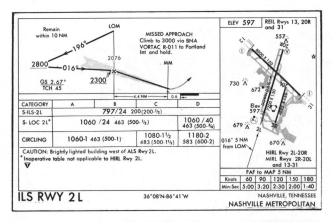

		LOM		ELEV 597	REIL Rwys 13, 20R and 31		

Remain within 10 NM

196°

LOM

MISSED APPROACH
Climb to 3000 via BNA
VORTAC R-011 to Portland
Int and hold.

2800 — 016° — 2076

MM

GS 2.67°
TCH 45   2300

4.4 NM   0.6

CATEGORY	A	B	C	D
S-ILS-2L		797/24 200(200-½)		
S- LOC 2L°	1060 /24 463 (500-½)			1060 /40 463 (500-¾)
CIRCLING	1060-1 463 (500-1)	1080-1½ 483 (500-1½)	1180-2 583 (600-2)	

CAUTION: Brightly lighted building west of ALS Rwy 2L.
° Inoperative table not applicable to HIRL Rwy 2L.

ELEV 597

557
C1
V

730   672

Elev 597

679   016° 5 NM from LOM   670

HIRL Rwy 2L-20R
MIRL Rwys 2R-20L
and 13-31

FAF to MAP 5 NM

Knots	60	90	120	150	180
Min:Sec	5:00	3:20	2:30	2:00	1:40

ILS RWY 2L   36°08'N-86°41'W   NASHVILLE, TENNESSEE
NASHVILLE METROPOLITAN

Fig. 4-12. Part of an ILS Approach Chart for Nashville Metropolitan Airport.

details on the ILS will come later but the idea of descending on the glide slope should be covered here. Chapter 5 has information on rates of descent required to follow various glide slopes (2-1/2°, 2-3/4°, and 3°) at different approach speeds. Seeing in Figure 4-12 that the glide slope at Nashville is 2.67°, a little interpolation between 2.5° and 3.0° would show that you would descend at the following rates for different speeds in order to stay on the glide slope (rounded off to nearest 5 fpm). Incidentally, you can get a *rough* idea of the rate of descent required by adding a "zero" to the speed and dividing by 2 (110 knots + 0 = 1100, divided by 2 = 550 fpm).

Knots	Rate — Feet/Min.
90	425
110	520
130	615
150	705
160	755

These speeds given are for *no-wind* conditions or, more accurately, are ground speeds. If you are indicating 110 knots and are only moving down the glide slope at 90 knots, but are using the descent as required for the 110 knots speed, you'd run below the glide slope and run out of altitude well before getting to the field. If, on the other hand, you are indicating 90 knots on the approach and have a tailwind (and, say, a groundspeed of 110 knots), you'd better descend at that rate required for 110 knots, *or* you'll be above the glide slope and won't get down to make the field. Of course, the problem is that you don't always know the groundspeed, and it may be constantly changing in gusty wind conditions. Your job will be to maintain a constant indicated airspeed and set the power required to maintain the proper rate of descent. In smooth air and no-wind conditions, it seems that you should be able to set the power at a particular manifold pressure and maintain a constant airspeed and fly right down the glide slope — but that's not always the way it works. Looking at Figure 4-12, you see that the airplane descends nearly 1500 feet on the glide slope during the approach. (The "Decision Height" is 797 feet MSL and the descent starts

at 2300 feet MSL, a descent of 1500 feet on the glide slope, but the full discussion of Decision Height and other terms will be reserved for Chapter 8.) If you have set the manifold pressure, say at 13 inches at the start of the approach and then left the throttle alone, you'd have roughly another 1.5 inches of manifold pressure before you got to the minimum altitude. (The manifold pressure will pick up this amount in descending the 1500 feet.) True, in order to maintain a constant power in descending, you would have needed to add about 0.35 to 0.40 of an inch to compensate for the decrease of 1500 feet. In Chapter 3, it was noted that you had to *decrease* the manifold pressure about 0.25 of an inch for the sample airplane for each thousand feet of altitude *increase* in order to maintain the same power at a constant rpm. But the manifold pressure has increased about 1.5 inches, so you'll have a little over an inch more of manifold pressure than is correct for the rate of descent. You'll find that leaving the throttle completely alone on the ILS approach (or any similar approach) will result in trouble in maintaining the rate of descent required to stay down on the glide slope. If the power change is ignored, the airplane must be nosed down and ends up crossing the boundary at a *higher speed* than desired. Figure 4-13 shows what happens to the power-required and power-available curve in such an instance.

You subconsciously lower the nose to pick up the speed at which the required power deficit (rate of sink) is again obtained. You don't know what this speed is, but nose over until the rate of sink is proper. You may have to overdo it slightly for a while to get back down on the glide slope — this is, of course, assuming that you haven't touched the throttle since the beginning of the descent.

An airplane of the type you'll be flying (for a while at least) will be likely to make its best ILS approach at an indicated airspeed of from 90 to 110 knots. What is meant is that you are not allowed this variation, but that you're not likely to pick less than 90 nor more than 110 knots for the approach airspeed. A too slow approach speed delays other traffic and in turbulent air can result in marginal control. A too fast airspeed may result in excessive floating. This may not be too much of a problem for a combination of your type of airplane and airports with ILS approaches. But this is a poor habit to get into for a guy who may be flying jet airliners after a while.

For instance, for a Piper Aztec, a speed of 125 mph is often used for an ILS approach for several reasons:

1. This is 110 knots, a figure easily interpolated on Approach Charts (actually, it's 109 knots, but this is close enough for practical purposes).

2. It is a speed that allows for a good margin of control in turbulent air.

3. It is fast enough to help expedite traffic flow and yet not so fast that upon reaching ILS minimums of 200 feet and 1/2 mile (with the airport in sight, naturally) the power can be reduced to idle and the airplane landed without excessive floating.

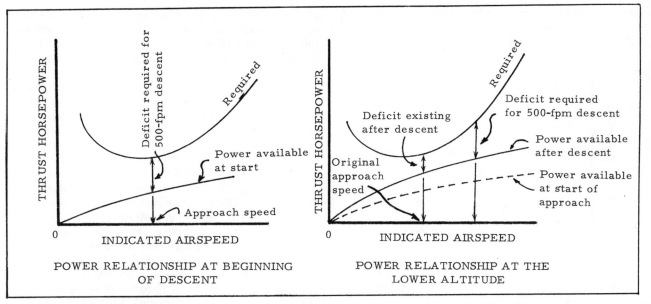

POWER RELATIONSHIP AT BEGINNING
OF DESCENT

POWER RELATIONSHIP AT THE
LOWER ALTITUDE

Fig. 4-13. Power-required and power-available curves for an airplane during an ILS approach. The throttle was set at the beginning and not changed. It's a vicious cycle, because if the pilot shoves over to a speed to pick up a 500-fpm rate of descent, the new airspeed will require a greater rate of descent in order to stay on the glide slope.

4. 125 mph is the maximum full-flaps speed, so that when visual contact is made with the field, the flaps can be extended without problems.

5. Because it is the max speed for full flaps, it is the top of the white arc on the airspeed indicator (Fig. 2-2) which makes a quick reference, or "how goes it," for an aid in your scan on the glide slope.

Figure 4-14 shows *one* technique, and you'll note that there is a period of level flight as you fly to intercept the glide slope. (Actually you may have about a couple of minutes to lose a hundred feet or so on some approaches, but for practical purposes, it would be level flight.) One procedure would be to fly this part of the approach at the final approach speed, using enough power to maintain a constant altitude at this speed in the clean condition. Then, as you approach the glide slope, extend the gear, maintaining the approach airspeed. For some airplanes of the type you'll be flying, extension of the gear is exactly what is needed to get the desired rate of descent of around 500 fpm.

Of course, the throttle must be retarded slightly as the plane descends, for the reasons cited earlier. The theory of Figure 4-14 is shown in Figure 4-15.

Other airplanes may require slight changes in power as the glide slope is approached, or it might be better to use partial flaps, as well as gear, all during the approach. For airplanes with fixed gear, the power will have to be decreased because on the level part of the approach it will already be in the "down and dirty" condition as far as the gear is concerned. Instead of increasing the horsepower required, you'll throttle back and *decrease* the horsepower available and get your horsepower deficit that way. Starting at the end of the level portion of the approach, you can set up a deficit horsepower for the required rate of sink.

Too many new instrument trainees try to fly only the glide slope instruments on the approach. This makes for a considerable amount of overcontrolling. They have no idea of what combination of power and airspeed is required to get the right rate of sink and try to outguess the glide slope needle by throttle

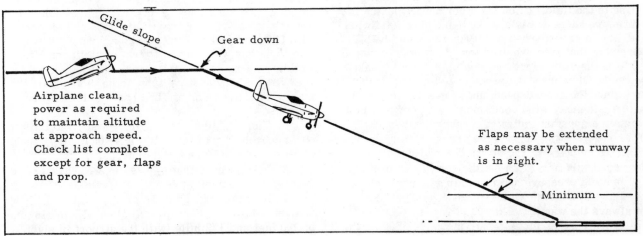

Fig. 4-14. As the airplane comes up on the glide slope extend the gear. The glide slope angle is exaggerated.

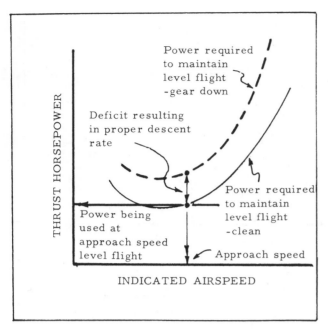

Fig. 4-15. Power-available and power-required curves for the two conditions shown in Figure 4-14.

jockeying and elevator flapping. The final result is a power variation from idle to full power and back to idle — with the glide slope needle going up and down like a band leader's baton. Figure 4-16 shows the instrument indications on an ILS approach.

The glide slope is only 1.4° thick. That is, from full-up to full-down deflection of the needle on the instrument means only 1.4° of travel through the glide slope. At 1/2 mile out, the 0.7° up or down means that you'd have a margin of about 35 feet up or down to keep the glide slope needle in the instrument.

Select your procedure, find a reference power setting for the descent, and use that as a quick and dirty figure. Obviously, that power setting will vary with conditions, such as variations in weight and density altitude, but you'll have something with which to start. Your instructor probably will have available all of the power settings for the various rates of climbs and descents that you'll be working with.

If you are flying a twin, practice some 500 fpm descents with one engine throttled back to zero thrust, so that you'll have a good idea of the power setting required with gear up or down. You might also shoot a few ILS approaches with a simulated engine failure.

Descents — T/S or T/C, Airspeed, and Altimeter

You should practice straight descents using the "partial-panel" instruments. The straight descent

Fig. 4-16. Instrument indications on an ILS approach. The attitude of the airplane as shown by the gyro is more nose down than would be expected in an actual situation. The airplane is slightly below the glide slope.

probably will be easier to handle than the climb because of the lack of torque. But if the heading gets out of control, it can still end up as a diving spiral.

The same general requirements of the partial-panel straight climb apply here. The needle or small airplane (and ball) is the primary directional and wing attitude indicator. If the indicator deviates to one side, deflect it (with coordinated controls) an equal amount to the other side for what you think is equal time. To try to chase what you think is the proper rate of descent with the throttle is a surefire way for everything to go berserk. If you forget everything else in this chapter, remember that even a barnyard power setting and airspeed to shoot for will keep the airplane from getting away if your attention is momentarily diverted — such as may happen on the glide slope when you have to report the outer marker, or the tower calls and gives information. If you are *only* flying the glide slope needle and using all kinds of power settings and airspeeds to do it, any distraction from the G/S needle would mean losing it — and the possibility of having to take it around for another try. With the right power and airspeed, you still might be off if distracted *but* not enough that you can't get things back on course.

Practice Maneuvers for Climbs and Descents

There are several good practice maneuvers that will help your transitions to and from climbs and descents.

Don't make the mistake some pilots make in overemphasizing the importance of the maneuvers. They are only intended to show you the fundamentals of instrument flying and are not an end in themselves. You can do beautiful practice patterns all day long and wouldn't be a foot closer to your destination. Don't just fly the maneuvers mechanically. *Remember what power settings it takes to get the 500* (or 250 or 1000) *fpm rate of climb or descent — and then use this knowledge*. Too many guys have the numbers for power settings and airspeeds for all kinds of flight configurations and maneuvers but forget this information under actual conditions.

There are all kinds of patterns, and you can make up your own. Figure 4-17 shows one that might be used to smooth out straight-ahead climbs and descents.

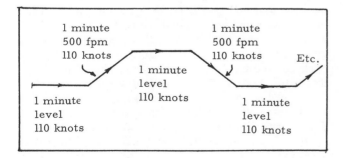

Fig. 4-17. A maneuver for smoothing out climbs, descents, and transitions thereto. Pick your own speed if you don't like that one.

The pattern in Figure 4-17 can be varied to fit your airplane. The descents can be made clean at first and then in approach configuration, or at least, in the configuration you plan to use for the ILS or other type letdowns. You can later use a 2-minute climb and descent at 500 fpm and also get a brief idea of a 1000-fpm change in altitude. Look at the power effects throughout the airspeed range, but don't waste your time practicing oddball and impractical combinations over and over. After you get proficient at full-panel patterns, practice the same patterns using T/S or T/C, airspeed, and altimeter. You might practice Figure 4-17 without any straight and level flight, going directly from climb to descent and vice versa. This will help you to pin down exact power settings.

Other good exercises under the hood would be to set up simulated traffic patterns at a safe altitude using gear and flaps (changing flap settings as might be done on approach) and maintaining constant altitudes or rates of descent as applicable. Practice missed approaches (go-arounds) adding climb power, pulling the gear up and flaps up in increments. This will help your scan. Also practice this with T/S or T/C, airspeed and altimeter to smooth out your procedures.

THE TURN

When you are under positive (radar) control, you'll be expected to make all turns standard rate, unless otherwise requested. This way the controller will be able to know, for instance, when to start you turning onto final to intercept the ILS. If you rack the airplane around one turn and sneak around on the other, the poor guy always will be trying to outguess you.

A standard-rate turn is 3° per second, or 180° per minute. For faster airplanes (jets), a half-standard turn rate is used (1.5° per second) so that the bank won't be too steep for good control. That's one thing you'll learn about instrument flying, particularly flying partial panel; the steeper the bank, the easier it is to lose control of the airplane. A good exercise for you in some of the clean, retractable gear airplanes in current use is to do some turns on partial panel at double, or triple, standard rate at about cruise airspeed. You'll find that at first it requires a great deal of concentration to maintain altitude and avoid a spiral at steep banks.

The thumb rule for angles of bank for a standard-rate turn given back in Chapter 2 still stands.

The airspeed used for the thumb rules is true airspeed rather than indicated or calibrated. But for practical use you can use indicated — for lower altitudes, anyway.

Knowing the angle of bank required will give you another check of the T/S or T/C. Before practicing basic instruments, check the calibration of the needle by setting up a standard-rate turn and checking as follows:

Set up the indication for a standard-rate turn in

either direction on the needle and ball, using a single needle width or a "doghouse" as required for your particular instrument. Turn for one minute, and then roll out. Find out the number of degrees turned in that time. If, say, 150° instead of 180° -- and you had held the proper needle deflection all the way — then you should hold about 1 1/5 times the indicated needle deflection, in order to get the required 180° in one minute — $\frac{180}{150} = 1\frac{1}{5}$. If you had turned 210° in the minute, you should then use 6/7 of the earlier needle deflection to get the proper rate of turn. To double check, turn the other direction and time again.

The reason for practicing timed turns is to have you able to do them almost unconsciously. You should be able to do them without actually having to use the clock. It's a good maneuver for insuring that the clock is included in your scan. In practical use, you'll have a specific heading to turn to and won't waste your time checking to see if your turn is exactly 3°, or 3.001°, per second. You'll have clearances to acknowledge, navigational aids to check, and other things that will be of more vital interest. You must be precise, but don't get so involved in one aspect of instrument flying that others are neglected. You could get so engrossed in making a perfect timed turn that you'd turn right past the heading needed to get on the localizer, or as given by the radar controller.

Suppose you were directed by the radar controller to make a 180° turn to the right. In an actual situation, you would set up a standard-rate turn and roll out on the proper heading. It's likely that you *won't* make a perfect timed turn and will come out off heading if you just followed the clock blindly. You would roll out on the required heading, whether a few seconds late or early. After all, that's the purpose of the turn; to get you to a specific heading, and a few seconds either way won't make any difference, whereas a few degrees would. A thumb rule for rolling out on a specific heading is to start rolling out at a point one-third the degree of bank. In other words, if you're in a 30° banked turn, start the roll-out about 10° before the desired heading is reached. It is unlikely that your standard-rate turns will be quite this steeply banked at this point of your training, as that would require a true airspeed of about 200 knots. As a round figure, 10° wouldn't be bad for a lead in rolling out for *all* standard-rate turns at cruise for airplanes of the type in which you'll be training.

As a little review of theory, Figure 4-18 shows the power required for a particular airplane to fly at a constant altitude (sea level) straight and level and in a 60° banked turn.

Point A in Figure 4-18 shows that a straight and level flight at 65 per cent power a certain speed (150 knots) results. In the 60° banked turn B, in order to fly at a constant altitude, a lower speed must result. Remember, a constant altitude results when the power available equals the power required. Of course, the power-required curves for banks of 1° to 59° would fall between those shown.

If, for some reason, you figured you also *had* to

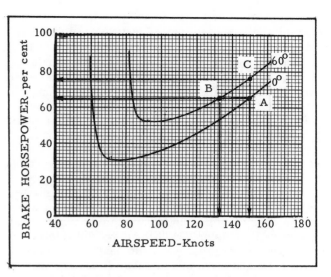

Fig. 4-18. Brake horsepower required to maintain a constant altitude for straight and level and a 60° banked turn. Brake horsepower is used because the pilot would be thinking of this in setting power.

maintain exactly the same airspeed in the turn as in straight and level flight, this could be done only by increasing the power available to the value shown by the higher line (which intersects the power-required curve top (Point C). Now you will be able to maintain a constant altitude at the greater airspeed. Did adding power increase your airspeed? Not at all. You could have increased the airspeed merely by relaxing back pressure and letting the nose drop. The added power let you fly at the higher airspeed *and still maintain a constant altitude.*

For the shallower banks used for standard-rate turns, the airspeed loss is negligible. For steep turns, it is a good idea to add power. This not only allows you to maintain altitude at a higher airspeed but also lowers the stall speed slightly. This spreads the two areas (stall and your flight speed) a little more than would be so otherwise and may help avert an unexpected stall in a tight situation.

Climbing turns are another matter. If you are turning while executing a maximum-rate climb, the climb rate will decrease slightly, and there's nothing you can do about it, as you are already using all of the legal power. For 500-fpm climbs, the decrease in climb rate due to the turn can be offset by slightly increased power (unless your plane is so underpowered that 500 fpm *is* the max rate of climb with climb power and wings level, so that there's no reserve). Where, for instance, 20 inches and 2400 rpm give a 500-fpm wings-level climb, in a climbing turn perhaps 20.5 inches might be necessary to maintain that rate. Look at Figure 4-19 at the maximum excess horsepower point, as shown for a wings-level at 500 fpm and a climbing turn at the same rate. Since thrust horsepower is the criterion for climb performance, Figure 4-19 uses THP and indicated airspeed.

At any rate, you should normally have no reason to exceed a standard-rate turn at any time in the climbing turn. The steeper the bank the less the

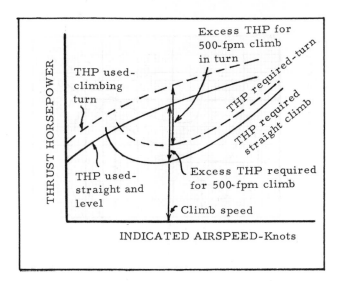

Fig. 4-19. Excess horsepower for the wings-level climb and the climbing turn. (Must be the same value for the same rates of climb.)

rate of climb for any given power setting. The same basic idea applies to the descending turn. If you've picked your speed and have a power setting for a wings-level descent of so many feet per minute, the turn should require more power to maintain that same descent. Where it took 12 inches for a 500-fpm descent in the wings-level clean condition (probably another 2 inches would be required with the gear down), the turn would require perhaps 12.5 inches for the same rate of descent. You will practice straight climbs and descents, plus climbing and descending turns, as a part of your introduction to basic instrument flying. Practically speaking, it won't make that much difference for you to worry about changing power in *shallow* climbing or descending turns. Such exercises as just covered are intended to help you become a more precise instrument pilot.

The Turn — T/S or T/C, Airspeed, and Altimeter

The timed turn will be of value when you have no heading indicator or equivalent. Again, a steeply banked turn — particularly using needle-ball, airspeed, and altimeter only — will radically increase chances of loss of control, so at no time make your turns steeper than standard rate.

In using the needle and ball or turn coordinator, some pilots say to lead with the rudder, deflecting the needle, then following with the ailerons. If this is the way you make all your turns — VFR and full-panel — then go ahead. If not, then don't throw in a new technique here. It is recommended that you use simultaneously coordinated controls so that the needle is deflected the proper amount and the ball is centered. If the ball is centered, the turn indicator will give an approximate picture of the wings' attitude. If the ball is centered and the indicator is deflected, the wings will be down in that direction. In instrument flying, in theory at least, the ball should *always* be centered — there are no maneuvers in normal, or predictable, IFR that would require slipping or skidding.

As you know by now, adverse yaw at the beginning of the roll-in may cause the needle temporarily to indicate a turn in the opposite direction. This is one of the reasons for leading slightly with rudder when using the turn and slip as a turn reference, as opposed to a turn coordinator. Since such action is expected, it shouldn't cause any confusion if you are slightly late in rudder action. In rough air, you'll have to average the needle swings to maintain a standard rate of turn and will be likely to overdo your roll-ins, roll-outs, and "averaging" of the needle swings at the beginning of your instrument training.

As a possible crutch in an emergency, another method of turning to an approximate heading is the use of the turn and slip and magnetic compass — using the "Four Main Directions" method.

In a shallow banked turn the compass is fairly accurate on the headings of East or West and lags by about 30° as the plane passes the heading of North and leads by about 30° as the plane passes the heading of South. The exact lag and lead will have to be checked for your situation, which will include latitude and other variables. For illustration purposes here it is assumed to be 30°. This "northerly turning error" affects the compass while the plane is turning.

Assume that you are flying on a heading of 060° and want to make a 180° turn. The desired new heading will be 240°. By turning to the left, you will have a cardinal heading (West) reasonably close to the new heading. A standard-rate turn is made to the left, but no timing attempted. You will be watching for West on the compass, and as the plane reaches that heading, you will start timing. The desired heading is 30° (270° minus 240°) past this, and the 30° will require ten seconds at the standard rate. The timing can be done by counting "one thousand and one, one thousand and two," and so forth, up to ten, at which time the needle and ball are centered (Fig. 4-20).

A turn to the right could have been made, realizing that the nearest major compass heading (South) will be 60°, or twenty seconds, short of the desired heading of 240°. In that case, you will set up a standard-rate turn to the right and will not start timing until the compass indicates 210°. Remember that the compass will lead on a turn through South and will be ahead of the actual heading by about 30°, as a round figure (Fig. 4-21).

When the 210° indication is given, the twenty-second timing begins, either by the sweep-second hand of the aircraft clock, your watch, or by counting.

There are several disadvantages to this system, the major one being that it requires a visual picture of the airplane's present and proposed heading and mental calculations. You may not have time or may be too excited for a mental exercise at this point. Also, in bumpy weather, the compass may not give accurate readings on the four main headings.

The main advantage is that reasonably accurate turns may be made without a timepiece, as any selected heading will never be more than 45°, or fifteen seconds, away from a major heading.

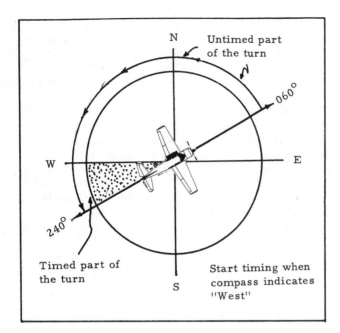

Fig. 4-20. A turn to the left using the four main directions method.

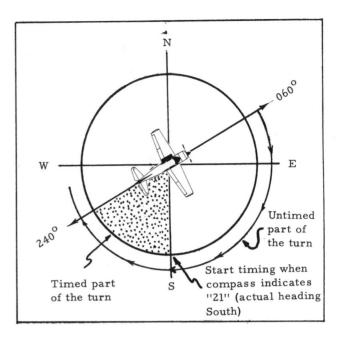

Fig. 4-21. A turn to the right using the four main directions method.

You should practice steep turns both full and emergency or partial panel. You'll see in such turns with the turn needle pegged that your scan will have to be well-developed or you'll end up in a spiral. It's also a graphic display of why you want to keep turns standard rate (and maybe even less when using the needle-ball, airspeed and altimeter).

PRACTICE MANEUVERS FOR THE FOUR FUNDAMENTALS

One maneuver that combines all the expected flight patterns is shown in Figure 4-22. You can make up others to suit your requirements. Do these using the full panel at first and, of course, under the hood with a safety pilot on board.

After problems on your timed turns and descents are pretty well ironed out, a realistic practice procedure is to simulate a holding pattern, as shown in Figure 4-23. At first, you'll do these without using a holding fix or electronic aids. Later in your training you'll practice holding over a VOR, radio beacon or intersection. Make sure the wind doesn't drift you over into the next county when practicing without electronic aids. Your safety pilot can keep an eye out for this.

For realism, you might make two level circuits and then descend 1000 feet (at 500 fpm) sometime during the next one. Make two more circuits and then descend another 1000 feet at 500 fpm. You'll note that the holding pattern in Figure 4-23 takes four minutes for a complete circuit. Descending at 500 fpm would mean that part of the descent would be in the straightaway and part in the turn. In an actual situation of being shuttled down from a holding pattern, such as would be the case for airplanes being

stacked over the approach fix, you will commence the descent immediately upon clearance from Air Traffic Control. You would not wait for the sweep-second hand to conveniently reach the twelve-o'clock position — you descend *immediately*. This means that a knowledge of the power setting required for a 500-fpm clean descent is important. You'll be holding in the clean condition (don't require extra power by having the gear or flaps down — if the gear is retractable, that is). You might also practice setting up the conditions of manifold pressure, rpm, and mixture, as would be used to conserve fuel during holding.

Practice holding patterns, using one-minute legs for both straightaways. You'll find that an actual holding pattern may require extending or shortening the outbound leg timewise to take care of wind for a one minute inbound leg. Figure 4-24 shows what might be the case.

Holding patterns as applied to an actual IFR situation will be covered later in this book. You should do some *practice* of them using only the needle-ball, airspeed, altimeter, and magnetic compass after becoming proficient using the full panel of instruments. You may ease into simple holding patterns, using a fix (VOR, etc.) at the end of these sessions.

Here are some maneuvers that can help build your confidence (in every case the manifold pressure gage and/or tachometer will be part of your scan):

Straight and level —

1. Cover the turn and slip, H/I and airspeed. Fly for two minutes, keeping the wings level with the attitude gyro and maintaining a constant altitude. Uncover the H/I and check your heading change (if any) after the two minute period. (Normal cruise — 65% power.)

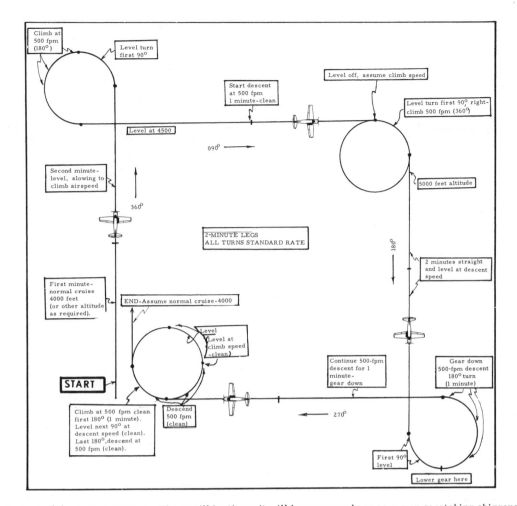

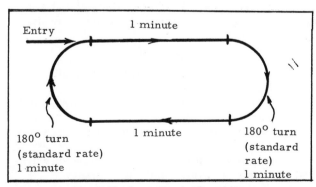

Fig. 4-22. A simple (?) practice pattern. There will be times it will keep you as busy as a man scratching chiggers but will help establish a good scan for your particular airplane.

2. Use only the H/I and altimeter to fly straight and level. You will find that the H/I is a very important aid in keeping the wings level. If the direction changes, a wing is down. You'll soon find you can use the H/I for wing attitude information almost as well as the attitude gyro. (Normal cruise.)

Note that the airspeed is less important for straight and level than for other maneuvers, but it can still be valuable as a check for trends of altitude loss or gain.

The vertical speed indicator will be seen as a good indicator of trends but, don't "chase" it.

Fig. 4-23. A practice holding pattern.

Turns —

If your airplane has a turn coordinator you'd substitute that instrument for the turn and slip in the maneuvers to be discussed.

1. Make a constant-altitude standard rate turn using attitude indicator, airspeed and altimeter only; setting up the bank as required to get a 3° per second turn for a particular airspeed (Normal cruise—65% power). Start out on a pre-chosen heading and make timed turns of 90°, 135°, 180°, etc. Uncover the heading indicator and check your accuracy after each turn.

2. Turn to predetermined headings using the H/I, airspeed and altimeter only. You can approximate the correct rate of turn by the rate of direction change. With a little practice you can set up a rate of turn that's close to standard rate. The main thing is to keep your scan going and not let the rate of direction change be too great (bank too steep). If this is the case, shallow the bank and get the turn back to a more reasonable rate. The point here is that in actual flight, if you have lost the attitude indicator and turn and slip (an unlikely loss-combination to be sure), a serious situation exists. The main idea is to keep the airplane under control and roll out on pre-determined headings. An exact standard turn rate is less important than those two requirements.

61

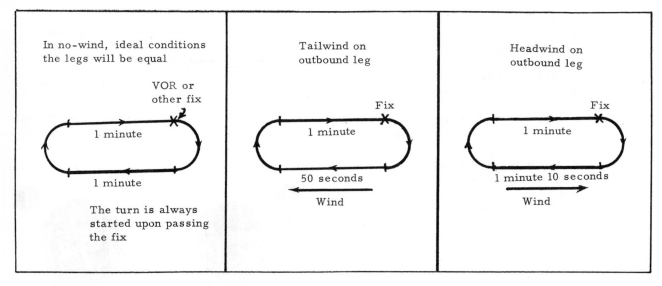

In no-wind, ideal conditions the legs will be equal

VOR or other fix

1 minute

1 minute

The turn is always started upon passing the fix

Tailwind on outbound leg

Fix

1 minute

50 seconds

Wind

Headwind on outbound leg

Fix

1 minute

1 minute 10 seconds

Wind

Fig. 4-24. Some possible variations of time for the outbound legs in the holding pattern for different wind conditions.

Practice the above at slow-flight speed and expected holding speed. Use the airspeed, and turn and slip only, for some turns.

Descents —

1. Make straight descents at 500 fpm using the heading indicator, airspeed, altimeter and vertical speed indicator. Use the power setting worked out earlier to get the 500 fpm descent at the chosen approach speed. Notice that the proper combination of power and airspeed gives the required rate of sink. In smooth air you can easily check the relationship of power, airspeed and descent by covering up the airspeed indicator. By setting power and carefully keeping the 500 fpm descent (with the elevators) you can uncover the airspeed indicator to find that the approach speed is being held. You'll see that by using the H/I, altimeter, vertical speed and power instruments, in smooth air you can make pretty accurate descents without airspeed, attitude gyro or turn and slip.

2. Do the same as above, using attitude gyro and then turn and slip or turn coordinator to replace the H/I. Check your heading (uncover the H/I) after a couple of minutes of straight descent.

You might cover the altimeter, also, during one of your descents and see what can be done with the H/I, power setting, vertical speed indicator and timing. (After a 500 fpm descent at a pre-chosen number of minutes — don't fly into the ground — uncover the altimeter and see how close you are to the correct altitude.)

Practice descending turns using the flight combinations just discussed.

Climbs —

1. Make climbs at recommended climb power at the recommended airspeed. Check the approximate rate of climb at a "medium" altitude (4000-6000 feet MSL for most of the U.S.). Say it's 700 feet per minute at the chosen altitude, power setting and airspeed.

Using the heading indicator, altimeter, vertical speed indicator, plus proper power, set up a straight climb (turn and slip, airspeed and attitude gyro covered). You can see that, by carefully (easy!) maintaining the expected rate of climb, the airspeed can be held reasonably close to the recommended value.

2. Make climbs and climbing turns using the airspeed and altimeter plus turn and slip or attitude gyro.

Summed up, you can practice the following maneuvers using *only* the flight instruments listed (assume that the power instruments are always available):

Straight and Level

1. Attitude indicator, airspeed and altimeter
2. H/I and altimeter
3. Attitude indicator, turn and slip and altimeter
4. T/S or T/C and airspeed indicator.
5. H/I and attitude gyro — use the vertical speed indicator as a trend instrument.

Turns

1. Level standard rate turns using attitude gyro (proper bank and timed turn) and altimeter to a pre-determined heading
2. Level standard rate turns using T/S or T/C and altimeter to a pre-determined heading
3. Level turns to headings using H/I and ASI. (Pitch up to decrease airspeed 3% from S&L value in standard rate turn.)

Descents

1. Straight descents — H/I, airspeed, altimeter and vertical speed (500 fpm)
2. Straight descents — T/S or T/C, airspeed and altimeter (500 fpm)
3. Straight descents — attitude indicator, vertical speed and altimeter
4. Descending turns to predetermined headings and altitudes — H/I, airspeed and altimeter
5. Descending turns to predetermined headings

and altitudes — attitude indicator (timed turn), airspeed and altimeter.

Climbs

1. Straight climbs — H/I, airspeed, altimeter and vertical speed indicator (max rate climbs)
2. Straight climbs — T/S or T/C, airspeed altimeter and vertical speed indicator
3. Straight climbs — attitude indicator, vertical speed and altimeter
4. Climbing turns to predetermined headings and altitudes — H/I, airspeed and altimeter
5. Climbing turns to predetermined headings and altitudes — attitude indicator (timed turn), airspeed and altimeter.

You don't want to spend hours of practice at these maneuvers, but they will show you the relationships between the various flight instruments and power settings. Your confidence in your ability to fly the airplane with some flight instruments out of action will increase radically.

RECOVERIES FROM UNUSUAL ATTITUDES

Turbulence or other outside factors could result in the airplane getting into such an attitude that control could be temporarily lost, as well as tumbling the attitude indicator and heading indicator. If this happens, you are left with the job of using the needle and ball, airspeed, and altimeter for immediate recovery.

Remember, for the average, less expensive heading indicators, the tumbling (or spilling) limits are 55° of pitch or roll; the attitude indicator will tumble at 70° of pitch and 100° of bank.

After these two instruments have tumbled, you'll recover with the needle and ball, etc., and reset the two that failed you. Naturally, all airplanes have caging and setting knobs on the H/I, but some airplane makes and models do not have a resetting knob for the attitude indicator. If it has tumbled, you'll have to fly *straight and level* for several minutes before it will re-erect itself. This is a decided disadvantage in turbulent air, and you may have to "needle-and-ball it" for some time. However, you'll find that getting the H/I back in action can be a great help.

The two most common results of loss of control are the power-on spiral and/or the climbing stall. The climbing stall usually turns into a power-on spiral and, sometimes vice versa, but you should catch it before this happens.

Because the attitude indicator could be tumbled in more radical attitudes, its information may not be trustworthy. If you are the kind of a guy who stares rigidly at the attitude indicator without cross-checking the other instruments — if the attitude indicator is the *only* instrument as far as you are concerned — then you're in for a harsh awakening when you try to fly a tumbled attitude indicator. Some beginning instrument pilots are so wrapped up in the attitude indicator that if through mechanical failure it

shows the airplane as rolling over on its back, the student would roll the airplane over on its back to "turn right side up" again, without bothering to cross-check other instruments.

The Power-on Spiral

Figure 4-25 shows what the partial-panel instruments would be showing in a power-on spiral — after the attitude indicator and H/I have tumbled.

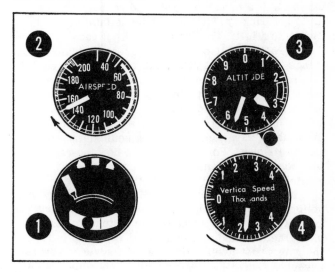

Fig. 4-25. A power-on spiral to the left. The H/I and attitude indicator are not in business.

Actually, it would take some doing to tumble the attitude indicator in the spiral — you'd have to be banked past vertical to do it — but perhaps the gyro tumbled earlier when you first flew into that thunderstorm and did the barrel roll.

Look at the indications of the instruments that are still functioning (Figure 4-25).

1. *Turn and Slip or Turn Coordinator* — The needle (or small airplane) will show a great rate of turn. The ball may or may not be centered.
2. *Airspeed* — High and/or increasing.
3. *Altimeter* — Showing a loss of altitude, probably rapid loss.
4. *Vertical Speed Indicator* — A high rate of descent.

It won't be any help for you to try to pull the nose up without leveling the wings. You can impose very high load factors at high airspeeds and could cause structural failure. The following steps are recommended:

A. *Reduce power.* This is particularly important for the fixed pitch prop, which may be turning up over the red line in the spiral. For the constant speed prop it will cause the blades to flatten and increase drag.

B. *Center the needle and ball or turn coordinator through coordinated and simultaneous use of the aileron and rudder.* As indicated by Figure 4-25, you would use right aileron and right rudder. Don't use any gimmicks; the instruments tell you the airplane is in a spiral dive. Under visual conditions, you would use coordinated controls in such a maneuver, so do it here. *Misuse of the aileron or rudder at high airspeeds can impose large twisting moments on the wings and/or can cause failure of the vertical fin by excess yaw.*

As the wings are leveled, it is most likely that the nose will start to rise sharply because of your original trim condition and/or because of back pressure you've added unconsciously. *Do not allow too quick a pull-up.* True, you have been, and are, going downhill at a prodigious rate and your every instinct is to get back *up,* but this can be as fatal as if you flew into a hill. You'll have to take it easy.

A *rolling* pull-out imposes greater stress on the wings than the straight one (all other factors equal). A g meter, or accelerometer, would show the same number of g's being pulled, but one wing will have more stress imposed on it than the "average," as shown on the g meter. If you are already pulling the limit as an "average," that wing could decide to go out on its own. For speeds *well* below the red line and for a nonviolent pull-up on your part, the rolling pull-out will expedite the recovery.

In an actual situation you'll tend to be jerky in your responses during the recovery and overcontrolling may be a factor. One problem students have in using the turn and slip in a spiral recovery is rolling the wings past level, particularly if adverse yaw is present. Okay, the answer is to be coordinated. Adverse yaw will be less a factor at high speeds than low and will be more of a problem on the recovery from an approach to a stall.

C. *Check the airspeed.* As the wings are being leveled, the nose will start to come up of its own accord. In fact, it's unlikely that any measurable amount of back pressure will be needed. You'll have to judge this for yourself, however.

In the recovery, an approximation of level nose attitude is reached when the airspeed makes its first perceptible change. In the spiral dive, the airspeed will either be increasing or will be steady at some high speed. As the nose moves up in recovery, the airspeed at some point will (a) stop increasing, or (b) start to decrease if it was steady. This hesitation or decrease occurs when the nose is approximately level. So, at this first sign of airspeed change, relax back pressure (or apply forward pressure, if necessary) to keep the nose from moving further upward.

Don't pull the nose up until cruise (or climb airspeed) is reached in the third major step of recovery. The nose will be so high that control will be lost (again). It's not inconceivable that you might end up on your back at the top of a loop.

Don't try to rush the airspeed back to cruise. It will soon settle down of its own accord if you stop the altimeter and have the power back at cruise.

D. *After the "rough" leveling is done by reference to the airspeed change, you will "stop the altimeter."* You'll be at some particular altitude when you make the rough recovery to level flight using the airspeed. *Try to keep it,* or at least, don't go below it (it's more likely that you will tend to climb above it

because of excess airspeed). Here you have two choices: (a) You can "pin down" that altitude and hold it as the airspeed eases back to cruise (assuming you have cruise power back on), or (b) you immediately can start a climb back to the original altitude as you recover. The second alternative is usually the best in that the loss of altitude may have been such that obstructions, such as mountains, television towers, etc., could be a hazard. Also, if you lost control while enroute on an IFR flight, you just might be down in somebody else's assigned altitude, and that's no place to be for any length of time. *Ease up to the airspeed and attitude; don't overdo it and lose control again.*

After the compass has settled down somewhat, set the heading indicator; even a quick and dirty setting will be better than nothing. You can make fine adjustments later. If the attitude indicator has a caging knob, cage it and uncage it again when the airplane is in straight and level flight as shown by the basic instruments. If the attitude indicator does not have a caging knob, the H/I will come in handy as an aid to the basic instruments in flying straight and level for the required period of time for it to re-erect itself.

Once you have regained control, you can make a climbing turn back to the original altitude and heading. Perhaps you'd better revise your estimate to the next IFR reporting point. You *have* been dawdling, you know.

Recovery from the Approach to a Climbing Stall and from the Stall Itself

The climbing stall often happens because the pilot was too eager to gain back altitude lost in the power-on spiral. Or maybe he neglected to catch that airspeed change when the nose was level and just kept pulling back until . . .

Being pessimistic (again), it will be assumed that you have lost the H/I and attitude indicator. Figure 4-26 shows what you would probably be seeing in such a case.

It's assumed that you'll enter this condition at cruise power. Reviewing the instrument indications:

1. *Turn and Slip* — May or may not be centered. However, it is likely that the needle will be deflected to the left, and the ball will show a left skidding turn because of "torque" effects. You could, however, be in the climbing *right* turn.

2. *Airspeed* — Decreasing, and probably doing so rapidly.

3. *Altimeter* — Altitude increasing or steady in last part of stall.

4. *Vertical Speed Indicator* — A high rate of climb. Because of the lag of the instrument, it is likely that it will still be showing a good rate of ascent even after the stall break has occurred.

The recovery technique will be as follows:

A. Relax the back pressure (or use forward pressure if necessary) until the airspeed stops decreasing — or, if it has been holding fairly constant,

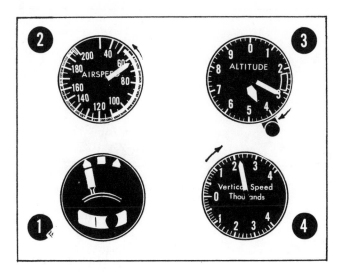

Fig. 4-26. Approach to a climbing stall. Attitude indicator and H/I are inoperative.

until it starts to increase. Try to stop the nose at the instant of the airspeed change. This works very well if the stall break hasn't occurred. If you've just reached the stall break, let the nose move down slightly past the point of this indication to assure enough of a nose-down attitude to avoid a secondary stall. In both situations, *apply full power as the nose is lowered.* This lowers the stall speed, hastens the recovery, and decreases altitude loss.

B. *Center the needle and ball or the turn coordinator.* If the stall has broken, make sure that the *ball* is kept centered if possible. It doesn't matter too much whether you are turning slightly (as shown by an offset needle or small airplane) during the recovery, *but* avoid yawing or skidding flight (as shown by the offset ball). A skid or slip can result in one wing stalling before the other, which could possibly lead to a spin. Try the unusual attitude maneuvers in VFR conditions without a hood. Notice that for most airplanes, as long as the ball remains centered at the break, problems of one wing or the other paying off first, with rolling tendencies, are minimized. *However, you should check the reactions of your particular airplane in this regard.* The problem is having enough rudder power at low speeds to keep the ball centered. In your VFR stalls, you'll notice in most cases that the airplane, at the break, tends to roll away from the ball. If the ball is to the right, roll, if present, will most likely be to the left (and vice versa). Notice that here the centering of the T/S or T/C is secondary to keeping the airspeed up. You can (and will) probably do Steps A and B simultaneously, as you progress.

C. *Use the altimeter to level off.* It's unlikely that the altitude *loss* will be the problem in this situation because you've been climbing at a goodly rate as the stall was approached. After you've checked airspeed, look at the altimeter. Then, rather than trying to stop the altimeter exactly there, perform as described in the first phase of the recovery. Allow about 100 feet of altitude loss during recovery to assure that you'll stay out of a secondary stall which

would cost even more altitude. Of course it's always best to recover without further loss of altitude. Perhaps you can do this in the trainer you're using, particularly if the break has not occurred. It depends on the particular airplane whether this can be done or not.

After the airspeed and altitude are under control, make the turn to the required heading. Get the H/I and A/I back in action as soon as possible to help in this. Adjust the power as necessary.

The primary objective is to get that airspeed back in a safe range, and you can make heading corrections later.

SPIN RECOVERIES

It's extremely unlikely that you will have problems with a spin. The modern certificated (normal and utility category) airplane is highly resistant to spins and, even if forced into one, will usually sneak into a diving spiral unless held into the spin. However, an illegally loaded plane could be a different matter.

The spin is an aggravated stall with autorotation. It's called autorotation because one wing is more deeply stalled and has less lift than the other. A rolling moment is produced which tends to maintain that stall condition and imbalance of lift.

The spin is usually the result of one wing being stalled before the other. This is why it's important to keep that ball centered during the stall approach and break, when you're on instruments. The normal procedure in practicing spins under VFR conditions (*don't,* unless you are in a utility--approved for spins-or acrobatic category airplane) is usually as follows:

You would be sure that you had plenty of altitude and the area was clear of other airplanes. Regulations require that the recovery be completed no lower than 1500 feet above the surface (3000 feet is better). Make sure that you are not in a control zone or Federal airway. (Take an instructor with you.)

You would "clear the area" by making a 90° turn in each direction, looking to all sides and particularly below you. Swallow that lump and wipe your sweaty palms one more time, then:

Pull the carburetor heat (if recommended for your airplane) and ease the nose up to do a straight-ahead, power-off stall.

Just before the break occurs, use full rudder in the direction desired to spin (as an example, to the left). This will yaw the nose to the left, slowing down the relative speed of the left wing and speeding up the right. A rolling motion is produced to move the left wing down, suddenly increasing its angle of attack past the critical point, and it stalls while the right wing is still flying. A definite rolling motion is produced, and the nose moves over and down to the left. If you relax back pressure, the maneuver would become a spiral, but you have the stick or wheel full back and are holding it.

In Figure 4-27 the spin is developing, and the indications show that the A/I and H/I have vacated the premises, so to speak.

65

Fig. 4-27. The instrument indications in the onset of a spin. The VSI here will be pegged *full down* as the spin develops. The H/I and attitude indicator have tumbled here.

The basic instruments (plus the vertical speed instrument) will indicate that the plane is in the spin by the following indications:

Turn and Slip or Turn Coordinator — The needle or small airplane will indicate the direction of rotation, but the ball in many airplanes will tend to go to the left side of the instrument (if the instrument is on the left side of the panel as shown in Figure 4-27).

Airspeed — Although you are descending at a goodly rate in the spin, the airspeed will remain low and fairly constant, although the value may oscillate to some extent.

Altimeter — This instrument will be showing a rapid loss of altitude.

Vertical Speed Indicator — Also shows that the airplane has a high rate of descent.

The clues that indicate the airplane is spinning must be judged together. There is rotation, as shown by the needle or small airplane. It could be that the needle and ball would indicate a *skidding turn;* don't use the ball but note what the other indications (needle or small airplane, airspeed, altimeter, vertical speed indicator) indicate. The facts that (1) the airspeed is low, (2) a rate of turn is indicated, and (3) a high rate of descent is occurring, lead to the conclusion that the airplane is spinning. The small airplane shows that the spin is to the left, so the following recovery technique should be used:

A. *Opposite rudder* (right rudder here) to needle or small airplane. This is not to imply that the needle is flown with the rudder and the ball with the ailerons, as is sometimes advocated. It's just that you've been told by the instruments that you're in a spin and are using the VFR mechanical technique.

You hold opposite rudder, until the nose stops moving. When this happens, *get off the rudder* (or neutralize); otherwise, you could start spinning in the opposite direction.

B. *Relax back pressure* or use brisk forward pressure as recommended by the manufacturer as soon as the rudder reaches its stop. (Some airplanes may require a *brisk* forward movement of the stick or wheel right after opposite rudder is used.) Get off the rudder when the rotation stops. *Check the airspeed.* As soon as it starts picking up, you are out of the spin and in a straight dive. The spin itself puts practically no stresses on the airplane, but a sloppy or delayed pull-out from the dive following the recovery does.

The first thought would be that you should perhaps keep the nose down until cruise speed is indicated before starting the pull-out to be sure to avoid another stall and possible spin. *However,* you would readily see in a VFR spin that, as soon as the back pressure is relaxed, the airspeed will pick up very rapidly (the nose of the airplane is practically straight down), so that any delay in the pull-out could cause you to exceed the red line speed, *plus* causing an excessive loss of altitude. So, as soon as the airspeed starts increasing, start applying back pressure *smoothly*. The airspeed will continue to increase during this process. Watch for the airspeed to stop increasing. That point, as discussed in the pull-out of the power-on spiral, shows that the nose is approximately level, and you should relax back pressure or ease on forward pressure to stop the nose in the level position.

C. *Check the altimeter* as soon as the forward pressure has been exerted (or back pressure has been relaxed). Use this instrument to level off.

Adjust power, heading, and altitude as necessary to get back to where you were when this fiasco started. Reset the attitude indicator and H/I as soon as possible. The following are added points to consider:

It's likely that instead of the power-off spin entry you would probably have cruise power on and would possibly get into it from some condition other than a wings-level, slow pull-up. (A climbing stall is a good place to start such a situation.)

Some airplanes will not stop rotating with application of rudder alone. The best all-around recovery is opposite rudder, followed immediately by brisk forward motion of the wheel or stick. When the needle, or small airplane, is centered (or more likely, "averaging" center as it flops back and forth) *then* neutralize the rudder and check the airspeed. This is the usual procedure for heavier airplanes because of their higher moments of inertia that need elevator input to supplement the opposite rudder. For many airplanes, closing the throttle(s) and making sure that the ailerons are neutral is the first step in recovery from an accidental spin. *Then* the rudder and elevators are used per the manufacturer's recommendation.

Again, it is *extremely* unlikely that you will ever get into a spin accidentally, either VFR or on instruments. Of course, it wouldn't be much help to you if people said, "Isn't it amazing; Ol' Joe spun in all the way from 9000 feet on IFR — first time I've heard of *that* in years."

INSTRUMENT TAKE-OFF

You've been making take-offs for some time and, probably, rightly feel that there's not much to be shown to you as far as this maneuver is concerned. However, during your instrument training, you'll be introduced to the instrument take-off, or ITO, as mentioned earlier.

Actually, except as a training maneuver, the ITO is not of too much value. If the weather is so bad that you must use the instruments to stay on the runway, you should think of an extremely good reason for not going. If the conditions are that lousy, how would you get back to the runway if trouble developed right after take-off? The ITO can be of practical use when the bad weather is strictly a local situation, such as early morning fog at a river bottom airport (and is CAVU everywhere else around). It could save your waiting for a fog only 100-200 feet thick to lift and burn off. But it's still a calculated risk, and you'd be banking on the fact that, in case of trouble, you could make it into another airport close by. For most cases, it's smart *not* to take off when ceiling and visibility are below approach minimums for that field, particularly if there are no fields with lower minimums nearby.

You may hear a number of "gimmicks" on setting the reference airplane on the attitude indicator before making an instrument take-off (or a take-off that will put you into instrument conditions right away). *Ignore them. Before take-off, set the small airplane as closely as you can to the actual attitude of the airplane at that time.* That way you will have a true picture of the airplane for take-off, climb, and cruise without having to change it later.

Setting the attitude indicator for the tricycle gear airplane is simple, in that the attitude of the airplane is generally considered to be level.

The tailwheel type is a different matter. If you are the scientific type, you could use a protractor and measure the angle the longitudinal axis makes with the ground, or better yet, some *Owner's Handbooks* contain this information. (For instance, the Beech Super 18 gives this figure as $11.5°$, assuming proper tire and oleo inflation.) During VFR or simulated instrument flight, you could, at a fairly low altitude (say, 2000 or 3000 feet), establish a high cruise. Maintain a constant altitude, and after a constant airspeed is established, set the small reference airplane (or the "pip" of the reference airplane — it depends on the type of instrument you have) on the horizon bar. If you don't rack the airplane around a lot before you land and don't hit the ground too hard, the small airplane should be at the right position when the airplane is sitting in the three-point position on the ground. You can note the number of horizon bar widths that the reference is above the horizon bar and use that in setting the instrument on the ground. The chances are that you could be accurate enough by setting the wings of the small airplane in line with some of the reference marks on the sides of the instrument face. Your instructor probably will already have the setting.

Instrument Take-off — Tailwheel Type

After you've completed the pretake-off check (to be covered later) and have received take-off clearance, taxi to the center of the runway and line up with the center line. Taxi forward slightly to straighten the tailwheel and lock it if the airplane has such a control. Set the H/I on the runway heading, and *make sure that it is uncaged.* Wake up the safety pilot.

Hold the brakes and run up the engine(s) to a setting that will aid in rudder control, then release the brakes as you apply full power.

Don't shove the power on abruptly but open it smoothly all the way. Your copilot may "back you up on the throttles" by putting his hand behind them to make sure they don't creep back. He'll be the lucky one to look out at the runway and will offer such helpful hints as "Dagnabit! Left! No, not so much" etc. until you are ready to clobber him — if you could spare the time from the gages. Here's the place during your training where you'll discover that even after quite a bit of experience with the old-fashioned type of heading indicator, you can still manage somehow to correct the wrong way for any heading deviations on the ITO. You and the safety pilot may have a brief scuffle as to who can exert the most pressure on the rudder pedals.

Here's where heading will be a most important factor. Maybe you've been a sloppy pilot and 5° of heading means nothing in the air. *But* if you are in the center of a 100-foot wide runway, a 5° error can put you off the edge before you get 600 feet down the runway.

In addition to the fact that you may want to correct the wrong way at first, there will be the problem of overcorrecting. The heading sneaks off, and you, thinking of that deep rich mud alongside the runway, decide that there'll be none of *that* and enthusiastically apply too much opposite rudder; and the take-off will resemble some of the first ones you ever made.

On some of the older tailwheel twins, it is sometimes necessary to use asymmetric power to help keep them straight on the take-off run before the rudder becomes effective (this is why you should add power before releasing brakes). This contributes to overcontrolling, gives the pilot much more to think about, and is definitely *not* recommended as a technique unless everything else fails. As far as using brakes to keep straight is concerned — *don't.* If your airplane is so tricky that brakes are required to keep it straight on the take-off run, well, you'd be better off to forget about instrument take-offs in it.

We've mentioned the precession effects resulting if the tail is raised abruptly, so don't shove it up. Allow, or assist slightly, the tail to come up so that the attitude of the airplane is that of a shallow climb.

It is assumed that your airplane will have a static system that has an outlet on the fuselage or at the pitot-static head. For some of the older and lighter trainers, the airspeed and altimeter (and vertical speed indicator, if available) had no static tubes but were open to the cabin. As speed picks up, the result is a drop in normal cabin (and instrument) static pressure. The airspeed will very likely read high, and the altimeter may show an altitude "jump," even though you can still feel and hear the tires rolling on the runway. Fortunately, this type of setup is going the way of the helmet and goggles.

Ground effect can induce instrument error by changing the airflow about the airplane. (This is speaking for the airplanes with static ports on the fuselage or a pitot-static tube.) The result is that the airspeed and altimeter will read *low.* This is, of course, a factor on the safe side. You may have noticed on VFR take-offs that as the airplane lifted, there seemed to be a sharp jump in airspeed and altitude. Ground effect may fool you into thinking the airplane is all set to climb out when that's not the case. Remember that the airplane performs better in ground effect, and things might not be so great when you get a few feet of altitude.

This is one time that you should assure yourself that you have sufficient flying speed before lift-off. In fact, it would be better to let the airplane lift itself off if you have established the proper attitude on the tail-up part of the run.

As the airplane lifts off, establish a rate of climb. Do not ease the nose over to try to pick up the best rate of climb speed right away — you could overdo it and settle back in. One problem, at this point, is that the forces of acceleration (you're picking up speed) work on the attitude indicator to give a slightly more nose-high reading than actually exists. You may lower the nose to get the "proper reading" and settle in. Of course, the attitude gyro can be used for wing attitude, but you would be better off to use airspeed and altimeter (and H/I) for proper climb *immediately* after lift-off.

Attain a safe altitude (at least 500 feet) before reducing power. You don't need the distraction of changing power when you're still getting used to the idea of being in the soup. Don't retract the gear under 100 feet above the surface. Establish the proper climb speed and follow your instrument clearance.

Instrument Take-off — Tricycle Gear

As in VFR take-offs, the nosewheel makes for more positive control and simplifies matters considerably. Naturally, you'll taxi out and line up with the center line and move forward a few feet to straighten the nosewheel. Don't use brakes for directional control.

As the airplane picks up airspeed, ease the nosewheel off and assume the normal take-off attitude. (Be prepared for the need for more right rudder as the nosewheel lifts.) For the light twin, it is recommended that the airplane not be lifted before V_{MC} (minimum control speed — single engine) is attained.

Don't rush *any* airplane.

The attitude of the airplane should be practically the same as that for the tailwheel type. As an example, take the Cessna 180 and 182 models; they have the same maximum weight, wing areas, airfoils, and stall and take-off speeds. In short, they are exactly alike except for the landing gear configuration, and as you know, the 180 is a tailwheel type and the 182 has a nosewheel. The attitudes at lift-off should be the same because, for all practical purposes, the airplanes *are* the same. You are interested, during the latter part of the run, in having the attitude which will allow the airplane to become airborne at the optimum time. If you keep the nosewheel on the ground or, in the tailwheel type, have the tail too high, you'll waste runway. On the other hand, the take-off will suffer if the tail is held too low on this part of the run. Start paying attention to the attitude indicator during VFR take-offs. What is best for a clear day is also the best for a day when clouds are overhead — assuming that runway conditions, airplane weight, and other variables are the same.

THE WELL-BALANCED PILOT

During the basic instrument part of your training you may first encounter the effects of "vertigo." (Turning or moving your head quickly while flying under the hood or on actual instruments is a good way to discover just what vertigo is.) If you've always associated vertigo with little old ladies who are a bit tiddly from sampling the cooking sherry, take a look at it from an instrument flying standpoint:

Under normal situations, most of your balance depends on sight (sure, the other senses help too). The feel of gravity, for instance, helps you know which way is "up" or "down" *under normal conditions*. But what about the artificial gravity created by turns, pull-ups, and vertical gusts? Your body doesn't know the difference unless sight helps to separate the "natural" from the "artificial" gravity. The fun houses in amusement parks are examples of how much the eyes have to do with equilibrium. The crazily angled walls make people forget the natural gravity force that all the "seat of the pants" pilots say they use.

So, your eyes can fool you too. Take autokinetic illusions, for instance. These five-dollar words mean that your eyes are telling you something that isn't true. The military used to have night fighter pilot trainees move into a darkened room where only one very small bright light was visible. Various pilots would be asked to describe the various motions ("up, now it's moving right," etc.). The lights were turned on and on the opposite wall was a permanently attached, stationary light. Without other references, the involuntary movements of the eyes gave the illusion of movement of the light.

Statements have been made by noninstrument pilots that they could fly in solid clouds without any flight instruments at all (or blindfolded — the choice of condition usually depended on how far the eve-

ning's festivities had progressed). There's no argument about that. They *can* fly in clouds without flight instruments — for quite a number of seconds sometimes (the length of time depends on how high the airplane was above the ground when the experiment started). Birds don't fly in solid instrument conditions. (Air Traffic Control records have no known case of a bird filing IFR, so this proves it.)

At any rate, the inner ear is the place where most of the *nonvisual* balance sensations originate. They (nearly everybody has two) react to forces and couldn't really care less about which way is "up" or "down" and are about as believable as the guest speaker at a Liar's Convention.

What does vertigo in flight feel like? It's a feeling that all is not well. The instruments are lying because your sensations tell you that you *couldn't* be doing what they indicate. There may be a struggle in your mind, but *always* believe the instruments over your own sensations when outside visual references aren't available. One instrument instructor noticed his instrument trainee leaning over to one side as he flew under the hood. The student was doing a great job, the instructor remarked — but why the Tower of Pisa bit? The student answered that he knew the blank blank instruments were right so he was flying by reference to them, but *he* felt more comfortable about the whole thing if he listed to starboard a little.

You should be prepared for a different sensation when flying actual instruments as compared to flying under the hood. First, flying actual instruments is easier in that you don't have to twist your head as much as when under the hood — you have a much wider field of vision. On the other hand, there is the psychological effect of being "committed." You're in the soup and in the system and have to go on with it; there's no taking off the hood and saying, "Well, I goofed on that last approach, didn't I?"

Sometimes the grayness (or blackness) outside creates the effect of the airplane rushing at greater than normal speeds, and at other times it seems that you are handing suspended and only the instruments' indications show that you are moving at such and such an airspeed and altitude. These effects can be disquieting if you encounter them on one of your first actual IFR flights with a load of passengers and you are the only pilot on board. This is a good reason for getting as much flying in actual conditions as possible during your training for the rating.

Something else to consider, too, is that turbulence could be bad enough so that you could have trouble reading the instruments — you'll be moving up and down so fast that they will be blurred and this could also tend to induce vertigo. It could also tend to induce hyperventilation, a condition caused by too deep and/or rapid breathing resulting in an imbalance between the carbon dioxide and oxygen. When people are scared, they tend to hyperventilate.

Vertigo can last a short while or an hour or more, depending on the situation and physical condition of the pilot. It's a rather weird sensation when first encountered, but take a deep breath and settle down to doing a good job with the instruments. The cockpits of some airplanes, it seems, are designed to induce vertigo, with widely separated radio equipment requiring twisting and/or quick head movements.

Your instructor may try to induce vertigo on one of your training flights to let you recognize it. Some pilots say they've never had vertigo and probably haven't.

The Medical Handbook for Pilots (DOT/FAA 1974) is a must for any pilot's library, instrument rated or not. It discusses in an interesting and easy to understand way such things as hypoxia, hyperventilation, alcohol, carbon monoxide, vertigo, fatigue, aging and other factors affecting pilots.

It goes without saying that drinking just before flying is stupid, but doubly so when flying instruments. The *Handbook* also brings out some points on drugs that would be of interest to you.

Fatigue has caused fatal accidents. The problem is that after fighting a long siege of weather enroute, you are usually at your physical worst when the most alertness is needed for the approach. Your senses may lie more easily because of this fatigue.

In short, when flying instruments, don't trust your own sensations at all; take the indications of one instrument with a grain of salt; confirm what all of the operating instruments combine to show.

SUMMARY OF THE CHAPTER

You should be able to have the airplane under complete control in any normal condition before attempting to add navigational and letdown work to the basic instrument flying. Control of the airplane through the instruments must become practically second nature. You have to be proficient in doing other things, such as calculations for a revised ETA, flying the VOR or other navaids, taking and acknowledging clearance, making voice reports, and getting weather information -- and the airplane won't be waiting for you.

You may not get to practice actual spin recoveries in your particular airplane; but that section was included so that if, under actual conditions, you get into a spin, the recovery technique might immediately come to mind.

Part Two

NAVIGATION
AND COMMUNICATIONS

Chapter 5

NAVIGATIONAL AIDS AND NAVIGATIONAL INSTRUMENTS

IT'S LIKELY that in your VFR flying you've been relying on radio navigation and have been keeping the sectional and WAC charts under the seat. However, you could always haul out a sectional or WAC chart and go on about your business should the omni receiver quit on you.

The following chapters are intended to tie in the gray areas of en route IFR navigation and the ATC (Air Traffic Control) system.

Too many neophyte instrument pilots get submerged in details of the flight and neglect to look at the overall picture, which, believe it or not, is comparatively simple.

You want to go from A to B: Under VFR conditions, you would plan your flight and go — and may or may not file a VFR flight plan — it's up to you. Okay, so the weather is below VFR minimums, but the purpose of the flight is the same. You are still trying to get from A to B. Because of the weather, you'll have to fly the airplane by reference to instruments. If you were the only person flying that day, you could, in theory anyway, hop in the plane and go — without worrying about other people — and could leave when you got ready and go the way you wanted to, after picking an altitude to clear all obstacles en route. However, there are fatheads who have the audacity to want to fly the same day and even in the same area *you* do. It's those other guys who make IFR a little more complicated.

Because separation between airplanes becomes a problem now that you can no longer "see and be seen," you'll have to follow predetermined paths,

and the ground coordinators (ATC) will be interested in knowing your altitude (and may assign you a different one than you requested). ATC may even send you to B by a different route than the one you prefer — all because of those other people who want to fly at the same time you do.

To keep closer tabs on airplanes, airways have been established so that pilots will follow known routes. In the beginning of cross-country instrument flying, these airways were established between the old low-frequency radio ranges and were named by colors. Red and Green airways ran generally east and west. Amber and Blue airways ran generally north and south. The word "generally" is used because, in its meanderings, a Red or Green airway might be running north or south for a part of the route; but the final result was that if you started at one end on a Red or Green airway, you would eventually end up west (or east, depending on the original purpose and direction of flight), albeit after a fatiguing and ear bending episode. Because navigation depended on aural signals, a two- or three-hour session of flying the LF/MF range could result in what was known among select medical circles as the "Dit-Dah Syndrome." (The pilot, on occasion, would swear that he could hear the dots and dashes for several hours after the flight was over.) If the introduction of VOR (VHF Omnirange) did nothing else, it at least cut down on the number of glassy-eyed pilots. (VOR airways are V-6, V-116, etc.)

Before going on, it might be a good idea to review the radio frequency bands:

Very Low Frequency (VLF)	10-30 kilohertz
Low Frequency (LF)	30-300 kHz
Medium Frequency (MF)	300-3000 kHz
High Frequency (HF)	3-30 Megahertz
Very High Frequency (VHF)	30-300 MHz
Ultrahigh Frequency (UHF)	300-3000 MHz

(Note: "Hertz" is used for "cycles per second.")
(Megahertz, kilohertz, etc.)

VOR — VHF OMNIRANGE

The VOR operates in the frequency range of 108.00-117.95 MHz, which puts it close to the middle of the VHF band (30-300 MHz per second).

They use even-tenths frequencies (108.2, 108.4, 109.0, 109.2, etc.). The odd tenths in that area (108.1, 108.3, 111.1, etc.) are ILS frequencies and will be covered later in this chapter.

The frequencies in the 112.00 to 117.95 MHz use all of the tenths, both odd and even (112.0, 112.1, 112.2 MHz, etc.). Some pilots have trouble at their first introduction to the radio in separating VHF and UHF from LF/MF, as far as reading the number in the *Airman's Information Manual* is concerned. The LF/MF frequencies are whole numbers (212, 332, etc.), while the VFR and UHF frequencies are always "point something" (112.3, 243.0). You would read the 112.3 as "one one two *point* three." You'll find that there won't be any confusion between the frequency bands after you start using them. The kilohertz is the lowest whole frequency measurement, so that 112.3 could not be LF/MF because this would be working with a part of a kilohertz. You won't be using any frequencies as low as 112 kilohertz, as none of the LF/MF air frequencies are that low. So, if it has a "point something," it's VHF or UHF.

Following are the VOR class designations and expected ranges of each, as given by the *Airman's Information Manual:*

Class of VOR, VORTAC, or TACAN	Operational Service Volume
T (Terminal)	25 nautical miles (NM) up to 12,000 feet MSL
L (Low altitude)	40 NM up to 18,000 ft MSL
H (High altitude)	40 NM from 18,000 ft MSL 130 NM from 18,000 ft MSL to 45,000 ft MSL Within the conterminous 48 states only, between 14,500 and 17,999 ft MSL (100 NM) 100 NM above 45,000 ft MSL

VOR Theory

The omni receiver in your airplane uses the principles of electronically measuring an angle. The VOR (ground equipment) puts out two signals; one is all-directional, and the other is a rotating signal. The all-directional signal contracts and expands 30 times a second, and the rotating signal turns clockwise at 30 revolutions per second. The rotating signal has a positive and a negative side.

The all-directional or reference signal is timed to transmit at the same instant the rotating beam passes *Magnetic North*. These rotating beams and the reference signal result in radial measurements.

Your omni receiver picks up the all-directional signal. Some time later, it picks up the maximum point of the positive rotating signal. The receiver electronically measures the time difference, and this is indicated in degrees as your *magnetic* bearing in relation to the station (Fig. 5-1). For instance, assume it takes a minute (it actually takes 1/30 of a second) for the rotating signal to make one revolution. You receive the all-directional signal, and 45 seconds later, you receive the rotating signal. This means that your position is 45/60 or 3/4 of the way around. (Three-fourths of 360° is 270°, and you are on the 270° radial.) The omni receiver does this in a quicker, more accurate way.

Since the VOR operates to give you the airplane's

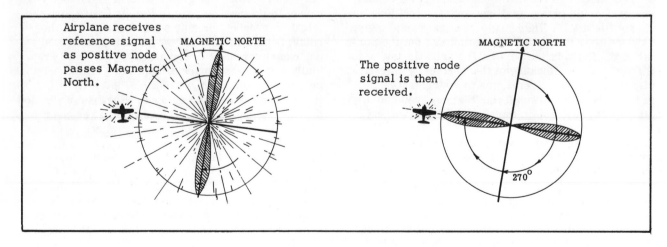

Fig. 5-1. VOR theory.

relative position to (or from) the station, based on Magnetic North, all directions on en route and letdown charts and the directions given relative to various navigation aids on charts are *magnetic*. This saves some extra figuring at a time when you might have your hands full.

The aircraft VOR receiver presentation is made up of four main parts: (1) *Frequency Selector*, (2) *Omnibearing Selector* (OBS) calibrated from 0 to 360, (3) a *Course Deviation Indicator*, a vertical needle that moves left or right and indicates the relative position of the airplane to the selected omni radial, and (4) a *TO-FROM Indicator* (Fig. 5-2).

Figure 5-3 shows a combination communications and navigation arrangement.

As the "radials" (there are 360 of them, or one for every degree) are fixed in reference to the ground, the airplane's heading has no bearing on what radial it is on. However, in tracking to or from a VOR it is best for the omnibearing selector to be selected to the *course* to be followed, so that the left-right needle reads correctly. Figure 5-4 shows the radial idea.

Airplane A doesn't have to select 050 on the OBS, but it's a lot more reasonable to do this; otherwise, the needle would sense incorrectly. The needle is sensing correctly when it is pointing in the proper direction to get you on the selected radial. Turn and fly "toward the needle." If the OBS of Airplane A were set at 230 and it drifted to the right of course, the needle would be deflected to the right, implying that a right turn would be needed to get back on the original course — which is incorrect.

To show that the heading of the airplane has nothing to do with the indications in the cockpit, take a look at Figure 5-5. The combination of needle, TO-FROM, and OBS merely gives you the airplane's position relative to the station.

Fig. 5-2. VOR receiver components. The small window would indicate "TO," "FROM" or "OFF" as applicable. The "ID" selection (pull) allows a clearer pick-up of distant stations' Morse code identifiers. (*Narco Avionics.*)

In Figure 5-5A, the pilot will get the indications as shown if he tunes in the VOR and sets the OBS to 050. Of course the needle would soon leave the center in the cases where he is flying *across* the 230 radial, but at the particular instant of tuning, the indications would be as shown. The VOR and the cockpit indications are straightforward. They merely combine to state that the airplane is to fly a magnetic course of 050° to get TO the station. What your particular heading is at that instant, the VOR and your cockpit equipment couldn't care less. Although as soon as the airplane leaves the selected course, the needle will let you know about it.

In Figure 5-5B, the VOR and cockpit equipment

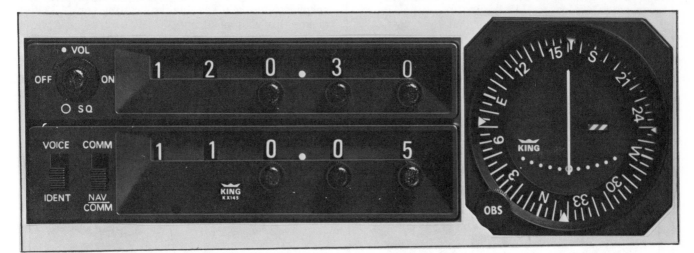

Fig. 5-3. A "one and a half" set consisting of a two-way communications (the "one") and a navigation and voice receiver (the "half"). The upper part of the control panel is a transceiver (set on 120.30 MHz) with a squelch control and the bottom is the NAV receiver (110.05 MHz). When the Voice/Ident selector is on IDENT, the selected VOR/LOC station voice and identification tone are both coupled to the cabin speaker or headphones. On VOICE the identification is eliminated but voice transmissions are still received. The COMM/NAVCOMM selector is used for transceiving and when NAV/COMM is selected the equipment acts as a NAV receiver using the frequency selected. By keying the mike, the selected COMM frequency may be used. After the key is released the NAV receiver goes back into action. Each dot on the fact of the indicator represents deviation from the selected radial, or centerline of the localizer, of 2° and 1/2° respectively. (*King Radio Corp.*)

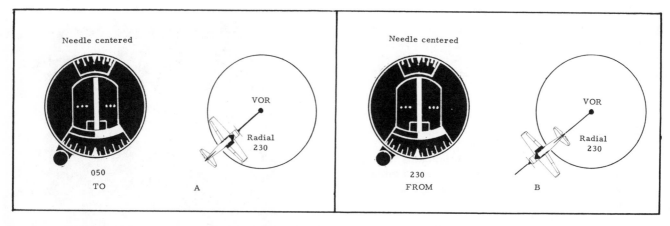

Fig. 5-4. Both airplanes are on the 230 radial, but A is inbound (course 050°) and B is outbound (course 230°).

are also doing their job. The airplane is on a bearing of 230° from the station, and that's all the equipment is supposed to tell. If you're just passing by, then it's only a matter of interest that your airplane is at that instant on a bearing of 230° magnetic from VOR "X."

Figure 5-6 shows what happens to an airplane that has the set tuned into a certain VOR and is trying to track inbound on the 230 radial (050° course). He has drifted to the right because of a northerly wind, and the indications are as shown.

The needle is indicating that the selected radial is to the left. In Figure 5-7, he is in the same situation, except that for some reason he set in 230 on the OBS but plans to track into the station on a course of 050°. The needle seems to be giving a bum steer.

The omni equipment in the airplane is still doing what it was designed to do — the pilot was the one who was out to lunch. If he suddenly rotated the airplane

to the heading he had set up on the OBS, the selected radial or course *would* be to the right (Fig. 5-8).

The cockpit indicators tell that you are south of the selected line of flight to the VOR.

Using the VOR for Cross-bearings

The idea covered in the last section can be used to check whether you've passed a VOR intersection or not (there is another method also, which will be covered shortly). Look at Figure 5-9.

You are flying along V-116, minding your own business, feeling satisfied that you've done your duty as an instrument pilot, having earlier given ATC an ETA for WIN, and everything is OK. (Use a lot of initials in your talk around the hangar; it'll drive VFR pilots crazy PDQ.)

Anyway, ATC asks, in a tone of voice that implies you have been extremely remiss, if you have

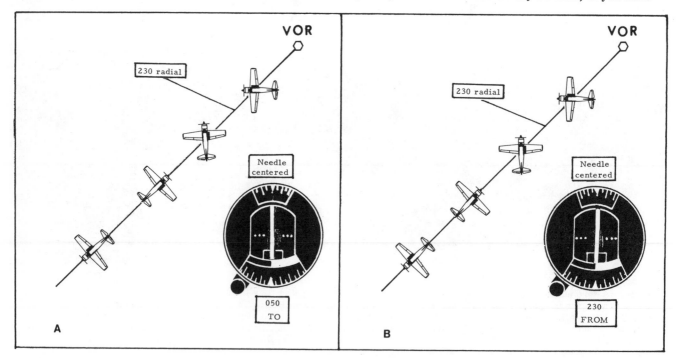

Fig. 5-5. The heading of the airplane has nothing to do with the cockpit indications. In A the OBS has been set to 050 and the TO-FROM indicator says TO. In B the OBS has been set to 230 and the TO-FROM indicator says FROM.

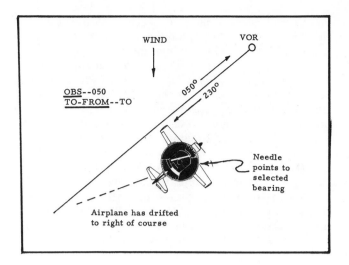

Fig. 5-6. The airplane has drifted to the right of course and the needle points toward the selected bearing.

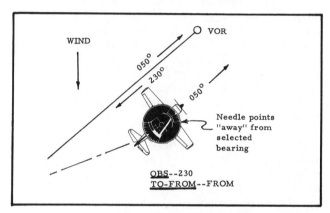

Fig. 5-7. The indications of the cockpit equipment for the same airplane position and heading but with the OBS set on 230, or reverse to the course to be flown.

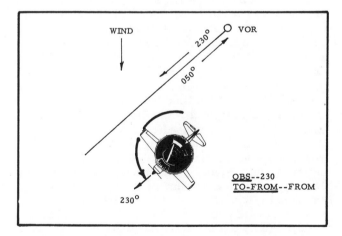

Fig. 5-8. The selected radial is to the right. Note that the needle did not move within the instrument, or it still has the same relative position to the pilot.

passed Hector intersection. Looking at the chart, you see Hector and figure that you are somewhere in that area but can't be sure whether you've passed it yet. (Your mind is going like a squirrel in a cage, trying to remember if previous clearances had

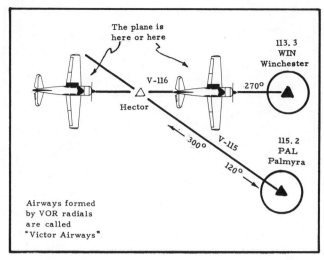

Fig. 5-9.

mentioned anything about reporting at Hector. Forget it; at this point you'd better find out whether you've passed it or not — and as soon as possible.)

You tune in PAL (Palmyra VOR) on the other VOR receiver — and *identify it*. Notice that the 300 radial of PAL intersects your airway, forming the intersection in question. You set up the OBS to 120; the bearing TO the station (from Hector) and the cockpit indications look like those shown in Figure 5-10.

Set 1 shows that you are right on course to Winchester, as you are supposed to be, but what about Hector intersection?

In your *mind's eye*, turn the airplane (at your present position) to a heading of 120°, and look at the needle on Set 2. It's to the left, so the radial making up the intersection is to the left — and you *have not passed it* (Fig. 5-11).

The needle is "pointing" to the radial in question. If you had set in 300, the same answer would be given (you haven't reached the intersection yet), but the presentation would be a little different, as shown in Figure 5-12.

In Figure 5-12, the needle is still "pointing" to the reference line (radial).

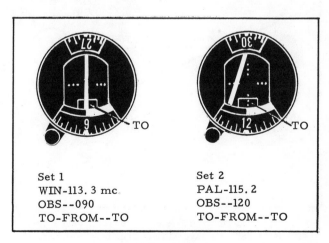

Fig. 5-10.

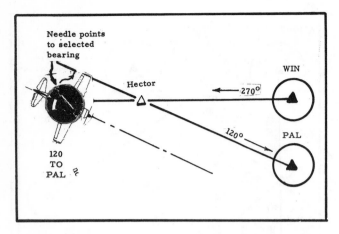

Fig. 5-11. The needle indicates that the radial is to the left on the imaginary heading, hence ahead of the airplane on the real heading.

One tip for this technique: If the station being used for the cross-bearing is "ahead of the wing tips" of the airplane, as in the illustration shown; use the TO indication (turn the OBS to the bearing that would give a "TO") and then look at the needle and "turn" the airplane in your mind. If it's behind the wing tip (dashed VOR and radial), use the cross-bearing that would give the "FROM" indication. This would mean that you would not have to "turn" the airplane so far in your mind.

If you had passed the intersection, the needle would show the opposite indications (to the right in Figure 5-11 and left in Figure 5-12).

Perhaps you don't like the idea of "turning" the airplane in your mind, so a quicker way would be always to set the actual radial for the cross-bearing VOR, or the bearing to the intersection FROM the station. If the VOR's relative position to your heading and the needle indication jibe, then you haven't reached the intersection (the cross-bearing VOR is to your *right* as you fly the course line — whether ahead or behind — and if the needle is to the *right*,

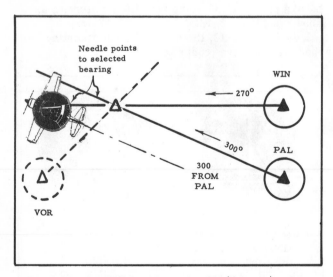

Fig. 5-12. The OBS has been set at 300 (Palmyra) and the TO-FROM indicator says FROM. The airplane is "turned" to a 300° heading.

you haven't reached the intersection). Naturally, if the VOR is to the left and the needle is to the left, you haven't reached the intersection, either.

So, tune the cross-bearing VOR and rotate the OBS to the outbound bearing from the VOR to the intersection as given on the chart. If the actual position of the VOR to your course line and the needle position match, then you haven't reached the intersection. If they don't match, you've passed it (Fig. 5-13).

Of course, it's easiest to have Distance Measuring Equipment in the airplane, and just read whether you have reached the intersection. (If the intersection is 17 miles on this side of the VORTAC ahead and the DME is indicating 20 miles, you still have 3 miles to go.) DME will be covered shortly.

Time-to-Station Work

One problem you may have in practicing for the flight test is the estimation of the required time to fly to a VOR off a wing tip. You'll find that, for a certain period of time, flying "perpendicular" to the course to the VOR, a certain time longer is required to reach the station after turning inbound to the station. (You recall from your trigonometry that the ratio of two legs of a right triangle can be found readily if the angle is known.) You will fly through an angle of either 10° or 20° and will have a fixed multiple of the time required in either case (6 and 3 respectively). Figure 5-14 shows the principle involved.

Suppose you want to find your distance (time) to the station in Figure 5-14.

You would tune the VOR and center the needle with the TO-FROM indicator showing TO. Figure 5-14A shows your relative position to the station. You would then turn the airplane to the nearest heading that would put that bearing 10° ahead of the wing tip. In the example, your original heading is 045°, so a right turn is in order. You would turn right to a heading of 090°. (You would see that this would put the station at a *relative* bearing of 080°, or 10° ahead of the right wing tip.) You would fly the 090° heading and make a note of the time that the needle was centered. Then, you would rotate the OBS to get a new bearing 180° TO. (The needle is no longer centered because you aren't at the newly selected radial yet.) You would hold the heading of 090° until the needle is centered again (Point B) and make a note of the time. The elapsed time multiplied by 6 would give the time required to fly to the VOR from Point B. If it took 1 minute and 20 seconds to fly from A to B, the time required to fly to the station would be 6 × 1 1/3 or 8 minutes. If you timed the flight from A to C (20°), the multiplier would be 3. (It should take 2 minutes and 40 seconds, and 3 × 2 2/3 = 8 minutes to the station.) Note that the multiplier and the number of degrees flown are a function of the number "60." If you fly only 10° of arc, the multiplier is 6 (6 × 10 = 60). If you fly 20° of arc, the multiplier is 3 (3 × 20 = 60). In theory, if you flew 30° of arc, the multiplier would be 2 (2 × 30 = 60), but 20° is plenty, and 10° works

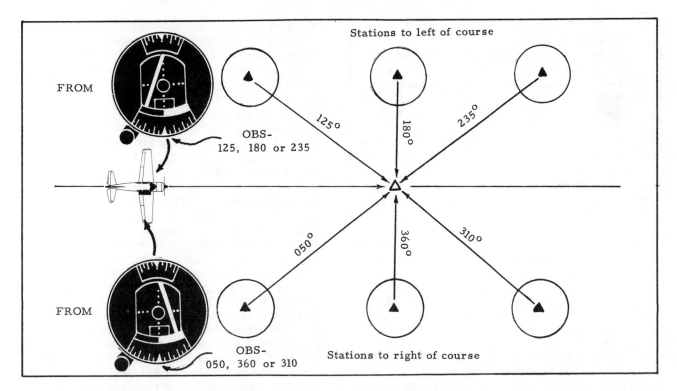

Fig. 5-13. After setting up *outbound* bearings from the cross-bearing VOR's, if the relative position of the needle and station match, the intersection has not been reached.

closely enough and saves time in actual practice. Or you may prefer to remember: "For each 10 seconds it takes to make a 10° change — you are 1 minute from the station." (If at a great distance out you may use "each 5 seconds for 5° equals 1 minute from the station.")

In no-wind conditions the estimate of time to the VOR should be reasonably accurate, but with a wind, the inbound leg could be affected. Wind from any direction or velocity at your altitude will affect the accuracy of the estimate.

You can actually track around a VOR and approach the station from any given direction. Under no-wind conditions the airplane (in theory) would remain a constant distance from the VOR — this is assuming that you made perfect corrections for the different compass deviation errors for different headings and that you flew those headings right on the button. There are VOR/DME approaches that use this basic principle of tracking around the station,

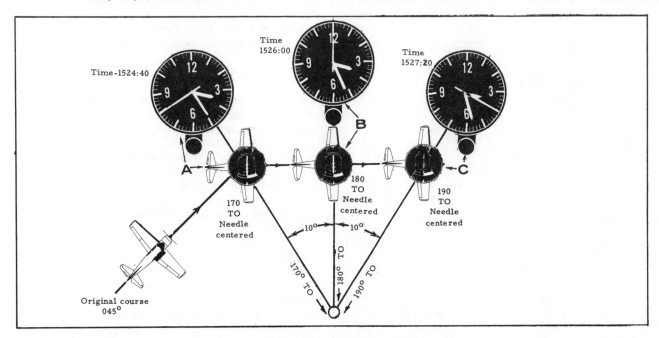

Fig. 5-14. Principles of time-to-station calculations. The "10°" angles have been exaggerated.

79

and the DME is a great aid in assuring that the proper radius is maintained.

In this day of radar, the idea of tracking around a station using only VOR (no DME) is academic. But as a training maneuver, it can serve in keeping you alert in flying the airplane and doing calculations at the same time. Looking back at Figure 5-14 at Point C, the heading would be changed 20° to the right and the OBS set to 210. When the needle centers, another 20° right turn is made and 20° is added to the OBS selection. This is done until the airplane is on the desired radial for turning in. (It may work out that you'll have to cut one of the segments short when the radial is reached.)

The amount of needle lead for a turn into the station depends on your distance from it. Normally 5°, or when the needle is deflected about one-half to one side (to the left in this case), is used as a lead. However, it really depends on the rate of needle movement, or your distance from the VOR. If you were 60 miles out, 5° would mean a 5-mile lead for a 90° standard-rate turn. For the airplanes you'll be flying, this will be too much.

You can check groundspeed enroute by cross-bearings, and probably have been doing it in your VFR flying. The drawback to this is a lack of accuracy in the airplane and/or ground equipment. As will be covered later, you could have up to 6° error in your airplane equipment, as checked, and still be OK for instrument flying.

You may also check the time to a station by turning either way 90° to your course and finding the time required to fly through 5°, 10°, or 20° of arc, as shown in Figure 5-15. (If using 5°, the multiplier is 12.)

The disadvantages of this are obvious. Supposedly, you are on an assigned Victor airway. This requires that you leave the center line of the assigned airway and mosey off into other peoples' airspace, particularly if you are some distance from the VOR (Victor airways extend 4 nautical miles out from the center line, or are 8 nautical miles wide). This technique could cause a great deal of interest down at the ARTCC radar console. For Victor airway navigation in the United States, you'll be a lot better off to take a cross-bearing, get an approximate distance to the station; and if you haven't had a groundspeed check, use your true airspeed and estimated wind to make a time estimate to the station. (VOR radial errors should be just about the same for only a 10° difference in selection.)

Another time check that may be used, though not particularly practical for operations on assigned airways, is called "double the angle off the bow," a term more descriptive when applied to the Automatic Direction Finder (ADF) than to the VOR — but the principle is the same in both cases. Basically, you will fly two legs of an isosceles (2 equal sides) triangle.

Assume you are flying to a VOR on an inbound bearing of 080°. Figure 5-16 shows the procedure. You may make the turn in either direction.

You would make a 10° turn to either the left or right (*left* in this case) and reset the OBS 10° to the *right* (or 10° greater — 090°). You were originally tracking on an inbound bearing of 080°, as shown by Figure 5-16. With a crosswind, you were holding some heading other than the inbound bearing. At any rate, change the *heading* 10° (to the left in this case). Fly that heading until the needle centers on the OBS selection of 090°; that is the completion of Leg 1. In theory and in no-wind conditions, Leg 1 should be the exact length (and time) as Leg 2. When the needle centers, the time into the station is the same as that required to fly the first leg. You can see, for instance, that a wind perpendicular to your original course could make a slight difference in the times on the two legs.

VOR Receiver Check

Part 91.25 of the Federal Aviation Regulations covers the tolerances of the airborne omni equipment as follows:

§91.25 VOR Equipment Check for IFR Operations.

(a) No person may operate a civil aircraft under IFR using the VOR system of radio navigation unless the VOR equipment of that aircraft —

    (1) Is maintained, checked, and inspected under an approved procedure; or

    (2) Has been operationally checked within the preceding 30 days, and was found to be within the limits of the permissible indicated bearing error set forth in paragraph (b) or (c) of this section.

(b) Except as provided in paragraph (c) of this section, each person conducting a VOR check under subparagraph (a)(2) of this section shall —

    (1) Use, at the airport of intended departure, an FAA operated or approved test signal or, outside the United States, a test signal operated or approved by appropriate authority, to check the VOR equipment (the maximum permissible

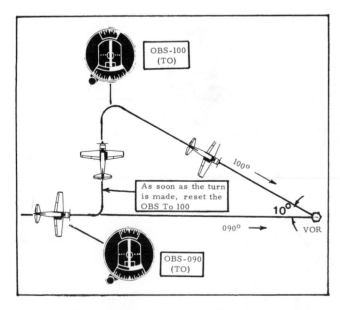

Fig. 5-15. Another time-to-station calculation method.

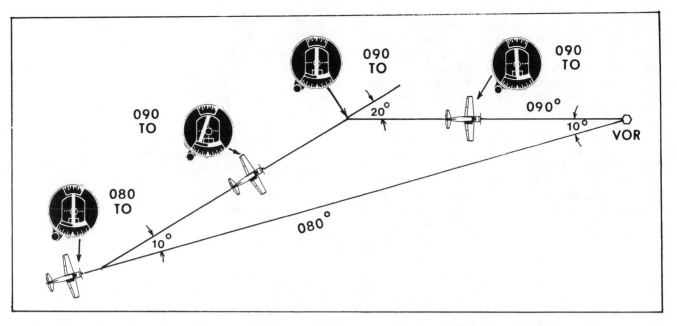

Fig. 5-16. The "double the angle off the bow" method of estimating time to the station.

indicated bearing error is plus or minus 4 degrees);

(2) If a test signal is not available at the airport of intended departure, use a point on an airport surface designated as a VOR system checkpoint by the Administrator or, outside the United States, by appropriate authority (the maximum permissible bearing error is plus or minus 4 degrees);

(3) If neither a test signal nor a designated checkpoint on the surface is available; use an airborne checkpoint designated by the Administrator or, outside the United States; by appropriate authority (the maximum permissible bearing error is plus or minus 6 degrees); or

(4) If no check signal or point is available, while in flight —

(i) Select a VOR radial that lies along the centerline of an established VOR airway;

(ii) Select a prominent ground point along the selected radial preferably more than 20 miles from the VOR ground facility and maneuver the aircraft directly over the point at a reasonably low altitude; and

(iii) Note the VOR bearing indicated by the receiver when over the ground point (the maximum permissible variation between the published radial and the indicated bearing is 6 degrees).

(c) If dual system VOR (units independent of each other except for the antenna) is installed in the aircraft, the person checking the equipment may check one system against the other in place of the check procedures specified in paragraph (b) of this section. He shall tune both systems to the same VOR ground facility and note the indicated bearings to that station. The maximum permissible variation between the two indicated bearings is 4 degrees.

(d) Each person making the VOR operational

check as specified in paragraph (b) or (c) of this section shall enter the date, place, bearing error, and his signature in the aircraft log or other record.

In addition, if a test signal radiated by a repair station, as specified in paragraph (b)(1) of this section, is used, an entry must be made in the aircraft log or other record by the repair station certificate holder or his representative certifying to the bearing transmitted by the repair station for the check and the date of transmission.

Note that the tolerance for ground checking is $\pm 4°$ and that for an airborne check is $\pm 6°$. The *Airport/Facility Directory* lists VOR receiver airborne and ground check points for various facilities.

The VOR test facility (VOT) transmits a test signal which gives the VOR user an accurate method of testing his receiver(s) on the ground. The airports having a VOT have the frequency listed in the *Airport/Facility Directory* with the other information pertaining to VOR check points.

When the receiver is tuned to the proper frequency and the needle centered by use of the OBS, it should indicate $0°$ when the TO-FROM needle indicates FROM, or $180°$ when the TO-FROM indicator says TO. The deviation from these figures is the error of the aircraft equipment. Some VOT's are identified by a continuous series of dots, while others use a continuous 1020 cycletone.

An RMI/VOR receiver (covered later in the chapter) will indicate $180°$ on any OBS setting when using a VOT.

The ground station accuracy is generally $\pm 1°$, but roughness is sometimes present, particularly in mountainous terrain. You may observe a brief left-right needle oscillation, such as would be expected as an indication of approaching the station. Always use the TO-FROM indicator as an assurance of passing the station, rather than assuming you're there when the needle suddenly pegs to one side.

A problem that sometimes occurs with the airborne equipment is that at certain prop rpm settings the left-right needle (or, more technically, the Course Deviation Indicator) may fluctuate as much as $\pm 6°$. A slight change in rpm will straighten out this problem (which, incidentally, can occur in helicopters as well). If you are having this sort of trouble, try the rpm change before casting aspersions on the veracity of your set (don't wreck it until you check it) or the ground station.

The Airport/Facility Directory contains the latest information on possible VOR problems. For instance, you could find that the Buckeye (Ariz.) VOR is unusable below 7000 feet MSL, beyond 40 nautical miles, and in the arc of radials 320°-345°, on account of reduced coverage.

VOR Identification

Always identify the station either by its Morse code identifier or by the code *and* automatic voice identifier. *Don't* identify, for instance, the Palmyra VOR just by hearing (on a VOR frequency) somebody in an FSS saying something about "Palmyra radio, etc." Remember that many Flight Service Stations operate several remote VOR's, and none may carry the name of the controlling facility.

If the VOR is down for maintenance, the code, or code and voice, is removed. The VOR receiver may indicate periodically. But if you don't hear the continuous identifier, don't trust it.

Remember, too, that being VHF, it only operates line of sight. If you are too low, it will have a warning flag showing.

It's suggested you get copies of the *Airman's Information Manual* and carefully read the section that goes into detail on the use of the VOR. In fact, if you plan on flying IFR, you should subscribe to that publication. It will be discussed in more detail in a later chapter.

VOR Receiver Antennas

In order to know some of the characteristics of your VOR receiver (and other radio equipment), you should know the locations of the antennas and which item of electronic gear uses which antenna. For instance, for airplanes with two transceivers, one of the communications antennas may be on top of the fuselage and the other on the bottom. When the airplane is sitting on the ground, the set with the bottom antenna may be blanketed out and unusable, as far as contacting ground control is concerned. The top antenna may be in a bad spot for communications directly over a facility. Rocking your wings may help or hinder.

Figure 5-17 shows some VOR receiver antenna types and probable locations on an airplane.

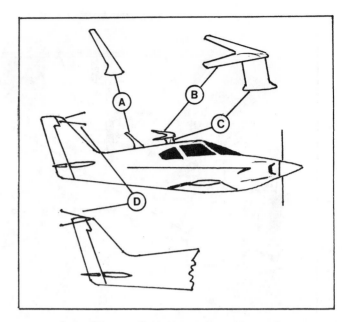

Fig. 5-17. Types of VOR and Communications antennas. (A) Broad-band communications antenna (118-136 MHz). (B) VOR/LOC antenna is combined with a (C) broad-band communications antenna. (D) VOR/LOC antennas. There are too many different models and types to show them here, but you should know which of the antennas on your airplane go to what radio and discuss with your instructor the possible weaknesses of each.

DISTANCE MEASURING EQUIPMENT — DME

This equipment, like the VOR, has done a great deal to take the headache out of instrument flying. When DME first came out and comparatively few were being used in general aviation airplanes, it pleased the pilot no end to be able to say that he was "exactly 17-1/2 miles east on Victor 116." Now DME is used by many airplanes, and controllers no longer fall off their chairs at such position pinpointing. (They've been watching you on radar, anyway.)

DME is a UHF facility, operating in the range from 962 to 1213 MHz. While the ground equipment is normally located at the VOR site (this is not the case for certain military installations), it's a separate piece of gear. Basically, it works this way:

When you select the station on your DME, the airborne equipment sends out paired pulses at a specific spacing (interrogation). The ground station equipment wakes up and transmits paired pulses back to the aircraft on a different frequency and pulse spacing. The time required for the round trip is read as distance (nautical miles) from the aircraft to the station. (As the pulses are moving at the speed of light — 161,000 nautical miles per second — you can imagine about how little time it takes to make a round trip of, say, 20 nautical miles.)

VORTAC is a combination of two facilities, VOR and TACAN (Tactical Air Navigation). TACAN, used by the military, provides both distance and azimuth information on UHF frequencies for the aircraft with the proper equipment installed.

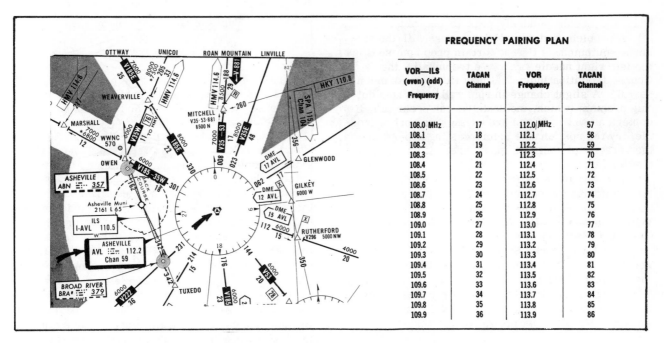

VOR—ILS (even) (odd) Frequency	TACAN Channel	VOR Frequency	TACAN Channel
108.0 MHz	17	112.0 MHz	57
108.1	18	112.1	58
108.2	19	112.2	59
108.3	20	112.3	70
108.4	21	112.4	71
108.5	22	112.5	72
108.6	23	112.6	73
108.7	24	112.7	74
108.8	25	112.8	75
108.9	26	112.9	76
109.0	27	113.0	77
109.1	28	113.1	78
109.2	29	113.2	79
109.3	30	113.3	80
109.4	31	113.4	81
109.5	32	113.5	82
109.6	33	113.6	83
109.7	34	113.7	84
109.8	35	113.8	85
109.9	36	113.9	86

Fig. 5-18. VORTAC frequency pairing plan.

Your equipment is able to pick up the distance information part of TACAN, but you'll use the VOR for azimuth information.

For each VOR frequency at VORTAC facilities there is an associated TACAN channel for distance information. For instance, all VORTAC's with the VOR portion having a frequency of 112.2 MHz have TACAN channel 59 (and that UHF frequency) associated with them. Current DME equipment uses this idea, and the DME frequency selector is set up as 112.2 rather than having a separate channel selector operating in the 962-1213 MHz (UHF) range. When you select 112.2 on the DME, you are actually selecting the proper UHF frequency and don't need to know what it is (Fig. 5-18).

On the Enroute Low Altitude Chart, the facilities having VORTAC are indicated as shown in Figure 5-18.

Basically speaking, on the National Ocean Survey chart, if a channel number is given with the VOR frequency box, then you can use DME equipment with that VOR facility (it's a VORTAC). Of course, you can also learn this by looking at the symbol in the center of the VOR rose as is shown in Figure 5-18. The quickest way is just to see if a channel is given. Each Ocean Survey chart has a legend which explains the symbols, and this will be covered later.

The DME equipment in the airplane measures the distance *direct* to the station or gives the *slant range*. If you are flying at an altitude of 6080 feet above the VORTAC, your distance indicator will never show less than a mile, even though you pass directly over the station, because you never get closer than 1 nautical mile (6080 feet) to it. Figure 5-19 shows this idea.

You will note in Figure 5-19 that at a distance of 5 miles, the lateral (or ground) distance is 4.9 miles. At a 2-mile indication, the lateral distance to the station is about 1.72 miles. Okay, so an error exists, but *you and the controller will always base reports and clearances on what the distance indicator says*. The error becomes very small at distances greater than 10 miles and is ignored. (The error also depends on altitude above the station, as well as the distance out.) *Don't* depend on the DME for station passage information; use your VOR TO-FROM indicator.

Some current DME equipment has the capability of reading off the distance from the selected VORTAC (in nautical miles), minutes to the station and ground speed. Figure 5-20 shows a DME receiver with these capabilities. You would select which of the three (miles, minutes or knots) you desired. This

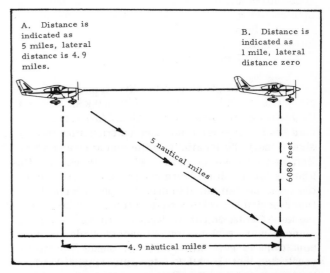

Fig. 5-19. The DME measures the distance direct to the VORTAC (slant range). The airplane at Point A would report his position as 5 miles (even if he calculated his exact ground distance).

Fig. 5-20. DME panel mounted, control box and indicator. This can be used with RNAV. (*Narco Avionics*.)

equipment can be used with Area Navigation (RNAV), to be discussed later in the chapter.

The DME uses a small antenna which weighs about one-fourth of a pound. It's located underneath the fuselage, usually in the center section area (Fig. 5-21). The identification is a single-coded one every 37-1/2 seconds, showing that the DME (only) is working. No ID, no working.

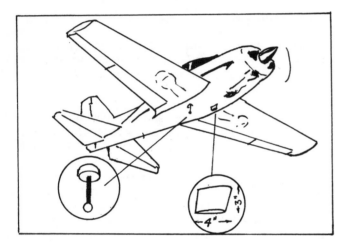

Fig. 5-21. DME antenna types.

LF/MF NAVIGATION AIDS

The radio beacon is a nondirectional ground facility that operates in the LF (30-300 kHz) and MF (300-3000 kHz) bands. (The old LF/MF ranges had four fixed legs — the radio beacon transmits in *all* directions.) The majority of the radio beacons are assigned a frequency in the 190-535 kHz range. Unlike the VOR, which gives geographic information, the radio beacon and airborne equipment give the station's position *relative* to the aircraft. You'll be using an ADF (Automatic Direction Finder) in the airplane to work with the radio beacon. Under certain conditions (precipitation static or thunderstorms), you'll find that the MDF (Manual Direction Finder) is more valuable, as will be covered later.

The ADF/MDF equipment can be used on radio beacons, compass locators, and certain types of the old LF/MF ranges.

Types of Ground Transmitters

Compass Locator

The compass locators are a part of the Instrument Landing System, and a particular ILS may have a compass locator at either, or both, the two (middle and outer) markers.

Figure 5-22 shows reduced-size ILS Approach Chart for Toledo Express Airport at Toledo, Ohio (TOL) with the LOM (Locator — Outer Marker) indicated. The frequency is 219 kilohertz and the identification is "TO" Morse code (continuously). If there were a compass locator at the middle marker (LMM) its identification would be "OL" and the LF/MF frequency would naturally be different from the LOM. Note that the LOM uses the first two letters of the three letter identifier for Toledo, and the LMM would use the second two letters. The LMM is being phased out at most locations except where it might be necessary for safety due to a local condition. Because the compass locators are used as a part of the ILS approach, they don't need to have a great range. They normally have a power output of less than 25 watts. Don't expect to receive one accurately

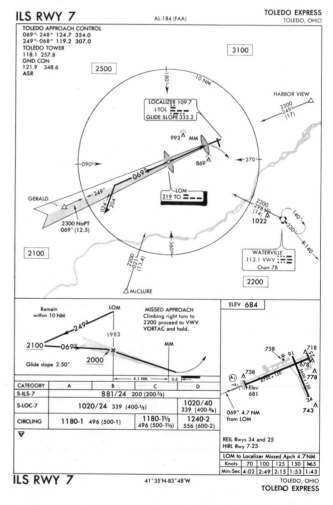

Fig. 5-22. A compass locator as depicted on an ILS Approach Chart.

at a distance over 15 miles. (The compass locator is considered to be a special class of radio beacon.) At some locations, higher power radio beacons, up to 400 watts, are used as outer marker compass locators. These generally carry Transcribed Weather Broadcast Information (discussed in Chapter 6). The complete ILS will be covered later in this chapter.

Radio Beacon

The term "radio beacon" is used for those non-directional LF/MF navigational or approach facilities *other* than compass locators — even though the LOM or LMM is a radio beacon also.

The radio beacon comes in three classes:

MH Facility — Power output less than 50 watts (expect up to about 25 miles of accurate reception under normal atmospheric and terrain conditions).

H Facility — Power output greater than 50 watts but less than 2000 watts. Expect about a 50-mile range, or less in some locations.

HH Facility — Power greater than 2000 watts (75 miles) and continuous identification in three-letter code as is the case for MH and H facilities. The radio beacons with voice facilities cut out the code during voice transmissions. You can get voice transmissions, except on those having the letter "W" (without voice), included in the class designation (*Airman's Information Manual*) or on the Enroute Chart or Approach Chart where the frequency is underlined if no voice is available.

Commercial Broadcast Stations

Commercial broadcast stations may be used for enroute ADF work, and a few semiofficial letdowns at uncontrolled airports may use them. The commercial station is more powerful but has the disadvantage of not giving continuous identification. At your home airport, you would recognize the announcers' voices and also get information from local advertising, such as when the announcer mentions "Schwartz's Livery Stable on Gitchygoomy Street." Basically, the commercial broadcasting station is a fine VFR homing aid and is good for picking up music or news, but it's not to be considered an IFR navigation aid.

The Loop Principle

You know that most portable AM radios (550-1650 kHz), such as you carry (or used to carry) on picnics with that good-looking member of the opposite sex, have directional properties. The reception was better (on the radio, that is) if it were turned in a certain direction.

The loop operates on the principle that the minimum reception is found when the plane of the loop is set perpendicular to the station. The maximum reception is obtained when the plane of the loop is in line with the station. Figure 5-23 shows a loop in the two positions.

Notice that the null or *minimum* reception area is much narrower (and, hence, would give more directional accuracy) than the *maximum* reception position. The null position of the loop is the one used in radio navigation. One way of remembering which position of the loop is null is to think in terms of most of the signal "slipping through the hole" — giving the minimum reception.

In earlier times, airplanes had a fixed loop antenna which allowed them to "home" into a station or fly to the station by pointing the nose at it. In a crosswind, a curved path always resulted (Fig. 5-24).

The loop could be accurately used on an LF/MF range only if the particular range had a continuous carrier wave or special extra antenna at the station.

There is the problem of 180° ambiguity with the old fixed loop and with the MDF. This impressive term means simply that, even though you have a null setup, you still don't know in which direction the station is located.

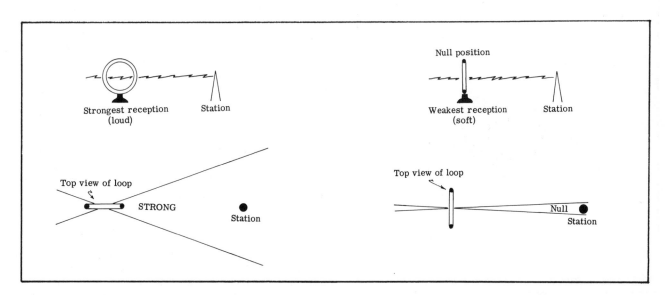

Fig. 5-23. The maximum and minimum reception loop positions.

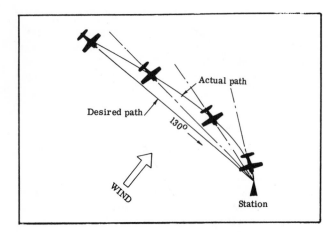

Fig. 5-24. Back in the olden days of the fixed loop and radio ranges, the airplane could only "home on the range."

The Manual Direction Finder

Suppose that things had just progressed to the point where MDF is being used by everybody and ADF hasn't been heard of yet. You're flying and are *temporarily disoriented with respect to the pre-planned route* (lost) and want to locate yourself with the MDF:

You tune in and *identify* a radio beacon and turn the loop until an aural null is found. The loop position indicator, or relative bearing indicator, shows that null is at a relative position, as shown in Figure 5-25.

The station can be in either one of the directions shown. What's your next move? Flip a coin? Eeeny Meeny, etc.? Or should a turn be made in the nearest direction of the null line? (A left turn of 60° in this case.) Either one of the three methods just mentioned should give you a 50-50 chance of heading for the selected station — but that's not good enough odds

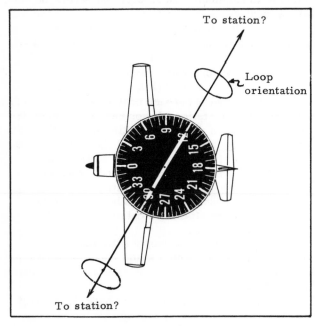

Fig. 5-25.

for your situation, so you use the scientific approach.

Turn the loop so that the null positions are through each wing tip, and turn the airplane to the right (which would be the quickest way to get the wing-tip null, as you can see in Figure 5-25), until the null is found again. Now the station is off one of the wing tips — but which one (Fig. 5-26)?

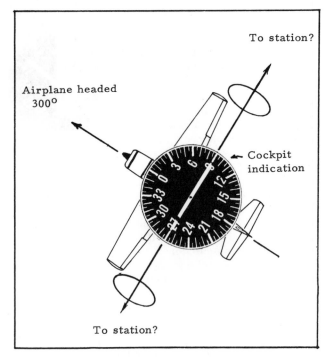

Fig. 5-26.

Note the heading required for the wing-tip null — and hold it. As you fly at a 90° angle to a line to it, the station will gradually move behind the wing. Your impression will be that the aural tone will start up again, and remain as you fly. When this happens, manually rotate the loop until you pick up the null again. Figure 5-27 shows what happens when the station is to the right.

If you had to move the right side of the relative position indicator back (or clockwise) to find the null, the station is to the right and vice versa. In this case, it is to the right. Shown by Figure 5-27, as an example, the station has moved back 20° behind the right wing tip so that the relative bearing to the station is 110°. For the example magnetic heading of 300°, the magnetic heading to the station is 300 + 110 = 410° (or, rather, 050° would be more like it). You could start the turn to the right and, meanwhile, turn the loop indicator to the 0°-180° bearing position as shown by Figure 5-28. It's a very good idea to time this process. If you fly through 10° of null travel, this time multiplied by 6 gives approximate time to the station. For 20°, multiply by 3 as mentioned in the section on the VOR.

If there were no wind, the problem would be simple. You'd hold a heading of 050°, the null would stay put, and you would fly directly over the station. With a crosswind, keeping the nose pointed at the

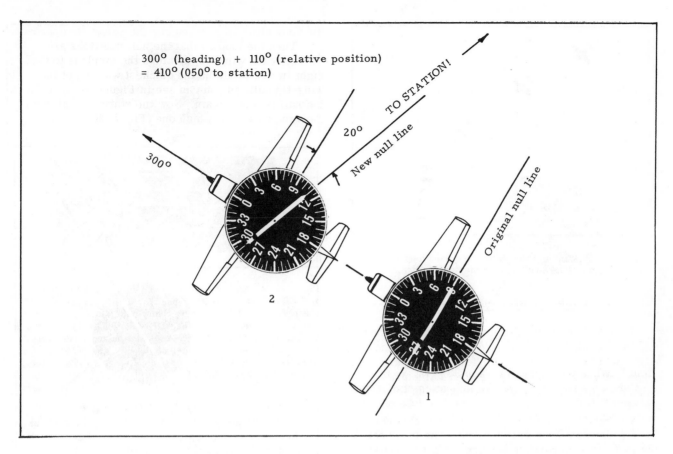

300° (heading) + 110° (relative position) = 410° (050° to station)

20°

TO STATION!

New null line

300°

Original null line

2

1

Fig. 5-27. As the plane continues on the 300° heading, the station (null) moves behind the wing.

station (homing) would result in the curved path of the guy with the fixed loop back in Figure 5-24. You can move your loop and so can "track" into the station.

Suppose, as an example, you have the exact winds at your altitude and work a wind triangle on your computer. You find that a 10° left correction is necessary for your airspeed on the 050° course. You

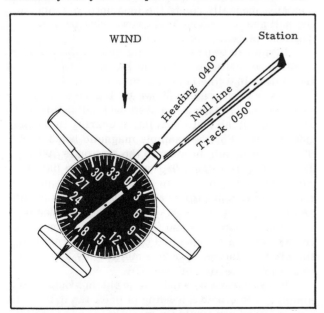

WIND

Station

Heading 040°

Null line

Track 050°

Fig. 5-28. Correcting for wind (tracking).

would turn *left* to the required heading of 040° and turn the loop indicator 10° to the *right*, as shown in Figure 5-28.

Assuming that you hold the heading of 040° and the wind remains the same, you should have a null all the way to the station.

In VFR flying you use the MDF (and ADF) as an aid only to help locate an airport and depend on visual sighting when you get fairly close in. In the majority of VFR situations you'll see the airport long before reaching the station. In IFR work you don't (normally) have the ability to see landmarks and must depend on your radio navigation equipment.

Assume that you are IFR, tracking in on the 050° course (MDF), and approach control wants you to report passing the station. If you keep the loop on the nose-tail position used for tracking in, you'll have a nice null all the way in — and past the station — and you will fly off into the rising sun, as the station sinks slowly over the horizon behind. What do you do? Turn the loop 90° to the position it was on the inbound track about two minutes before your estimated time to the station as obtained originally (Fig. 5-29).

It's best to turn the loop a little early, rather than a little too late which would mean missing the station passage and another delay while you oriented yourself again.

Tracking inbound, you should have the correct heading pretty well set up and so shouldn't be too badly off by holding it for a couple of minutes "in the

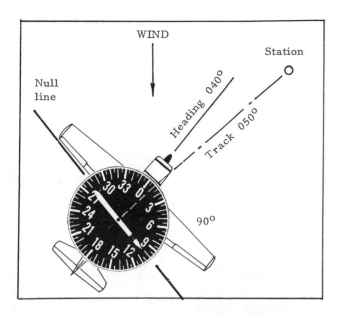

Fig. 5-29. At an estimated two minutes before reaching the station set the null indicator 90° to the enroute setting.

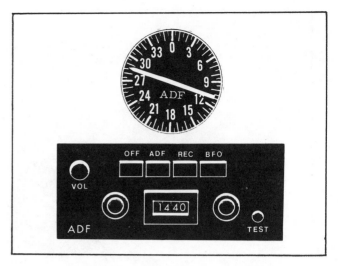

Fig. 5-30. An ADF console and indicator. The REC (Receiver) or ANT (antenna) position is used for tuning and clearer AM reception. The BFO (beat frequency oscillator) is used for stations outside the U.S. The ADF selection gets the needle in action. IDENT selection increases station acquisition range and helps the reception of distant station signals (Morse code identifier). The ADF receiver normally has a range of reception of from 190 to 1750 kHz.

blind." When you turned the loop 90°, you got the maximum signal again, and the indication of a station passage will be the return of the null or loss of aural signal for a short period. Note that a station passage can be indicated whether the airplane passes directly over the station or not. (The station will "pass through" the null line.)

On the problem of orientation just covered, you might use a rule of thumb for the point of turning the loop for station passage. If your original estimate is 12 minutes, rotate the loop 2 minutes before you expect to get there. In other words, use one-sixth of the inbound time as a lead time (for an estimate of 18 minutes, turn the loop 3 minutes early). It's best never to use less than 2 minutes, unless you have a very good idea of your groundspeed.

The MDF is less affected by thunderstorms and precipitation static than the ADF. Most higher priced radio compass installations allow the pilot to select the MDF setting and manually rotate the loop as necessary. It would be unusual in this day of radar, ILS, and VOR to have to use the MDF, but it was covered as an introduction to the ADF idea. Figure 5-30 shows an aircraft radio compass installation.

Automatic Direction Finder

The next step in the development of airborne direction finding equipment was the ADF. The problem of 180° ambiguity was solved — the needle always points to the station. (Of course, this depends on whether you have the set tuned properly, or whether a big lusty thunderstorm happens to be closer to the station than you are. The latter consideration, however, has been somewhat less of a factor with later equipment.) Always tune for maximum signal when tuning in a station, either by ear or by using the max signal indicator (older type equipment). Newer ADF equipment has decreased the pointing-to-the-

thunderstorm tendency, but it is still present in varying degrees. The ADF can be used for tracking around a station to approach from a certain direction, as discussed for the omni.

Basically, the idea consists of turning the airplane until the ADF needle is pointing either 080° or 280° relative (depending on which way you want to track around) and holding the resulting heading until the needle moves back 20° (say, to 100° relative) then making a 20° turn to the right. This puts the needle back at the 080° relative position, and you fly the new heading until the needle is again back to the 100° relative position, and the step is repeated. You can hold up traffic for quite a while doing such an action, and for practical purposes this technique is no longer used. (It's guaranteed to make the radar controllers feel surrounded.) This is strictly a practice maneuver which will get you in the habit of "thinking ADF." In turning into the station, allow yourself about 5° as was mentioned for the omni.

You can also use the "double the angle off the bow" method in checking the time from a station. Head toward the station so that the needle indicates a 000 relative bearing; note the heading on the H/I. Turn right or left 10° and fly this new heading until the relative position of the needle has doubled. In other words, when you turn that 10°, the needle will be offset that amount. Note the time. Fly the new heading until the needle is offset 20° from straight ahead and check the clock. The time it took for this to occur is the number of minutes you are out from the station when the needle hits the 20° mark. Turn and track inbound.

Tracking around the station and the "double the angle, etc.," are highly unlikely to be required in an actual situation of flying IFR in the continental United States. However, they are good exercises for getting used to thinking in terms of ADF and MDF.

The use of ADF (or MDF), other than for homing or simple tracking, is probably the toughest part of your training in getting ready for the flight test. There are no real shortcuts to learning the procedure in the air. (You can probably work ADF problems very easily with paper and pencil at a desk where you won't have to worry about flying the airplane and other such minor distractions.) If you can visualize your heading and relative position (geographically) with respect to the station, the problem becomes much easier.

ADF Tracking To and From a Radio Beacon

In flying IFR, you could receive a clearance from approach control such as: "N3456J IS CLEARED FROM ZILCH INTERSECTION DIRECT TO THE OUTER MARKER FOR AN ILS APPROACH, REPORT AT THE OUTER MARKER, OUTBOUND." Your situation could be like that of Figure 5-31.

Homing, as shown by the dashed lines, would result in your hitting the localizer at too great an angle and would result in overshooting (unless you were really sharp) and possible overcorrections. This is no way to start an ILS approach.

The numbered points you see in Figure 5-31 will be covered in more detail in the illustrations and text which follow.

The best technique would be, upon reaching Zilch intersection, to turn directly to the compass locator at the outer marker (which you have already thoughtfully tuned in and *identified*). Check your H/I when the ADF needle is at the 000° position — this is your course to the locator. Take a cut, as you feel necessary, to correct for wind (say, as an example, 15° — it's a strong wind). The ADF and heading situation should look like that shown in Figure 5-31.

Okay, so you cross the LOM (Point 2), and the needle swings 180 (from the 015° to the 195° relative position). You would now want to track outbound on a 090 course on the localizer, so that you could make a procedure turn and get back to the business of the approach. For the sake of an illustration in flying outbound ADF on a particular track, take a look at Figure 5-32, which covers the area of the LOM in more detail than is given by Figure 5-31. (The numbers correspond to those in Figure 5-31.)

As soon as you cross over the outer locator (and it will be assumed that you crossed directly over it), you'll turn to the right. As close in as you are and with the wind as it is, an angle of 15° would be a good one for the example. So in order to intercept a line oriented 090° at a 15° angle from your present position at Point 3, you'll have to use a heading of 105°. Here's where the whole secret of using the ADF is found. *Don't* memorize numbers, but visualize your position relative to the radio beacon and the localizer. If you are flying at a heading of a 15° angle to the required line (the localizer) and it is to your right, the ADF needle will be 15° off tail to the

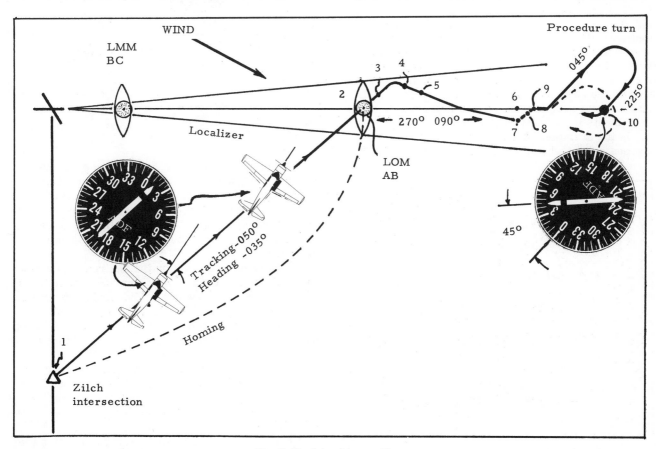

Fig. 5-31. A tracking problem.

89

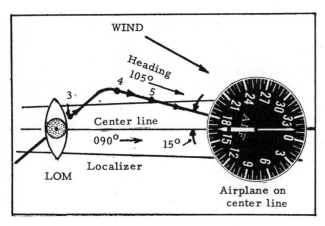

Fig. 5-32.

right when you intercept the desired track.

Note that when you first turn, the ADF needle moves from a tail position (Point 3) to an approximate wing-tip position (Points 4 and 5). As the plane flies toward the line of intercept, the needle moves farther back to the "tail" (180°) position. When the needle is the same angle off the tail as that of the intercept angle, well, you're on the line. As shown by Figure 5-32, the needle would be pointing 165° relative. As a check of this number racket, your heading is 105°, the relative bearing of the LOM is 165°, so the course to the station is 270°—or you are 090° from it. If you turn the airplane so that the needle is right on the tail, the H/I should indicate 090°.

The present method of marking the ADF azimuth dial (from 0°-359° relative) sometimes results in confusion and makes for ponderous work. Most pilots when working ADF don't think in terms of the numbers on the instrument face but look at the number of degrees the needle is offset from a prime position, such as the nose, tail, or either wing. There's nothing like computing the course to a radio beacon on instruments in turbulent air (Fig. 5-33).

In the example, you should lead the turn a few degrees so that when the plane is turned to the 090°

Fig. 5-33. ADF work can be extremely interesting at times.

heading, it will be centered on the localizer. (The amount of lead would depend on the angle of approach and groundspeed.)

Tracking outbound from the LOM entails the same general idea as tracking inbound. The idea is to set up a crab and check the results. Basically, the idea is that if the airplane is maintaining its track, the amount of crab angle as shown by the H/I should be equalled by the outset of the ADF needle from the 180° position. (For the sake of the example assume a 15° wind correction is necessary and you are holding a heading of 075°.)

In other words, if at any time you turned to the outbound course of 090°, the needle should fall exactly on 180° on the ADF relative bearing indicator. (You can "turn" the airplane to the 090° in your mind.)

So you're tracking outbound, holding the 075° heading you started with, and look down to see that the ADF needle has moved *from* the magic 195° relative (or 15° to the left of the tail). It's now 20° from the 180 position, or 200° relative (Fig. 5-34).

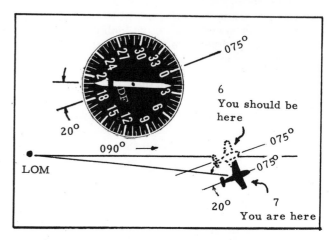

Fig. 5-34. The airplane has drifted from the required bearing. (Also see Figure 5-31.)

As Figure 5-34 shows, you have moved down to Point 7 (you should have been at 6 at this point), and assuming that your H/I (and magnetic compass) are correct, you have been *under*correcting for wind. In other words, you haven't turned to the left to move the needle; therefore, the airplane must have moved off to the right.

You want to get back on the desired course so will make a turn to the left. Turn 30° further left so that the heading is now 045°. You would note that by turning the airplane left 30°, you also "turned" the relative bearing needle another 30°, so that the relative bearing of the LOM is now 230°, or 50° off to the left of the tail.

You are now headed back to the required course of 090° at an angle shown by Point 8 in Figure 5-35. *When you are back on course, the relative bearing of the needle should be 45° off the left of the tail, or indicating 225°.*

Basically, the ADF needle always indicates the straight line to the radio beacon (assuming proper tuning, no thunderstorms, etc.), and you can "move" its relative position by turning the airplane or by the wind's moving the airplane's position relative to the station (or both).

As you get back to the course, the needle will indicate as shown by Point 9 in Figure 5-35. After you are on course, turn back to the right with an estimated correction for wind. As a "cut" of 15° didn't hold you before, then a cut of 20°, or a heading of 070°, should be tried.

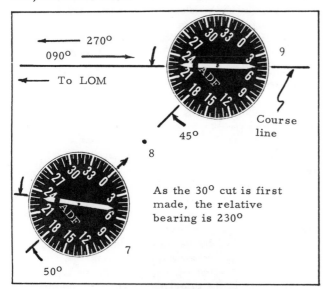

As the 30° cut is first made, the relative bearing is 230°

Fig. 5-35. Getting back on course. Although the airplane is headed at a 45° angle to the course, its path will be at a slightly shallower angle because of the crosswind. However, the rule still stands that it will be on the required bearing when the combination of its *heading* and the relative bearing combine to give the proper number.

On an ILS you would have the localizer as an aid. But this example is just using the ADF, as would be the case for the outbound portion of an ADF approach or tracking outbound from a radio beacon on a predetermined course — the principle is the same. For one thing, on an ADF *or* ILS approach, you would be tracking outbound in most cases for only two to three minutes, and we've played rather fast and loose with the amount of drift and correction angles used. (If the deviation indications and corrections likely to be seen under actual conditions were shown in the drawings, it would be a tough proposition to see that anything was happening.) The amount of cut used to return to the course depends on how fast you left it (the faster you left it, the greater the cut) and how far you are from the station. To cover all possibilities would require too many pages.

Figure 5-36 shows an ADF approach for Winnebago County Airport at Oshkosh, Wisconsin. This approach uses the outer marker locator (LOM) for the primary navaid (see the profile). Winnebago County Airport has an ILS approach also, so that the LOM is doing double duty.

The procedure turn symbol is shown with the outbound (224°) and inbound (044°) headings. You will note that the first part of the procedure turn is a turn

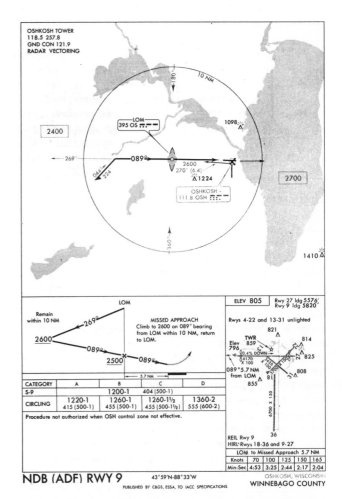

Fig. 5-36. An ADF approach to Winnebago County Airport, Oshkosh, Wisconsin.

to the left (which is considered standard) and at 45° to the outbound course of 269°. Reversing the course in the procedure turn is done by making a 180° turn *away* from the station — in this case, a *right* turn is made from the heading of 224° to 044° to get back to the course line of 089° *to* the LOM. But, getting back to the "canned" example (Fig. 5-31).

It's time to make a procedure turn, so you turn the airplane 45° to the left to a heading of 045°.

Normally, you will turn to a *heading* of 045° to the outbound leg and not worry about the wind. That heading would be flown for one minute, and a right turn of 180° would be made (to 225° in the example). With the wind situation as shown in the example, or with a strong headwind on that outbound part of the procedure turn, you might just expect to be very close to either the localizer center or prechosen inbound course very shortly after completing the 180° turn. (Or you may have to continue the turn on around past the inbound heading under extreme conditions.) Look back at Figure 5-31 at the dashed lines to see what *could* happen.

Under no-wind conditions, the second leg of the procedure turn is shorter in both time and distance, as shown by the solid lines in Figure 5-31. For a strong wind from the southwest, you might expect a pretty good period of straight flight getting back to

91

the localizer or letdown course. This is where "see-ing" your situation and recognizing clues as to what's been happening are helpful. If you've been correcting to either side of the localizer (outbound), you can visualize what will probably happen to your procedure turn. It seems that some pilots correct on the localizer for a strong wind very well but are completely nonplussed when the procedure turn seems to turn to worms. Valuable time and distance are wasted on the final approach leg while they try to get organized again.

At the completion of the 180° turn, the ADF needle will be showing less than a 45° deflection from the nose. So you'll fly that leg until the needle moves back to that relative position at Point 10. You are headed at a 45° angle to the prechosen line to the station. When the needle indicates that relative bearing, you are on that line. Of course, you will start the turn slightly before the ADF needle has slid all the way back to a relative bearing of 045°, to allow for the distance required in the turn. How much lead you'll allow will depend on how fast you are approaching the prechosen line. In theory, any-way, if you start the turn for Point 10 with the proper amount of lead, the ADF needle will point dead ahead (000°) when you roll out on the inbound heading of 270°. For a crosswind, you would set up an esti-mated correction or could roll out of the turn with a correction already set in. You would then track in-bound using the technique covered earlier. The pro-cedure turn discussed here has been the old fashioned "one minute out and a standard rate turn" type since

this gave a better look at the wind problems. The ar-row for the procedure turn as shown in Figure 5-36 is the symbol indicating that the pilot may use the type of reversal procedure he chooses and the old type is no longer "mandatory." Other reversing procedures will be discussed later in this book.

The altitude situation was not covered in this example. You would naturally be letting down at various points, but this will be discussed in later chapters.

ADF/MDF Antennas

Shown in Figure 5-37 are several types of ADF/MDF antennas.

INSTRUMENT LANDING SYSTEM (ILS)

The ILS is the backbone of the approach aids and allows the pilot not only to fly a precise course to the runway but also to fly a precise descent, which allows for lower landing minimums than for VOR or ADF approaches. Figure 5-38 shows the ILS com-ponents.

Localizer — Course information.
Glide Path — Descent information.
Marker Beacons — Range or distance information.

In addition, the compass locators mentioned ear-lier and special approach lighting will be a part of the system.

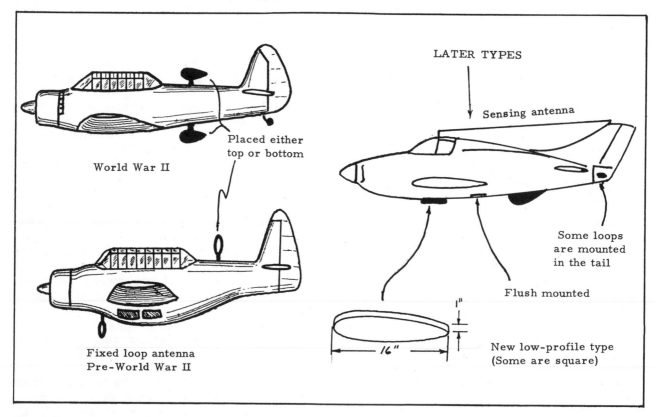

Fig. 5-37. A cross section of fixed and ADF/MDF antenna types. Some of the new low-profile antennas have "fixed loops" (the loop is "turned" electronically). The sensing antenna is a fixed "reference" antenna.

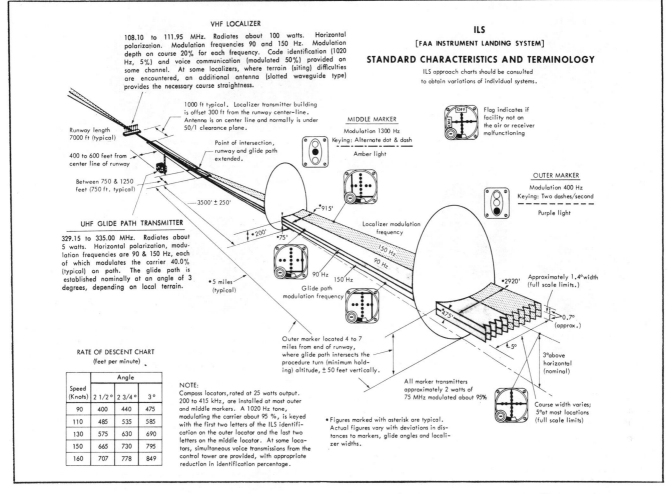

108.10 to 111.95 MHz. Radiates about 100 watts. Horizontal polarization. Modulation frequencies 90 and 150 Hz. Modulation depth on course 20% for each frequency. Code identification (1020 Hz, 5%) and voice communication (modulated 50%) provided on some channel. At some localizers, where terrain (siting) difficulties are encountered, an additional antenna (slotted waveguide type) provides the necessary course straightness.

ILS approach charts should be consulted to obtain variations of individual systems.

1000 ft typical. Localizer transmitter building is offset 300 ft from the runway center-line. Antenna is on center line and normally is under 50/1 clearance plane.

Runway length 7000 ft (typical)

400 to 600 feet from center line of runway

Between 750 & 1250 feet (750 ft. typical)

Point of intersection, runway and glide path extended.

MIDDLE MARKER
Modulation 1300 Hz
Keying: Alternate dot & dash
Amber light

Flag indicates if facility not on the air or receiver malfunctioning

OUTER MARKER
Modulation 400 Hz
Keying: Two dashes/second
Purple light

—3500' ± 250'—

UHF GLIDE PATH TRANSMITTER

329.15 to 335.00 MHz. Radiates about 5 watts. Horizontal polarization, modulation frequencies are 90 & 150 Hz, each of which modulates the carrier 40.0% (typical) on path. The glide path is established nominally at an angle of 3 degrees, depending on local terrain.

•915'
•200'
•75'

Localizer modulation frequency

150 Hz
90 Hz

90 Hz   150 Hz

•5 miles (typical)

Glide path modulation frequency

•2920'
•475'

Approximately 1.4° width (full scale limits.)

0.7° (approx.)

5°

3° above horizontal (nominal)

Course width varies; 5° at most locations (full scale limits)

Outer marker located 4 to 7 miles from end of runway, where glide path intersects the procedure turn (minimum holding) altitude, ± 50 feet vertically.

All marker transmitters approximately 2 watts of 75 MHz modulated about 95%

RATE OF DESCENT CHART
(feet per minute)

Speed (Knots)	Angle		
	2 1/2°	2 3/4°	3°
90	400	440	475
110	485	535	585
130	575	630	690
150	665	730	795
160	707	778	849

NOTE:
Compass locators, rated at 25 watts output. 200 to 415 kHz, are installed at most outer and middle markers. A 1020 Hz tone, modulating the carrier about 95 %, is keyed with the first two letters of the ILS identification on the outer locator and the last two letters on the middle locator. At some locators, simultaneous voice transmissions from the control tower are provided, with appropriate reduction in identification percentage.

• Figures marked with asterisk are typical. Actual figures vary with deviations in distances to markers, glide angles and localizer widths.

Fig. 5-38. The Instrument Landing System components (*Airman's Information Manual.*)

Localizer

The localizer transmitter is located at the far end of the ILS runway and operates on the *odd* VHF frequencies between 108.0 and 112.0 (or, more properly, from 108.10 through 111.95 MHz). You remember that the *even* frequencies in that frequency range are used by VOR's.

The localizer signal emitted is adjusted to produce an angular width of between 3° and 6°, as necessary to provide a linear width of approximately 700 feet at the runway threshold. Five degrees is considered "standard."

The transmitter sends two signal patterns, one modulated at 90 cycles per second (cps) and the other at 150 cps. When the airplane is at a position where these patterns have an equal signal strength, it is on an extension of the center line of the runway, and the localizer needle (the same needle used for VOR work) is centered. If the pilot is on the proper heading to keep the needle centered, the airplane will remain on the line down the center of the runway.

The 150 cps (or blue) area is to the right of the center line for the airplane approaching on the "front" or normal course. (Obviously, this leaves the 90 cps or yellow sector on the left.) The localizer course extends on in the opposite direction, and this portion may be used for a "back course approach" (Fig. 5-39).

For the airplane approaching on the back course, the blue sector will be on the left. You'll note on your omni head that the blue is on the *left* side of the face and the yellow is on the right. *The needle always indicates the color of the sector the airplane is in. If the airplane is on the front course, you correct toward the needle as would be expected.* Look at Airplane A in Figure 5-39. It's over in the blue sector, and the needle shows this. The needle also indicates that the center line is to the left. Airplane B is in the yellow sector, and the needle gives this news (and indicates that a correction to the right is needed). Airplane C is right on the center line (at least, temporarily).

Looking at the back course, Airplane D is to the left of the center line and in the blue sector. The needle is to the left. But *on a back course inbound, you correct opposite to the needle.* Airplane E is to the right of the center line (in the yellow, as the needle indicates), and he would have to correct *left, away* from the needle. Airplane F is, of course, right down the center.

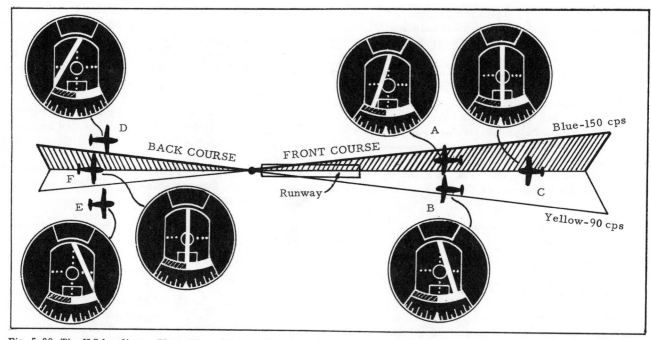

Fig. 5-39. The ILS localizer. The setting of the omnibearing selector has no effect on the needle indications when the set is tuned to a localizer frequency. However, many pilots set up the published inbound course on the OBS as a quick reminder of the base course when on approach. You might also remember that a left needle indication is in the blue sector, a right indication in the yellow.

The airplanes on the front course (A, B, and C), should they choose to fly on past the runway down the localizer, would still correct *into* the needle. If the airplanes on the back course flew on past the airport, they would still have to correct *against* the needle. The needle gives the straight story on the color sector, whether front or back course, but you only correct *into* it if you are flying the airplane on the localizer using the front course magnetic bearing. Some airplanes have special equipment (Course Director) to set it otherwise. It's doubtful that your trainer

will be so equipped, but check on it. Figure 5-40 shows the plan views for front and back course approaches for the Greater Southwest International (Dallas-Fort Worth) Airport. The localizer identification consists of a three letter code preceded by the letter I (··) transmitted on the localizer frequency (I — GSW for Greater Southwest and I — TOL for Toledo Express back in Figure 5-22). Some few chosen locations have voice facilities so that the tower may give advisories to the pilot on approach.

The localizer is only 5° wide, or 2.5° on each

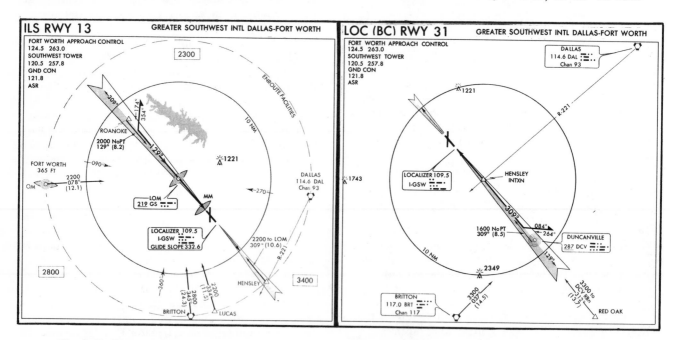

Fig. 5-40. Plan views of front and back course ILS approaches for the Greater Southwest International Airport.

side of the center line. The needle, when deflected completely to the side of the deviation indicator, indicates that the airplane is 2.5° (or more) from the center line. You remember that for most omni heads a full deflection of the needle to either side meant that you were 10° (or more) from the selected radial. One of your biggest problems in starting work with the ILS is this sensitivity of the needles on both the localizer and glide slope (particularly the glide slope). You can consider the needle to be approximately four times as sensitive on the localizer as it was for omni, so watch those corrections — don't overdo it!

The localizer antenna on the airplane is the same one used for the VOR (Fig. 5-17).

Glide Slope (or Glide Path)

The glide slope transmitter is UHF (329.15 to 335.00 MHz), and 40 channels are available for use. Each glide slope channel is associated with a particular localizer frequency, as shown by Figure 5-41. Older airborne equipment had a separate selector for the glide slope. On newer crystal controlled equipment, it's just a matter of turning the G/S power switch on and selecting the proper localizer frequency, and the G/S is automatically tuned in.

ILS			
Localizer mHz	Glide Slope mHz	Localizer mHz	Glide Slope mHz
108.10	334.70	110.1	334.40
108.15	334.55	110.15	334.25
108.3	334.10	110.3	335.00
108.35	333.95	110.35	334.85
108.5	329.90	110.5	329.60
108.55	329.75	110.55	329.45
108.7	330.50	110.70	330.20
108.75	330.35	110.75	330.05
108.9	329.30	110.90	330.80
108.95	329.15	110.95	330.65
109.1	331.40	111.10	331.70
109.15	331.25	111.15	331.55
109.3	332.00	111.30	332.30
109.35	331.85	111.35	332.15
109.50	332.60	111.50	332.9
109.55	332.45	111.55	332.75
109.70	333.20	111.70	333.5
109.75	333.05	111.75	333.35
109.90	333.80	111.90	331.1
109.95	333.65	111.95	330.95

Fig. 5-41. Localizer/glide slope frequency pairings.

The glide slope (or glide path) transmitter is situated 750 to 1250 feet in from the approach end of the runway and 400-600 feet from the center line. Whereas the localizer can be used from both directions (at some airports, obstructions make a back course approach unfeasible), the glide slope is a one-directional item at present.

The glide slope works on basically the same idea as the localizer (except that it is oriented differently) in that the center of the glide slope is found at the area of equal signal strength between 90 and 150 cps patterns.

The glide slope extends about 0.7° (7/10°) above and below its center. On the omni head, a full deflection of the glide slope needle represents this

amount (0.7°). (If you think the localizer is going to be sensitive as compared to the omni, wait until the first time you get in close on the glide slope!)

The glide slope normally is between 2.5° and 3° above the horizontal, so that it intersects the middle marker at about 200 feet and the outer marker at about 1400 feet above the runway elevation. (Don't get the idea that a particular ILS glide slope varies from 2.5° to 3° — it stays the same, but that represents the extremes of the various installations in the country.)

There are false courses (reversed in sensing) at angles much greater than the usual (2.5° to 3°) path (Fig. 5-42).

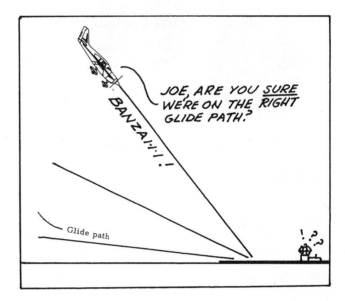

Fig. 5-42.

Nulls occur above the glide slope which result in the flag on the omni head showing (as well as the centering of the glide slope needle). You'll notice this particularly when flying over the airport with the glide slope tuned in.

The glide slope signal flares from 18-27 feet above the runway so don't plan on using it all the way down to touch-down.

If you are making a back course approach, don't expect to get glide slope information. If the glide slope needle is acting like it knows what it's doing in such a situation, *ignore it*. You'll have to depend on VOR cross-bearings or other aids for knowing when to descend.

Glide Slope Antennas

Figure 5-43 shows typical glide slope antenna installations. Most older types are exposed while newer ones may be installed behind a fiberglass nose cone. More than one greenhorn has tried to tow the airplane by the older type glide slope antenna.

Marker Beacon

The marker beacon is a VHF navigational aid with a frequency of 75 MHz. The airway fan marker

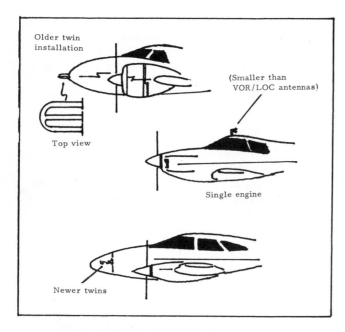

Fig. 5-43. Types of glide slope antennas.

was a valuable aid during the early days of IFR and the LF/MF range and was designed to give the pilot a position location. The fan markers were keyed to indicate on which range leg they were located as well as their relative position on that leg.

The first leg of the radio range past the north N quadrant was the number one leg and transmitted single dashes. The next marker out on that leg was identified by two dots and a dash. (The airways fan markers have a power output of about 100 watts.)

The Z-marker or station locator was an aid in locating the "cone of silence" on the old LF/MF ranges.

The airborne equipment consists of a three-light aural and visual system, as shown in Figure 5-44. The airway fan markers and Z-markers are modulated at 3000 Hz (cycles per second) and trigger the white light. The Z-marker gives a steady aural tone and light, and the airway fan markers give the coded signal plus the white light.

The outer and middle markers are low-powered elliptical markers (up to 3 watts power output). The

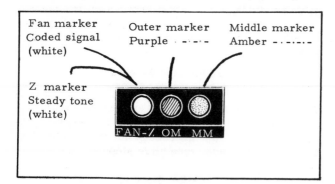

Fig. 5-44. Fan, A, outer and middle marker aural signals and the color signals as would be indicated on the marker beacon light panel.

outer marker, located at from 4 to 7 miles from the approach end of the runway, is keyed at two dashes per second and is modulated at 400 Hz. A purple light and aural tone are indicated. The middle marker is modulated at 1300 Hz and triggers an amber light and an alternating dots and dashes aural signal. It's located 3500 feet, plus or minus 250 feet, from the end of the runway.

Several selected larger airports have installed DME facilities to be used with the ILS localizer, so that distance information will be available throughout the approach. Possibly in the future, the OM and MM will be phased out.

At some airports having "Category II" minimums (a special lower-minimums approach) an inner marker is used to indicate the point at which the airplane is at the "Decision Height" (land? or execute a missed approach?) on the glide slope between the middle marker and the landing threshold. The IM is modulated at 3000 Hz (cycles per second) and identified with continuous dots keyed at six dots per second. (Replaces FAN-Z marker spot on the panel.)

Marker Beacon Antennas

Figure 5-45 shows three types of marker beacon antennas.

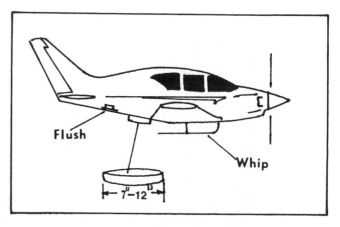

Fig. 5-45. Marker beacon antennas.

TRANSPONDER

Basically, a transponder is an airborne "radar transceiver." It picks up the interrogations of the ATCRBS (Air Traffic Control Radar Beacon System) and transmits or "replies." There are actually six modes or divisions of radar beacon services available. Each mode is a function of pulse spacing.

Mode Application

1 Military (IFF)
2 Military (IFF)
A/3 Common (ATC)
B Civil (ATC)
C Civil (altitude)
D Civil (unassigned)

At first, there were only 64 codes available on Mode A/3, but now 4096 codes are available for use with that mode.

Modes 1 and 2 are military tactical modes, and Mode A/3 was just discussed. Mode B has been designated a civil air traffic mode, but it will not be used in the United States. Mode C is used for automatic altitude transmissions. Mode D has been established, but its use has not been specified.

Mode C will be discussed in the next chapter from the controllers' standpoint, but *you* should have an idea of what aircraft equipment is required. Basically, altitude reporters speak to ground radar stations through the aircraft's transponder, letting the ATC computer know your altitude relative to standard pressure at 29.92 inches of mercury. The computer converts this to known local pressure and displays the precise altitude on the scope. (There will be some sample Mode C displays in Chapter 6.) Encoders are required in Group I TCA's (Terminal Control Areas) and are most useful in providing positive altitude separation in IFR flying.

There are two basic types: (1) the encoding altimeters that combine a normal barometric altimeter with a built-in altitude digitizer and (2) "blind" encoders which operate independent of the aircraft's altimeter as Figure 5-46 shows.

A transponder will accept only the interrogator signals of its mode. After it accepts the interrogation, the transponder will transmit a coded reply. The ground station is "tuned" in to receive this re-ply (after all, he told you to reply on this code), and you no longer are wandering in the wilderness.

When you file an IFR flight plan, you would write in after the aircraft type, for instance, a slant and the letter "B." This tells ATC that you have a DME and a 4096-code transponder with no altitude encoding. The flight plan designators are as follows:

/X no transponder
/T 4096-code transponder, no altitude encoding capability
/U 4096-code transponder, with altitude encoding capability
/D DME, no transponder
/B DME, 4096-code transponder with no altitude encoding
/A DME, 4096-code transponder with altitude encoding capability
/M TACAN only, no transponder
/N TACAN only, 4096-code transponder with no altitude encoding
/P TACAN only, 4096-code transponder with altitude encoding
/C RNAV, 4096-code transponder, no altitude encoding
/F RNAV, 4096-code transponder with altitude encoding
/W RNAV and no transponder

The suffix is only to be used on the *flight plan* and with the *aircraft type*. When you've done this, you've passed the word to ATC. There's no need of

KEA 125: 20,000 ft. unit with internal altitude encoder.

KEA 126: 35,000 ft. unit with internal altitude encoder.

KE 127: 20,000 ft. blind encoder for use with existing altimeter.

KAE 128: 30,000 ft. matched set standard altimeter and blind encoder.

Fig. 5-46. Encoding altimeters and blind encoders. The system is linked to the transponder to send altitude information to the Air Traffic Control Radar Beacon System (ATCRBS). There will be more detail on this in the next chapter. (*King Radio Corp.*)

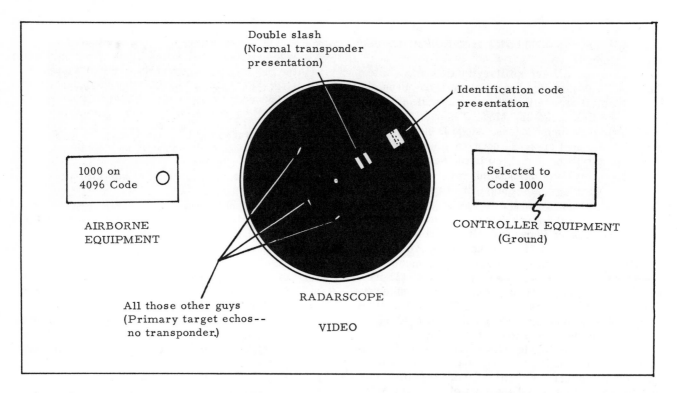

Fig. 5-47. The combination of your transponder and the radar controller equipment settings makes the airplane show up as a more outstanding target.

mentioning it further, either on the rest of the flight plan or over the radio.

Some airborne equipment has only a 64-code capability, so you can see that there could not be that many combinations of pip presentations possible to display on the screen. He tells you what he wants, dials in that code, and you do the same, with the result that you will show up on the screen as shown in Figure 5-47.

The presentation shown in Figure 5-47 is called a "double slash" presentation. If there is a considerable amount of traffic in your area, the controller may elect to have you show up as a single slash.

Maybe your transponder is of the older type with only a 64-code capability. If so, you would select Code 10 (Code 1000 is based on the 4096-code system) (Fig. 5-47).

If there is question as to your identification, the sector controller may have you "Ident," which means that you will press the "Ident" button on your transponder and your pip will "fill in the space" between the two slashes. Your target will stand out from all the rest, as shown by Figure 5-47.

In earlier times, separate radar equipment was necessary for the use of the IFF (Identification — Friend or Foe). But now the primary radar (which picks up reflected signals) and secondary surveillance radar (which receives transmitted airborne transponder or IFF signals) are presented on the same screen. The Mark X (IFF) transponder is being phased out because of the limited number of codes available.

Figure 5-48 shows the basic idea of the synchronized primary and secondary radars. Note that the interrogator transmits on a frequency of 1030 megahertz, and your transponder replies on 1090 megahertz *for all codes*.

The VFR code is 1200 for any altitude.
Loss of Communications — 76(00)
Emergency — 77(00).

Notice that each number is the basic code. With the increase in traffic, the Center may have several airplanes climbing (for instance) at low altitudes,

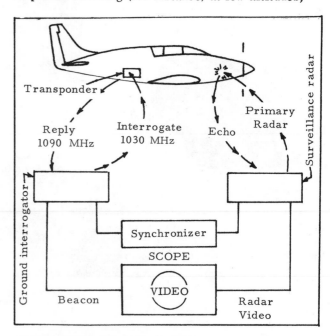

Fig. 5-48. Basic primary (echo return) and secondary (transponder reply) radar system.

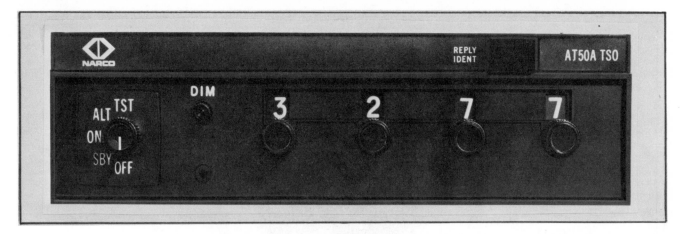

Fig. 5-49. A transponder control panel. (*Narco Avionics.*)

and with the greater number of codes available, several codes in the 1000 range may be used (1010, 1020, etc.).

If you are flying VFR (even not on a flight plan) and want your presence known to the Center or approach control so they can vector their traffic out of your way, you'd better keep squawking code 1200 (or other, as applicable) as you fly. Don't "squawk" an IFR Code unless you are really IFR.

The emergency code is 77(00), and all secondary surveillance radar sites are ready to receive this one at all times. (You remember that the other codes are chosen by the controller at will, but he must keep this one standing by.) Code 7700 practically lights up the entire radar screen (and fires off red lights), and you definitely won't be overlooked.

There is even a transponder code to use if you are being hijacked. It's best not to publish it and you should check with your local facilities (FSDO, tower, etc.) for this information.

Figure 5-49 shows the control head of a transponder. The transponder antenna is similar in appearance (and location on the aircraft) to the DME antenna.

RADIO MAGNETIC INDICATOR (RMI)

The Radio Magnetic Indicator (hereafter to be known as the RMI) is aimed at simplifying the problems associated with VOR and ADF work.

Basically, the RMI combines the airplane's heading information with VOR and/or radio beacon input. Figure 5-50 shows the presentation of a two-pointer RMI head.

The pointers may both be hooked up to two VOR receivers or two ADF's or, as shown in Figure 5-50, one of each. The VOR and ADF needles in Figure 5-50, under these conditions, will always indicate the magnetic bearing to the station(s) tuned (automatic type).

You can see in Figure 5-50 that the airplane is headed 030°, the magnetic course to the VOR is 350° (or, as shown by the tail of the needle, you are on a magnetic bearing of 170° from the VOR), and the

magnetic course to the radio beacon you've tuned in is 080° (no adding of heading and relative bearing to get 080° — you just read it off).

As far as VOR work is concerned, there is no TO or FROM indication. If you want the bearing, just tune the VOR and a built-in servosystem takes care of it. The head of the omni arrow is TO, and the tail is FROM. If you want to take cross-bearings, you tune the omni and read the tail of the pointer — or pointers, if you have omni hooked up to both of them. You could then read the magnetic bearing FROM the station (or stations). You would track outbound, using the "tail" of the particular pointer as necessary.

AREA NAVIGATION

Area Navigation (RNAV) equipment in the aircraft allows the pilot to "move" a VORTAC and fly directly to these "waypoints" rather than having to fly directly to the "real" VORTAC. Some line

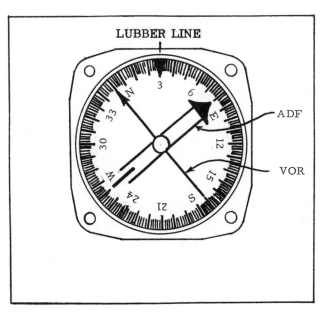

Fig. 5-50. The radio Magnetic Indicator indications.

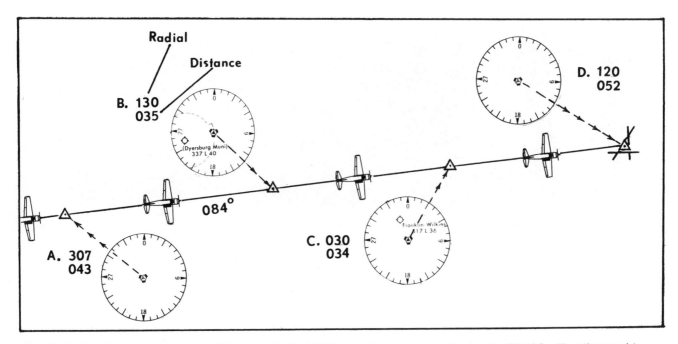

Fig. 5-51. Area Navigation equipment allows straight line VOR navigation away from the actual VORTAC. The pilot tuned in VORTAC (A) and "moved" it along the 307 radial a distance of 43 nautical miles to put in on his route. At (D) he has "moved" the VORTAC to the destination airport (120 radial, 52 nautical miles) and can track directly to it. The distance to a waypoint may be read from the DME indicator of the aircraft. The pilot tracks directly to or from the phantom stations in this example. The aircraft's course is 084° magnetic. Some equipment will indicate lateral and longitudinal distances from the selected waypoints. You have to be careful in setting up your own "airways" because of reception and obstruction clearance problems.

computer systems can move a VORTAC up to 67 miles in any direction from its actual position.

Basically, the system allows the pilot to set up the desired position of the phantom station (or way point) by dialing into the computer the bearing and distance from a "real" VORTAC; the VORTAC is then "moved" to that location. Figure 5-51 shows how an airplane may fly direct from waypoint to waypoint and finally move a VORTAC to a destination airport not so equipped.

VISUAL DESCENT POINT

The VDP is a specified point on the final approach of a non-precision approach (no glide slope), electronically marked by either a DME fix or a 75 MHz marker beacon. It defines a definite point to which the pilot should proceed at or above the minimum descent altitude. It also identifies the point in the approach from which a normal descent (about 3 degrees) can be made to the touchdown point on the runway *if* the pilot has visual contact with the runway, light, etc. A VASI system is normally associated with the VDP. If the VDP is reached and the ground isn't seen, a missed approach is likely.

Throughout this book, various new aids will be mentioned; but the primary purpose here is to cover the fundamentals of VOR, ADF, etc., so that you may, with only a short period of instruction on use of the new aid, go on about your business. The aim is not to copy the avionics manufacturers' operating manuals. Look at the equipment in one of the bigger airplanes on the field, and fly the plane if you have a

chance. For further information, check the listing at the end of the book for avionics (and other) manufacturers' addresses.

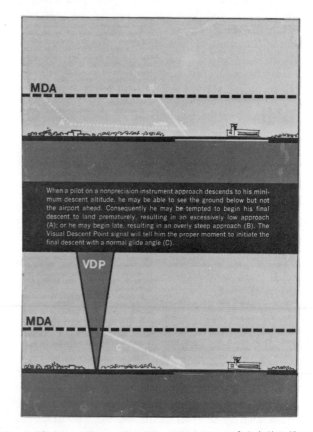

Fig. 5-52. Visual Descent Points. *(FAA General Aviation News.)*

100

Chapter 6

COMMUNICATIONS AND THE CONTROL OF AIR TRAFFIC

WHEN YOU were first flying out of that uncontrolled field, you were on your own and kept separated from other aircraft by the eyeball method. This worked fine because there was very little transient traffic, and you could keep up with everybody anyway. Things were so quiet that everyone would quit hangar flying and rush out of the airport office to watch a transient airplane land. There would be much discussion as he approached as to where he could be from. (You would know as soon as he got down, but it was part of the program to throw out conjectures.) Later, as transient traffic began to pick up and radio equipment was available for the airplanes, a Unicom, or aeronautical advisory station, was set up in the office. It was just that — an advisory station — and woe betide the guy who started thinking he was a controller in the tower at Washington National Airport and issued take-off or landing clearances to all and sundry within range. Unicom was a further step; you could communicate with approaching and departing airplanes, and it helped you keep up with the increased traffic in and near the field. Unicom got to be pretty much old hat, and you'd been using the radio like a professional — around your home airport.

Then came the day when you had to fly some parts for Joe, the operator, over to the field at Whitesville where they had a TOWER. Well, you kind of sweated that one on the way over and practiced your lines until you had them cold and, of course, got mike fright and called "ZEPHYR ONE TWO THREE FOUR PAPA, THIS IS WHITESVILLE TOWER,

OVER" — instead of the other way around — and wound up drenched with sweat by the time you had landed and taxied in. (And, if you recall, there was a little problem contacting ground control, so there was a period of limbo when you weren't talking — or listening — to anybody.) Well, they didn't arrest you and even let you taxi back out and take off when you got ready to go home. Going into controlled airports is routine now. You even know the controllers personally and visit the tower and drink coffee with them in the airport restaurant. In fact, after you'd done it a few times, you felt more comfortable going into a controlled airport than landing at an uncontrolled airport unfamiliar to you. You figured that at the controlled field somebody was helping you keep up with other traffic, even if the responsibility for safety was still yours.

Then you flew into St. Louis (or Memphis or Chicago) and used approach control for the first time to get traffic information. You also later used VFR traffic advisory (radar), both for approaching and departing from big airports. For en route service, you've been using the Flight Service Stations for some time, so that's no problem.

Probably at this stage of your career, you've used all of the facilities available to you except the Air Route Traffic Control Center (ARTCC), but it would be a good idea to review a little to tie it all together. The purpose of this chapter is to cover the communications areas generally; the specifics will come later.

FLIGHT SERVICE STATION (FSS)

Flight Service Stations have the prime responsibility for preflight pilot briefing, enroute communications with VFR flights (and IFR flights, too, if necessary), assisting lost VFR aircraft, originating NOTAMS, broadcasting aviation weather, accepting and closing flight plans, monitoring radio NAVAIDS, participating with search and rescue units in locating missing aircraft and operating the national weather teletype systems. At some locations FSS's take weather observations, issue airport advisories, give airman written examinations and advise Customs and Immigration of transborder flights.

Selected Flight Service Stations will provide Enroute Flight Advisory Service (EFAS) on a frequency of 122.0 MHz. This service was implemented on the West Coast, and now covers the 48 contiguous states. (See the *Airport/Facility Directory* for locations.)

THE TOWER

Local Control

Local control is the function that pilots think of as "the tower." The operations of the tower are broken down into three main divisions or "positions."

Local control has jurisdiction over air traffic within airport traffic areas. (*Airport traffic areas* are five statute miles in radius from the center of a *controlled airport* and extend from the surface up to, but not including, 3000 feet above the surface.)

A *control zone* also extends in a 5-mile radius from the center of the airport but may have legs or extensions for IFR approaches or departures. Control zones extend upward from the surface to the base of the Continental Control Area (14,500 feet MSL) and are thought of more in terms of IFR operations.

The tower is considered to control the traffic pattern entry and the pattern itself, including take-offs and landings. The local controllers are in the glassed-in part of the tower, as their control is dependent on visual identification of aircraft for take-offs and landings. The local control (tower) frequencies are in the *Airport/Facility Directory*.

Ground Control

Ground control regulates traffic moving on the taxiways and those runways not being used for take-offs and landings. Ground control will coordinate with the tower if you have to cross a "hot," or active, runway. Ground control is, naturally, on a different frequency than that of the tower. You can imagine what radio clutter would result if some pilots were asking for taxi directions while other pilots were calling in for landing instructions.

For simplification and summary of local (tower) and ground control duties: Local control has

jurisdiction of aircraft in the process of landing and taking off. This includes aircraft while in the pattern and *on* the active runway. Ground control is used for ground traffic at the airport *other than on the active runway during the take-off or landing process*. The ground controller will be in the tower beside the local controller. (In some cases, the same man may talk to you in both capacities, but on a different frequency, of course.)

You may get your IFR clearances on the ground control frequencies, but this will be covered later.

Approach Control

Approach control may be a radar or nonradar setup. For a nonradar facility, the approach controller is in the glassed-in portion of the tower with the other two positions. His primary duty is to coordinate IFR traffic approaching the control zone, but he may coordinate IFR and VFR traffic during marginal VFR conditions, as well as coordinating VFR traffic at busy terminals even in CAVU conditions.

Approach controllers using radar are usually in an IFR room which is located at the tower building but not necessarily in the glassed-in portion. Approach controllers work directly with the tower, and their duties are rotated between tower and IFR room positions.

A simplified summary of facilities is given at the end of this chapter.

Clearance Delivery

At the less busy airports, you'll be given your instrument clearance on the ground control frequency, usually after you reach the warm-up area. (You'll be notified that the clearance is forthcoming.) However, at the airports with heavy IFR traffic, this could cause a great deal of clutter on that frequency, so a special frequency called "clearance delivery" is set up. It's just as the name implies, a frequency used strictly for pretake-off IFR clearances. You'll be told to contact clearance delivery on such-and-such a frequency. This may be done (as before) at the warm-up spot or at the ramp before taxiing at certain locations. *Don't* be taxiing out on clearance delivery and taking a clearance when you should be listening to ground control and minding the store. If you have a trustworthy copilot, he might put on a set of earphones and be dealing with clearance problems to save time, while you and ground control are working to get you to the active runway in one piece. (Taxiing into a large, expensive, immovable object while copying a clearance is hardly an ideal way to start a flight.) The clearance delivery frequency is given with the other airport data in the *Airport/Facility Directory*. Clearance delivery has nothing to do with the direct control of air or ground traffic.

Departure Control

You've used VFR traffic information in departing, so, this won't be entirely new to you — except that now you'll be flying by references to the gages.

Shortly after you've taken off on your IFR flight, the tower (local control) will say something like: "THREE FOUR PAPA CONTACT DEPARTURE CONTROL." (You will have already been told the frequency by ground control or clearance delivery, as applicable.)

You: "BLANK DEPARTURE (CONTROL) THIS IS ZEPHYR ONE TWO THREE FOUR PAPA, OVER."

Departure Control: "ZEPHYR THREE FOUR PAPA, BLANK DEPARTURE, RADAR CONTACT. TURN LEFT HEADING ZERO ONE ZERO, MAINTAIN FOUR THOUSAND, REPORT LEAVING THREE (THOUSAND)." You would acknowledge and carry out their instructions. The radar may vector you off airways to a particular reporting point (such as an intersection) or may give you a heading to join your course (airway). At any rate, it sure makes it a lot easier for you than moving out on your own. Other traffic may require that they vector you at what seems to be a roundabout way to get started, but it's for the best.

After you've gently been led from the "group grope" around the airport, you'll be told to contact another facility (usually the Air Route Traffic Control Center), and the frequency will be given, which you will acknowledge and use.

AIR ROUTE TRAFFIC CONTROL CENTER

When you filed your IFR flight plan at the Flight Service Station, you set off a beehive of activity. Your plans, hopes, and dreams (and flight plan) for the next few hours go into the hands of the Air Route Traffic Control Center for your area.

The United States (the 48-states portion) is divided into 20 areas, each the responsibility of a particular ARTCC (Atlanta, Memphis, Los Angeles, etc.). Figure 6-1 shows the location and relative size of the Memphis Center area.

The job of each Center is to coordinate IFR traffic within its area and alert the next Center of your approach to *its* area. In nearly all areas, if you're not flying at too low an altitude, they will be able to monitor your flight by radar all the way. In fact, your flight is so well monitored that you may be quietly chided at being "four miles south of course," or may hear other ego shattering statements that

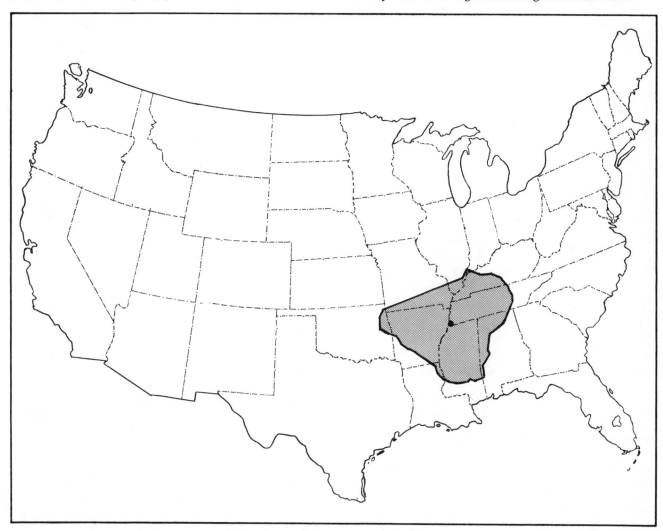

Fig. 6-1. The approximate size and location of the Memphis Air Route Traffic Control Center Area.

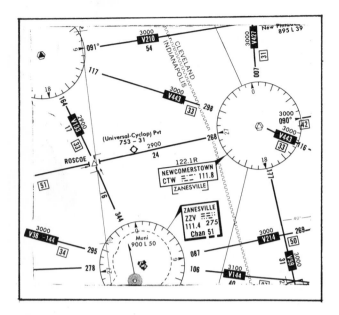

Fig. 6-2. The boundary between two Center areas as depicted on an Enroute Chart.

blare out over the cabin speaker for your passengers to hear. Figure 6-2 shows the boundary between two Centers as depicted on an Enroute Chart.

As the U.S. is broken down into Center areas, so are the Centers divided into Sectors.

Figure 6-3 shows a simplified look at the Memphis Center boundary and the Sectors within this area.

The heavy black line is the route of an example instrument flight to be discussed in this and later chapters.

Figure 6-4 is the Center area showing the Sectors, 5 areas of responsibility (Central, North, East, South, and West) plus airways and other details. Each Sector has its own controller Sectors and assigned VHF and UHF frequencies.

Looking at Sector 23 as an example, it shows that it is the low altitude Sector with GRAHAM VOR included (GHM-L) and the VHF/UHF frequencies are 125.85/381.4 MHz respectively. The information in the box (D-32 and A-57) are dial codes for use between Sectors and other Center business. The numbers are for the D (Manual) and A (Assistant) control positions, which will be covered shortly.

Suppose you are filing an IFR flight plan from Memphis International Airport to Nashville Metro Airport via V-54S, V-7 below 18,000 feet, the route as shown by the heavy line in Figures 6-3 and 6-4. (Sure, going the straight route via V-16 would be the best way, but, for this example and the sample trip later in the book, the V-54S, V-7 route will give a better chance to look at some extra details.)

Figure 6-5 is part of the basic layout of the Sector positions of the Memphis Center. The High Altitude (FL 240 through 330) and Ultra High Altitude Sector positions (FL 35 through 600) aren't shown in detail. Take a look at the low altitude Sector positions (surface to include FL 230), and look back at Figure 6-3 at your route on the trip to Nashville.

You'll be flying through Sectors 8, 13, 9, and 23 and these are the controller positions that will be working you.

You'd file your IFR flight plan with the FSS by phone or in person. It's sent to the 9020 Computer complex at the Center by teletype. A master copy is kept by the computer. The "A" control position for each Sector involved in your flight (see Fig. 6-5) will have its strip coming off of the teletype. Since you will be going into Sector 8 first, the computer will activate the flight strip printer at that station and "run off" your first strip.

The computer also sends a departure strip to the appropriate Approach Control (or, if you prefer, the departure control) when the Flight Plan is received (30 minutes before ETD), and tells APC the airport of departure (not always the big one with the tower), requested altitude(s) and route of flight. The computer assigns a discrete Beacon Code for the full trip (departure, enroute *and* arrival) if the computers (ARTCC and ARTS) can coordinate. It may be that you'll have to use a new code for the departure or destination airport if they aren't tied in with the Center computer.

When the IFR flight plan gets to the 9020 Computer, it analyzes it for route, departure area and will buy any altitude you put on the flight. It will reject erroneous routes and airspeeds, so if you put in your flight plan that you're going from Memphis to Nashville via some airway that runs north and south along the West Coast and plan to do it in a Cessna 150 cruising at a TAS of 500 knots, the computer will sneer and spit it out. It will reject all "errors" except those controller prerogatives such as cruise and climb rates.

ARTS (Automatic Radar Terminal System) facilities are tied into the Center computers and give automatic acquisition when the aircraft leaves the runway and ARTS passes the departure time to the Center computers. This is known as a Silent Departure Clearance. There will be no verbal communications between the APC and Center. Approach (departure) control will talk to the *pilot*, however.

If the pilot doesn't want a SID (Standard Instrument Departure — which will be covered in more detail in Chapter 8) he will so indicate on the remarks portion of the flight plan, and the departure strip will have typed on it "NO SID." You'll be given verbally in detail the procedure in getting away from the airport and on your way. For instance, clearance delivery might give you a clearance such as:

NOVEMBER 3456 JULIET CLEARED TO THE NASHVILLE AIRPORT. MAINTAIN 3000 UNTIL CROSSING THE 310 RADIAL OF HOLLY SPRINGS VORTAC (ETC.) AS FILED. MAINTAIN 5000. MAINTAIN RUNWAY HEADING FOR RADAR VECTOR. DEPARTURE CONTROL FREQUENCY 124.15.

If, after you've filed, you decide on a different route to Nashville (or maybe even changed your destination) contact the FSS and they will pass the word to the Center. Your clearance, in order to avoid confusing it with the earlier flight plan, would

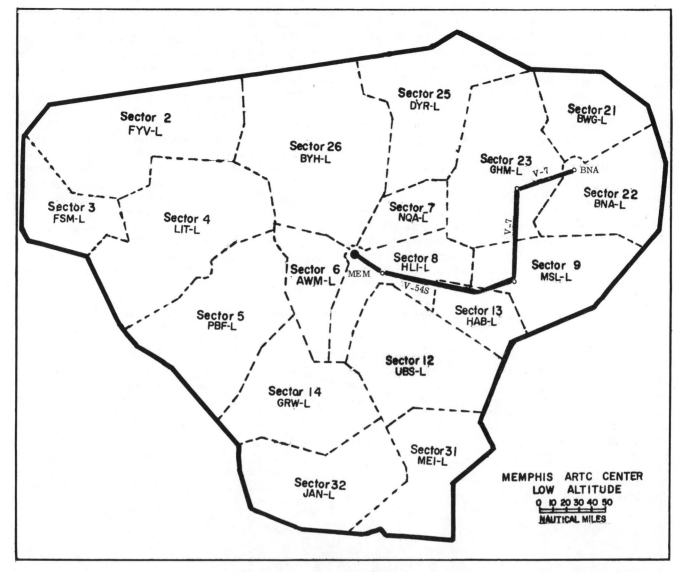

Fig. 6-3. A simplified look at the Sectors in the Memphis Center Area. The heavy black line indicates the route of the sample IFR flight used in this book.

contain the full route structure (no "cleared as filed"). Incidentally, the computer will hold your flight plan for 2 hours after your estimated departure time; after that you must refile.

Figure 6-6 shows that through coordination with the Center, departure control may climb you before reaching its boundary.

Radar handoffs are completely automatic if the aircraft is on course. This includes ARTS to Center, Sector to Sector, and Center to Center. There's no chatter between controllers, but the computers "talk" to each other. If the aircraft is off course when crossing a boundary, the pertinent controllers have verbal communications.

The computer automatically computes groundspeeds when the aircraft passes reporting points and updates the proper Sectors via a readout device, therefore the times are extremely accurate.

Note in Figure 6-5 that each Sector has different stations:

R (Radar Controller) — He follows you on radar

and makes handoffs. He assures separation and gives clearances. Under normal circumstances he's the number one man of the Sector.

D (Manual Controller) — He should be able to maintain contact when (or if) the radar is inoperative. He keeps up the strips and follows the actions of the aircraft so that if radar contact is lost he can immediately move in to control. If the computer radar, with its electronic data-following process, fails he would move to keep up with the traffic by voice (and position reporting) only if necessary.

A (Assistant Controller) — The Assistant Controller mans the flight strip printer and removes flight strips. He checks route changes, enters flight plans filed in the air and will get weather for pilots who request it.

Note again in Figure 6-5, that the A positions are shared in most cases by two sectors (A7 performs the Assistant Controller duties for Sectors 7 and 8).

Still looking at the Center Sector floor plan, you'll see BU 8, BU 6, BU 4, etc. These are back-up

105

radio positions to be used as needed. The "C's" noted at various intervals are control positions for coordination between sectors, but with the use of the computer, these positions are being phased out.

Other important stations are shown in the layout:

DSO (Data Systems Officer Position) — The DSO keeps the computer going and coordinates between the electronics technicians and controllers.

A/C (Assistant Chief Controller Position) — The Assistant Chief on duty is in charge of the control rooms.

MC (Military Coordinator) — This position is responsible for security procedures and mass movements of military aircraft.

SUP A (Super-assistant) — He can amend flight plans and he coordinates with Flight Service Stations and the military. This position is called the *Master Assistant's Position* and it makes sure that any flight plans sent from Flight Service Stations and military

facilities are in the proper format to be entered into the computer.

TMC (Traffic Management Coordinator) — He coordinates with major terminal facilities and may reroute traffic in case of problems in those terminal facilities.

Memphis Center has four remote radar antennas for the 9020 computer system in addition to the main complex at the Center. All radars are fed to the computer; it selects the best returns and feeds it to the Sectors. The R controller in a Sector normally doesn't know which radar antenna is in use by the computer.

There is also a broad band (non-computer controlled) radar available as back-up at each sector position and, if the computer radar fails, this can be used as in the olden days with the controller moving "shrimp boats" by hand across the now-horizontal screen. Shrimp boats are clear plastic markers with the aircrafts' N-numbers or airline flight

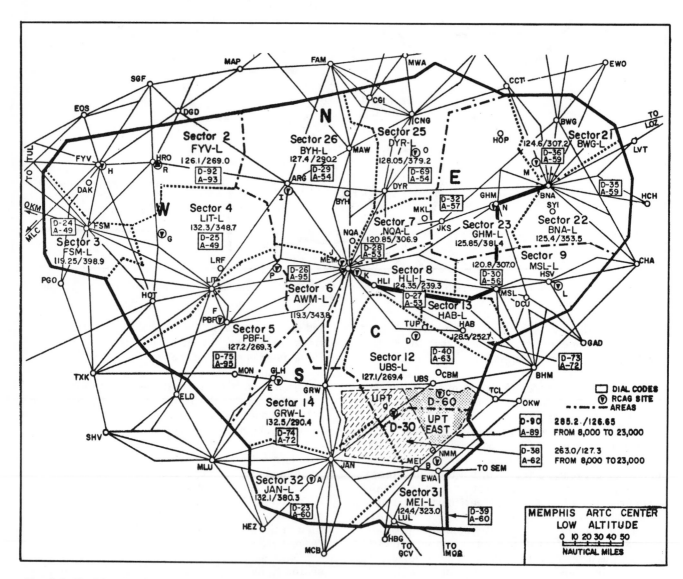

Fig. 6-4. The Memphis Center Area. The Sectors, airways, RCAG (Remote Communications-Air to Ground) sites and Sector frequencies are included. The shaded areas in the southeast corner of the Center Area (D-30 and D-60) are special areas connected with military training. Note the sample route to Nashville drawn as a heavy line.

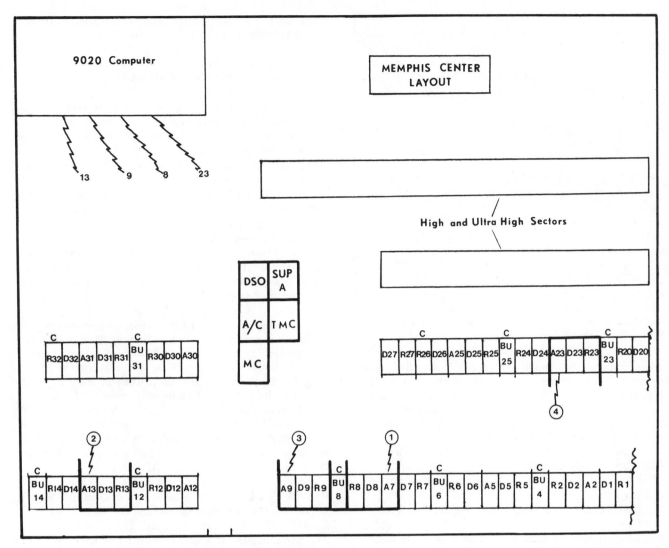

9020 Computer

High and Ultra High Sectors

DSO	SUP A
A/C	TMC
M C	

Fig. 6-5. The layout of the low-altitude Sector positions. Some positions have been left out to simplify the illustration.

numbers and altitudes as appropriate (Fig. 6-7).

So, if the 9020 computer system fails, the older radar (or broad band radar) and shrimp boats can be used. If both radars fail, the controllers will require that you go to the old fashioned pilot-estimates-and-position-reporting-operation. It could be that as

a further move you'd have to give your position report to the nearest FSS for direct line relay to the Center. This is the way IFR flying was done for many years before the good coverage of radar and direct contact with the Sector controllers. (Also the navigation was done by the four-leg low frequency

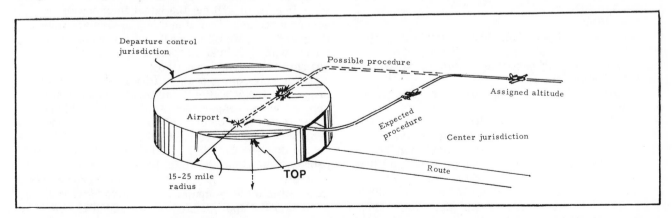

Fig. 6-6. Departure Control will usually hold you below a certain altitude until you are out of its area of jurisdiction. If traffic permits, departure will coordinate with the Center and you may be climbed to the assigned altitude sooner. The approach (departure) control area probably won't be a smooth round cylinder as shown, but may have corridors and projections.

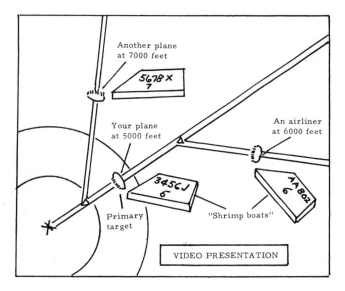

Fig. 6-7. The "shrimp boat," a clear plastic indicator marked with the aircraft number and assigned altitude, is moved with the target of the airplane by the Sector controller, if the computer radar system fails.

radio range as discussed in Chapter 5, so things are better now even in the worst case.) If the controller is operating in the blind he'll keep any crossing traffic 10 minutes apart, or have 1000 feet altitude separation. Soon a complete computer back-up system will be available at all Centers.

Okay, suppose you have to work directly with Flight Service Stations? It's not as convenient as working directly with the Center, but there's no problem.

If you had to make a position report to a FSS you'd use the PTA-TEN procedure. Suppose you're over Muscle Shoals VOR, your last verbal contact with the Center was when they told you that radar contact was lost. Let's say you can't talk on the Sector frequency because your com set doesn't go that high.

You could call Muscle Shoals radio on the proper frequency and after establishing initial contact you'd give the following information:

P — Position (Muscle Shoals VOR)
T — Time Over
A — Altitude
T — Type of Flight Plan. The Center knows full well that you are blundering toward Nashville, but Muscle Shoals doesn't have this knowledge so you'd tell them that you are IFR.
E — ETA at the next compulsory reporting point (Graham VOR)
N — Next succeeding reporting point after that estimated. You'd say "Nashville," which is the next reporting point after Graham, but wouldn't give an ETA for Nashville. Or, putting it together: MUSCLE SHOALS RADIO THIS IS ZEPHYR 3456J, OVER. (and, after contact): MUSCLE SHOALS AT 31, 5000 FEET, INSTRUMENT FLIGHT RULES, GRAHAM AT 55, NASHVILLE, OVER. (Muscle Shoals would confirm and probably give you an altimeter setting.

Your time over MSL, altitude and estimate to GHM would be called into the Center by phone.)

SECTORS AND STRIPS

Figure 6-8 shows sample strips received by the Sector teletypes for the route.

Strip 1: Looking at the far left section of the strip (A) the N-number of your airplane and equipment on board, (it's a Zephyr ZA-6 (/A equipment on board) with a filed true airspeed of 148 knots. This is Sector 8's strip (Ø8). The 7 (Ø7) indicates that it will go to the A7 position which assists Sectors 7 and 8. The 495 is your CID (Computer Identification) number for the entire flight and is printed in red. When a controller wants computer information about your airplane's actions in flight he would interrogate the computer by punching in this number and "asking" it what he needs to know. The Ø1 indicates that this is strip number one.

Looking at the next bit of information on this strip (C) it notes that you are leaving Memphis with a proposed time-off, say, of 2040Z. The small arrow (↑) notes that this will be a climb portion of your flight. Your requested altitude is 5000 (E). As indicated by (F) you'll be using a Moscow Departure (XPC) and your route will be from MEM via V54S to Muscle Shoals, V-7 to Graham and on to Nashville (BNA). Column (G) notes that your assigned transponder code is 4365 and that you are departing from Memphis International Airport.

Strip 2: Since you will penetrate Sector 13 next, its strip (coming out of that Sector's teletype) will arrive as soon as your departure time has been confirmed. (It will cover departure time and route.) The groundspeed in knots is worked by the computer as you fly (G163). It notes (B) that the *actual* time of departure from MEM was 2040Z and (C) gives an ETA to ALLSO intersection (30 miles this side of Muscle Shoals) as 2119Z and with estimate to MSL of 2131Z (E). The strips for each Sector do not come off the teletype simultaneously but are typed out at each Sector control position 15 minutes before the airplane penetrates the Sector. The Sector next to be in control will get the latest information on groundspeed and other items as the airplane enters its boundary. Most intersections now have five-letter identifiers and a pronounceable name, a great change from earlier practices.

For an example of the timing of the strips to the proper Sectors: a flight plan is filed and the strip appears at the proper Sector (in our example Sector 8 was the first to be entered) before the proposed departure time. If the pilot calls early (and it's suggested that you file at least 30 minutes before your proposed off-time) the flight plan is stored and the controller can call it up. When a departure time is inserted into the computer the required strips are printed at the appropriate Sectors and at the departure control position. Flights coming from other Centers are automatically fed to the Memphis

	A	B	C	D	E	F	G
①	N3456J ZA6/A T148 08 07 495 01		↑ [ ][ ] MEM P2040		50	MOSCOW XPC MEM V54S MSL V7 GHM BNA	4365 MM+ MEM
②	N3456J ZA6/A T148 G163 495 13 02	MEM 2040	21¹⁹ [ ][ ] ALLSO	50	MSL 2131	MEM MOSCOW XPC V54S MSL V7 GHM BNA	4365 HSV
③	N3456J ZA6/A T148 G162 495 09 03	ALLSO 2119	21³¹ [ ][ ] MSL	50		MEM./.GHM XTE GATE BNA/ 2215	4365
④	N3456J ZA6/A T148 G165 495 23 04	MSL 359 025 2140	21⁵⁵ [ ][ ] GHM	50		MEM./.GHM XTE GATE BNA/ 2215	4365
	A	B	C	D	E	F	G

Fig. 6-8. Flight strips for the sample flight from Memphis International Airport to Nashville via V-54S and V-7.

computer and the strips are printed 30 minutes before the Center boundary-crossing time. As the flight continues the Sector controllers will make notes on the strips.

Strip 3: The third strip will go to Sector 9 (check Figures 6-3, 6-4, and 6-5). The computer expects your groundspeed to be 162 knots and your ETA at MSL to be 2131Z. Note that as the flight progresses normally less information is given in detail. On strip 3 (Row F) the route information is shortened and the slash between MEM and GHM is a "tailoring symbol" denoting that part of the route has been deleted; the same has been done on strip 4. Both of those strips indicate that your flight will be picked up at Graham by Nashville approach control.

Strip 4: Strip 4 repeats much of the information given on the others, but looking across:

(A) Shows a groundspeed of 165 knots and Sector 23 gets the posting of this strip. (B) Indicates that the previous fix was Muscle Shoals and Sector 23 will assume responsibility at a point 359°, 25 nautical miles north of MSL at 2140Z.

(C) Notes that the ETA at GHM is 2155Z.

(D) Says 5000 is still the assigned altitude.

(E) (Blank)

(F) Notes that Graham Gate will be used and 2215Z is the ETA at Nashville. As (G) indicates you'll still be squawking 4365 on the transponder.

The destination approach control will get an estimate strip at the same time as the Sector joining the APC area (in this case, Sector 23).

The alphanumeric system has necessary information following the target on the radar screen. Assuming that you had an encoding altimeter (Mode C Transponder) as you were picked up by Sector 8 (and you're squawking 4365) your target would show

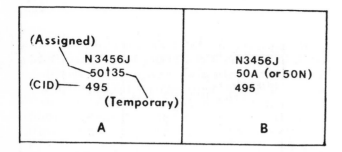

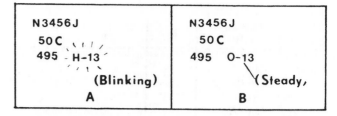

Fig. 6-9. (A) Symbols with a Mode C Transponder (encoding altimeter). The information by your return indicates your airplane's registration and that you are assigned 5000 feet MSL and are climbing through 3500 MSL.

(B) You have no altitude reporting equipment and have *reported* at 5000 feet (50A). If you haven't reported, or the controller hasn't punched in that you've reported, the symbol will be 50N (not reported).

up on the PVD (Plan View Display) as shown in Figure 6-9 (A).

Later as you've leveled off and at the boundary of Sector 8 and are going to be handed off to Sector 13, the controller in Sector 8 sees your alphanumerics

Fig. 6-10. (A) The computer is notifying the controller that it's handing your aircraft off to Sector 13 (B) The aircraft has been handed off automatically to Sector 13 (O-13) and Sector 8 is through with you except to give you a frequency change. There was no verbal communication between Sectors 8 and 13 in this case.

doing as shown in Figure 6-10.

The reverse would occur for an aircraft approaching Memphis Airport from the opposite direction at 5000 feet. The level portion of the flight would be in Sector 13 and with descent in Sector 8.

SCOPE SYMBOLS

Figure 6-11, taken from the *"Airman's Information Manual,"* shows an ARTS III Radar Scope

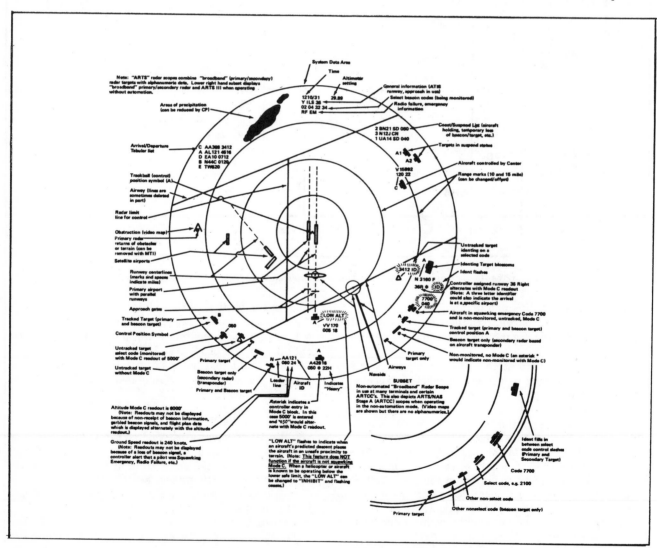

Fig. 6-11. ARTS III Radar Scope Alphanumeric Data. (*Airman's Information Manual, Part 1.*)

with various alphanumeric data. A number of radars don't have this equipment. The ARTCC and ARTS Computer facilities can "talk" to each other. The lower right hand shows presentation of the older type radar display.

Figure 6-12, also taken from the AIM shows the information available at a NAS Stage A Controllers Plan View Display when operating on the full automation RDP (Radar Data Processing) mode. When not in the automation mode, the display is similar to that shown in Figure 6-11.

It's too much for you to be able to take in all of the information given in Figures 6-11 and 6-12 at once. But you may want to use them as references to get a better picture of the controllers' displays and the radar information available to ATC and to the pilot through communications with ATC.

Memphis Center and MEM tower will have a Letter of Agreement concerning specific departure routes and arrival gates for the Memphis terminal area. You may be departed from the airport in a direction that's not precisely direct to your route. The Letter of Agreement sets out the best coordination between the facilities for best traffic sequencing in or out.

If you have a general idea of what the Center does with a flight plan, and what the Sector layout looks like, you'll feel more at ease with the system. One instrument pilot wondered why, just when he was having good communication with a controller, he was asked to switch to another Sector controller on another frequency. A short discussion of Sectors

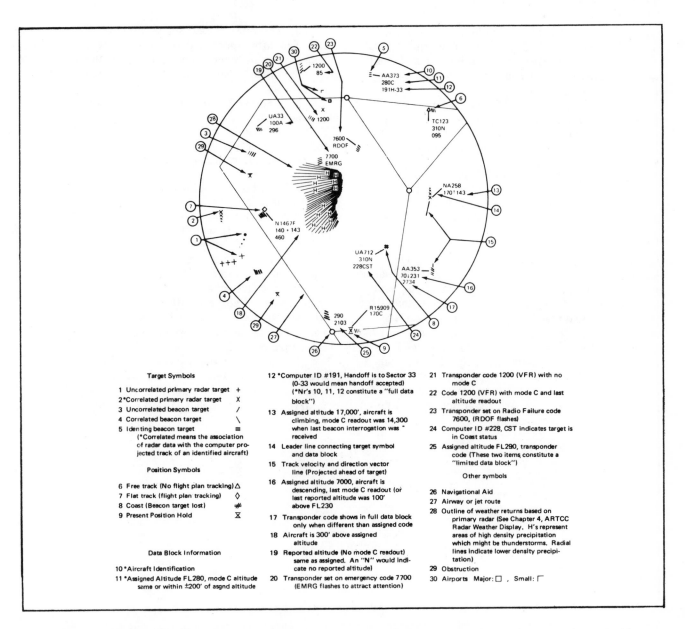

Target Symbols

1 Uncorrelated primary radar target +
2 *Correlated primary radar target X
3 Uncorrelated beacon target /
4 Correlated beacon target \
5 Identing beacon target ≡
  (*Correlated means the association of radar data with the computer projected track of an identified aircraft)

Position Symbols

6 Free track (No flight plan tracking) △
7 Flat track (flight plan tracking) ◇
8 Coast (Beacon target lost) ⧣
9 Present Position Hold ⊠

Data Block Information

10 *Aircraft Identification
11 *Assigned Altitude FL280, mode C altitude same or within ±200' of asgnd altitude

12 *Computer ID #191, Handoff is to Sector 33 (0-33 would mean handoff accepted) (*Nr's 10, 11, 12 constitute a "full data block")
13 Assigned altitude 17,000', aircraft is climbing, mode C readout was 14,300 when last beacon interrogation was received
14 Leader line connecting target symbol and data block
15 Track velocity and direction vector line (Projected ahead of target)
16 Assigned altitude 7000, aircraft is descending, last mode C readout (or last reported altitude was 100' above FL230
17 Transponder code shows in full data block only when different than assigned code
18 Aircraft is 300' above assigned altitude
19 Reported altitude (No mode C readout) same as assigned. An "N" would indicate no reported altitude)
20 Transponder set on emergency code 7700 (EMRG flashes to attract attention)

21 Transponder code 1200 (VFR) with no mode C
22 Code 1200 (VFR) with mode C and last altitude readout
23 Transponder set on Radio Failure code 7600, (RDOF flashes)
24 Computer ID #228, CST indicates target is in Coast status
25 Assigned altitude FL290, transponder code (These two items constitute a "limited data block")

Other symbols

26 Navigational Aid
27 Airway or jet route
28 Outline of weather returns based on primary radar (See Chapter 4, ARTCC Radar Weather Display. H's represent areas of high density precipitation which might be thunderstorms. Radial lines indicate lower density precipitation)
29 Obstruction
30 Airports Major: □ , Small: ⌐

Fig. 6-12. NAS (National Aerospace System) Stage A Controller's Plan View Display. (*Airman's Information Manual.*)

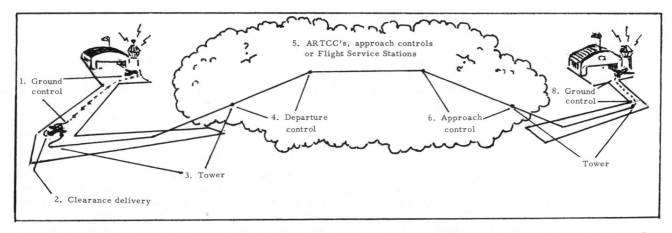

Fig. 6-13. A simplified view of facilities used on a "typical" IFR flight. At some airports you may contact clearance delivery for your clearance before taxiing. At others you will be instructed by ground control to switch to clearance delivery, or may get your clearance on the ground control frequency. The Center may be climbing or descending you in parts of the areas shown for departure control or approach control.

and Sector controllers' duties cleared things up. (*He should have visited the Center and found this all out when he started working on the rating. It's strongly suggested that you visit and get a good look at the operations.*)

Figures 6-3, 6-4 and 6-5 are examples of Sector numbers, boundaries or frequencies used. Those factors may change but the principle is valid as far as showing what happens to your flight plan at the Center.

The example flight in Part 4 will cover the Center/APC actions more from a pilot's standpoint.

As far as communications techniques are concerned, the *Airman's Information Manual* has some good tips.

1. *Listen* before you transmit. Many times you can get the information you want through ATIS or by monitoring the frequency. Except for a few situations where some frequency overlap occurs, if you hear someone else talking, the keying of your transmitter will be futile and you will probably jam their receivers causing them to repeat their call. If you have just changed frequencies, pause for your receiver to tune, listen and make sure the frequency is clear.

2. *Think before* keying your transmitter. Know what you want to say and if it is lengthy, e.g., a flight plan or IFR position report, jot it down. (But do not lock your head in the cockpit).

3. The microphone should be very close to your lips and after pressing the mike button, a slight pause may be necessary to be sure the first word is transmitted. Speak in a normal conversational tone.

4. When you release the button, wait a few seconds before calling again. The controller or FSS specialist may be jotting down your number, looking for your flight plan, transmitting on a different frequency, or selecting his transmitter to your frequency.

5. Be alert to the sounds *or lack of sounds* in your receiver. Check your volume, recheck your frequency and *make sure that your microphone is not stuck* in the transmit position.

6. Be sure that you are within the performance range of your radio equipment and the ground station equipment. Remote radio sites do not always transmit and receive on all of a facility's available frequencies, particularly with regard to VOR sites where you can hear but not reach a ground station's receiver. Remember that higher altitude increases the range of VHF "line of sight" communications.

7. Except for emergencies, avoid calling FSS stations at 15 and 45 minutes past the hour. Many stations have limited manpower and your call may interfere with SIGMET/AIRMET Alerts, etc.

Part Three

PLANNING THE
INSTRUMENT FLIGHT

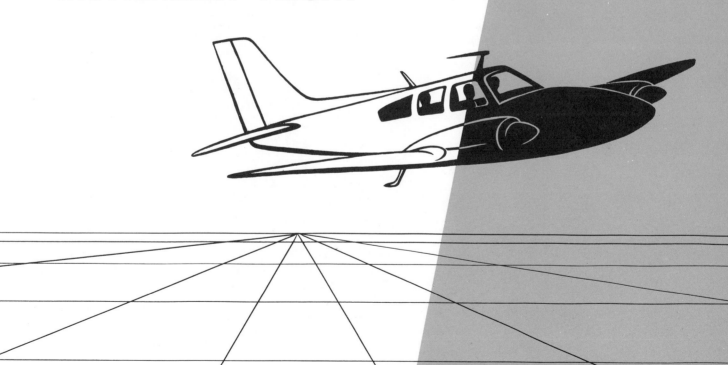

Chapter 7

WEATHER SYSTEMS AND WEATHER PLANNING

BEFORE WORKING out the navigation, you'd better check the weather to see whether you can go or not and get some wind information for estimating groundspeeds. In order to do this, weather systems and hazards and Weather Services available should be reviewed.

This book is not going into detail on meteorological theory. There are complete books — and good ones — dedicated to weather (see recommended texts at the end of the book). While it might be nice to know that the low ceilings at your destination airport were created by a Maritime Tropical or Maritime Polar air mass, it still doesn't alter the fact that certain conditions exist, and you'll have to cope with them or cancel the flight. In fact, you may not have access to information as to the type of air mass involved but *will* have ceilings, visibilities, temperatures, and other information that will tell you what to expect.

PRESSURE AREAS

You've been watching TV weather reports long enough to have gained a good idea of how pressure areas affect the weather (and studied it in getting your private and/or commercial certificate).

High-pressure areas, you've learned, *usually* mean good weather. Low-pressure areas *usually* mean less than good weather. Sometimes, the circulation around a high-pressure area (clockwise and out in the Northern Hemisphere) can pull warm

moist air into an area where it is cooled and condensed to such an extent that fog and/or low clouds are formed (Fig. 7-1).

The circulation around a Low in the Northern Hemisphere is counterclockwise and inward, caused by a combination of low pressure and the earth's rotation, just as the clockwise (and outward) circulation around a High is caused by the high pressure and the earth's rotation. The effect of the earth's rotation is called "Coriolis Effect" and is a good conversational gambit if nothing else is available.

Buys-Ballot's law states that, if an observer in the Northern Hemisphere stands with his back to the wind and sticks out his left hand, he will be pointing to the low-pressure area. However, local effects could be such (obstructions, etc.) that the observer is merely pointing to the left at some object of dubious interest. You'll do better to check weather information.

Lines connecting points of equal pressure are called *isobars*, as shown in Figure 7-1.

Elongated high-pressure systems are called ridges. The equivalent low-pressure shapes are called troughs. A "Col" is a neutral area between two Lows and two Highs.

FRONTAL SYSTEMS

Fronts in General

A front is a boundary between two air masses of different character. Although a front is considered

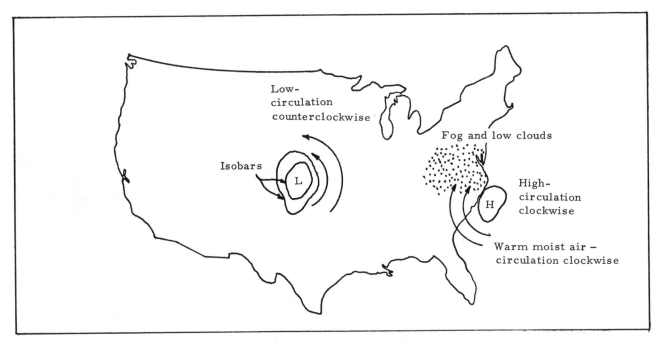

Fig. 7-1. Pressure areas and isobars.

to be a sharply defined line, it may be many miles in width. The more different the characteristics of the two air masses, the more defined the frontal zone.

Figure 7-2 shows a sample weather system with pressure patterns and frontal systems existing at a particular time.

Some weather is the result of circulation or local conditions, but most problems are caused by frontal systems. Let's examine the weather associated with the various types of fronts.

The Cold Front

Figure 7-3 is the cross section of a cold front, as indicated by A-A in Figure 7-2.

The cold front normally contains more violent weather than the warm front, and the band of clouds and precipitation is narrower. The faster the front moves, the more violent the weather ahead of it. Cold fronts normally move at about 20 to 25 knots, but some (called "fast-moving cold fronts" for obvious reasons) move as fast as 60 knots.

The slope of the front, as indicated in Figure 7-3, is exaggerated. Slopes of cold fronts vary from 1/50 to 1/150 and average about 1/80. The "top" of the cold air mass would be at 1 mile altitude at a position 80 miles behind the surface position of the front.

In the Northern Hemisphere, strong cold fronts are usually oriented in a northeast - southwest direction and move east or southeast.

As a typical cold front approaches, the southerly winds in the warm air ahead pick up in velocity. Altocumulus clouds move in from the direction from which the front is approaching. The barometric pressure decreases rapidly (Fig. 7-4).

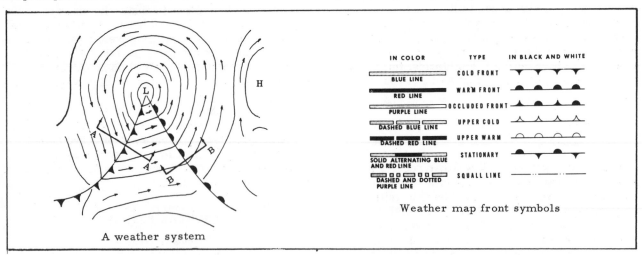

A weather system

Weather map front symbols

Fig. 7-2. Fronts and pressure areas as would be depicted on a current surface weather map.

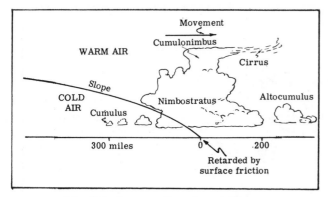

Fig. 7-3. Cross section of a cold front.

The ceiling will lower rapidly as the cumulonimbus clouds move in. Rain will occur and will intensify as the front approaches. After frontal passage, the wind will shift to westerly or northerly, and the pressure rises in short order. Rapid clearing (with lower temperatures and dew points) is the usual rule after the cold front passage. The surface winds are likely to be strong and gusty.

A slow-moving cold front may have a wider band of weather with lesser buildups if the warm air is stable and may have many of the characteristics of a warm front.

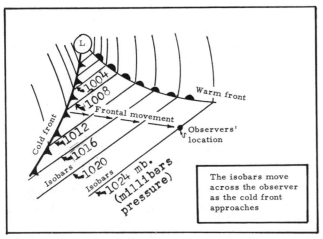

Fig. 7-4. As the cold front approaches there is a drop in pressure.

Squall Lines

Sometimes in front of a rapidly moving cold front, a solid line of thunderstorms develop. Such "squall lines" may extend up to 40,000 feet with isolated buildups to 60,000 to 70,000 feet. The squall line sometimes is found 50 to 300 miles ahead of the front and is aligned generally parallel to it.

Warm Front

The warm front normally has a wider band of less violent weather (it says here), and ceilings and visibilities are low. The warm front may hang around for days. More than one pilot (instrument-rated) has had to sit staring at the four walls of a

hotel room because practically half the country (his half) was below IFR minimums.

Figure 7-5 shows the cross section of a "typical" warm front, as shown by the B-B in Figure 7-2. (You are looking the way the arrows are pointing in Figure 7-2.)

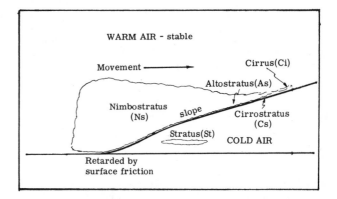

Fig. 7-5. Cross section of a warm front (warm air stable). Rain falls out of the warm air (clouds) and if the cold air is below 32°F, freezing rain results.

The slope of the warm front is about 1/100 as an average, but slopes may vary from 1/50 to 1/200. The warm front moves about one-half as fast as the cold front, and as the band of weather is much broader, the result is that it can be in the area a longer time.

In Figure 7-5 the warm air is stable, which means that stratus type clouds would be expected to predominate. If the warm air is unstable, clouds of vertical development may be found.

In the winter, freezing rain may be encountered if the cold air ahead of the front is below freezing.

If freezing rain is encountered, climb if possible. The air above the front line is warm, and the rain will be in its usual liquid form. (Air Traffic Control would be interested in your altitude changes if you are IFR or plan on getting into the clouds.) You can expect to have to let down through it and should make the transition as expeditiously as possible without overdoing it. Incidentally, you are required by the FAR to report the encountering of *any* icing, and freezing rain qualifies very well in this regard.

Occluded Front

The fact that the cold front moves faster than the warm front can manufacture a situation such as the occluded front. Figure 7-6 shows the cross section of "typical" cold and warm front type occlusions as shown at C-C in that figure.

Notice that by sliding under the cool and warm air, the cold air has created an upper warm-front condition. As the occlusion develops, the warm-front cloud system disappears, and the weather and clouds are similar to that associated with a cold front. The warm-front occlusion is less common than the cold-front type.

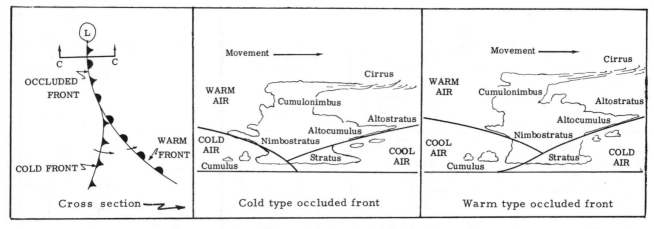

Fig. 7-6. Warm and cold type occluded fronts.

In this case, the air ahead of the warm front is colder than that behind the cold front. The cool air moves up over the more dense cold air. The surface weather would be similar to that of a warm front; but in flying through the occlusion during its initial stages, you might expect to encounter weather of both types of fronts, with thunderstorms within stratus cloud areas. As the development progresses, the severity of the associated weather decreases.

Stationary Front

Sometimes the pressures and circulation on each side of a front act in such a way as to stop the frontal movement. Such a front is naturally called a stationary front, which is as good a name as any. The weather associated with a stationary front is a milder form of warm-front type clouds and precipitation. The problem is that if the front bogs down, the weather can be from below average to unsatisfactory for several days until things get moving again.

CLOUDS

Before now your interest in clouds has been academic. You were mostly interested in (1) the heights of the bases, (2) the amount of cloud cover, and (3) whether various forms of precipitation could fall out on you. Now their internal characteristics will be of prime importance.

Clouds are broken down into families pertaining to their heights.

Low Clouds — The bases are found from the surface to 6500 feet.

Middle Clouds — The bases are found from 6500 to 16,500 feet.

High Clouds — The bases are found at heights of from 16,500 up to 45,000 feet.

Clouds With Extensive Vertical Development — The height of the cloud bases may be from 1000 to 10,000 feet. The cloud can, in extreme cases, extend up to 60,000 to 70,000 feet.

The clouds are further described by their form and appearance. The puffy or billowy type formed by local vertical currents are called "cumulus," and those formed of widespread (or fairly widespread) layers have the term "stratus" or "strata" somewhere in the name. Figure 7-7 shows some representative clouds of the various families.

A cloud with the term "nimbo" or "nimbus" in the name is expected to produce precipitation.

Flying near clouds of stratus type formation, you would expect fairly smooth air. Cumulus clouds, by their very nature, are the product of air conditions that indicate the presence of vertical currents.

Clouds are composed of minute ice crystals or water droplets and are the result of moist air being cooled to the point of condensation. The high clouds (cirrus, cirrostratus, and cirrocumulus) are composed of extremely fine ice crystals. When the water droplets become a certain size, rain results (or snow or sleet, depending on the conditions). Hail is a form of precipitation associated with cumulonimbus type clouds. It is the result of rain being lifted by vertical currents until it reaches an altitude where it freezes and is carried downward again to gain more moisture. The cycle may be repeated several times, giving the larger hailstones their characteristic "layers," or strata.

Turbulence can be found near these clouds, and hail can fall from the "anvil head" into what could appear to be a clear area.

Clouds may be composed of supercooled moisture. The impact of your airplane on these particles causes them to freeze immediately on the airplane.

Clouds are formed by moist air being cooled to the point of condensation, and this leads to the subject of lapse rates.

For dry air the adiabatic lapse rate is 5.5° F per thousand feet. (Adiabatic is a process during which no heat is withdrawn or added to the system or body concerned.) The normal lapse rate for "average" air is 3.5° F, or 2° C. The moist adiabatic lapse rate is the lapse rate produced by convection in a saturated atmosphere, such as within a cumulus cloud. At high temperatures, it will be 2°-3° F per thousand feet, and at low temperatures, it will be in the vicinity of 4°-5° F. The dew point lapse rate is about 1° F per thousand feet.

118

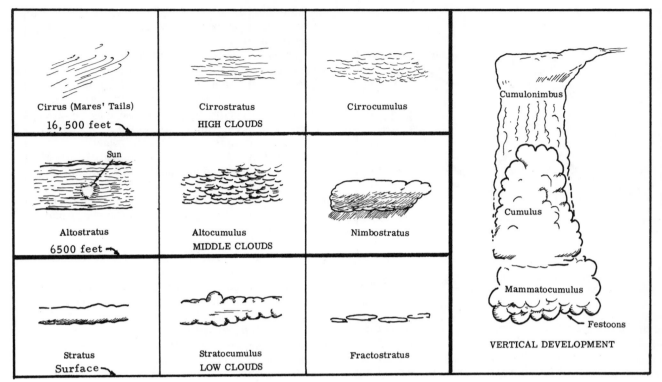

Fig. 7-7. Some cloud types.

For cumulus type clouds that are formed by surface heating, the base of the clouds may be estimated by the rate at which the dry lapse rate "catches" the dew point. (The dry lapse rate is 5.5° F, the dew point drop is 1° F per thousand, so the temperature is dropping 4.5° F faster than the dew point, per thousand feet.) Assume the surface temperature is 76° F and the dew point 58° F, a difference of 18° F: Dividing this number by 4.5, it is found that the temperature and dew point make connections at 4000 feet — the approximate base of the clouds. This only works for the type of cloud formed by surface heating and is not suitable for locations in mountainous or hilly terrain.

When clouds have a temperature of between 0° C and -15° C, they consist mostly of supercooled water droplets with some ice crystals. There's an old physics experiment in which distilled water is cooled very slowly and remains liquid below 32° F — until the container is jostled or other outside factors are introduced, then it freezes instantly. In the case of the supercooled water droplets in the cloud, *your airplane* is the outside factor. The shock of the airplane flying into the particles can cause them to freeze on the airplane, but in-flight icing will be covered in more detail later.

When the temperature within the cloud is lower than -15° C, the cloud is usually composed entirely of ice crystals.

As you have noted in your flying, the addition of nuclei to moist air can result in the formation of clouds or fog. Specks of dust or smoke form the center of the particles, and airports located in river bottoms near industrial plants are notorious for being socked in when everything else is good VFR.

When moist air is orographically lifted (moved up a slope), it may be cooled and condensed to the point where clouds are formed. If the warm moist air is unstable to begin with, the clouds that result may be well-developed cumuliform clouds.

HAZARDS TO FLIGHT

Thunderstorms

Thunderstorms constitute a real menace to the instrument pilot because he may fly into one with little or no warning. Airborne radar has been of great value in finding "soft spots" (or *comparatively* soft spots). Center radar can help too.

A big problem (*one* of your big problems) will be turbulence within the cell. Updrafts and downdrafts may cause structural failure of the airplane — or loss of control.

If you know you're about to penetrate a thunderstorm, you'll want to slow the airplane below the maneuvering speed or speed recommended by the manufacturer for such a situation. Secure all loose gear. Set your power to maintain level altitude at the recommended speed *before* penetration and *leave the power alone.* Fly a straight and level attitude. Don't, repeat, *don't* try to maintain a constant altitude (height). In attempting to keep at the same altitude, you may put extreme stresses on the airplane. You fly into an updraft, shove the nose over — just as you hit a violent downdraft. Remember that even near the cells the air can be extremely violent, and many a VFR pilot has found out about "sucker holes" when he tried to fly between cells.

The "altitude hold" portion of the autopilot should be off, or the airplane can be overstressed as the

119

Fig. 7-8. Flying through thunderstorms can be a "rending" experience.

equipment tries to do an impossible job of maintaining a constant altitude.

Maybe you haven't flown into severe or extreme turbulence and don't realize that you may be bouncing around so much that the instruments are very hard to read, and since you are IFR, this can be an interesting situation.

Precipitation static will be a problem for the LF/MF equipment in heavy rain or snow, so you might as well turn the volume down. Attempting ADF tracking in a thunderstorm or thunderstorm area can be one of the biggest wastes of time in your flying career. You'll be busy enough trying to keep the airplane under control.

Try to maintain as near a constant heading as possible. Pick a heading that should get you out in the shortest time — you wouldn't want to go through a squall line the *long way*.

Because of lightning flashes, it would be best to have the cockpit lights full bright during night penetrations. You'll lose your night vision anyway and shouldn't be looking out at this stage. Lightning has damaged airplanes on occasion, but this relatively rare possibility will be of secondary importance compared to what turbulence and hail can do to you.

To repeat a statement made in an earlier chapter: Never jeopardize control for voice transmissions. If ATC calls when you are hard put to just fly the airplane, tell them to wait; or if necessary, hang on to the controls and forget the mike and let *them* worry, also.

Icing

Area forecasts are a good source to check for possible structural icing along your route.

Carburetor Ice

You've probably had experience with carburetor ice and know that it is "not the heat (or lack of it) but the humidity" that is the big factor concerned. Carburetor icing can occur on warm days without a cloud in the sky.

The warm moist air enters the carburetor, where it is cooled by the combination of two factors: (1) the vaporization of the fuel, and (2) the venturi effect of pressure change through the carburetor. The temperature drop will vary but may be up to 72° F. If the resulting temperature is below freezing, ice forms in the venturi and downstream in the intake system. As a review of the indications of carburetor ice: For the airplane with a fixed pitch propeller, the rpm creeps off with no change in throttle position. Pulling the carburetor heat results in a still further drop in rpm — and that's where many pilots make a mistake; they don't leave the heat ON long enough but push it off with a feeling of wellbeing (no ice). Leave the heat on for at least 10 seconds. If there is ice, the rpm will pick up from its even lower setting. When you push the heat off, the rpm will pick up sharply, particularly if you'd been unconsciously easing the throttle forward to take care of the ice-caused power loss. If there is a lot of ice, it may cause a temporary roughness as the deluge goes through the engine.

A manifold pressure loss is the big indicator of ice for the airplane with a constant-speed propeller. The governor will mask the power drop (the blades flatten out to maintain the preset rpm). When heat is applied, the manifold pressure will drop further and will pick up *past* the MAP at which the heat was finally pulled, after the ice is cleared out and the heat is pushed off. Carburetor air temperature gages or inlet air temperature gages can be a great aid in preventing carb ice problems.

The use of full or partial heat depends on the airplane. Usually, for light trainer types, it's all or nothing. For bigger airplanes, particularly those with carburetor air temperature gages, partial heat is fine. Some of the bigger systems can cut down power by nearly 20 per cent when on full heat.

Excessive use of heat while taxiing can be bad. In most cases, the carburetor heat system is taking in unfiltered air; and dust, sand, and other foreign material can be sent through the engine.

If the airscoop becomes iced over, the alternate air system can save the day (Fig. 7-9).

The warm air, being less dense than the outside air, causes some loss in power, but that is much better than a total power loss.

Structural Icing

The airscoop icing situation just mentioned occurs when structural icing is the big problem. The windshield, wings, empennage, props, antennas, etc., will also be gaining weight and adding drag.

In most cases, the weight of the added ice will be a comparatively minor factor — the drag increase, thrust decrease (for prop icing), and lift decrease are the factors that cause the big problems.

Structural icing is broken down into two main

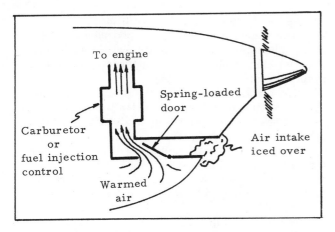

Fig. 7-9. Engine suction opens the spring-loaded door and allows the warm air from the engine compartment to enter the carburetor. Alternate air (or its equivalent) may be manually selected, also.

The accretion rate of structural ice depends primarily on (1) the amount of liquid water, (2) the drop size, (3) airspeed, and (4) the size and shape of the airfoil. If you fly into an area of icing, it would be well to remember that up to about 400 knots ice collection increases with speed. Above this, frictional heating of the skin tends to lessen the chances of the ice sticking.

The effects of icing on the airplane are all bad. Lift (for a given angle of attack) decreases, thrust falls off, drag and weight increase. The stall speed rises sharply.

If your airplane has a stabilator, you should be aware of the possibility that the airflow disturbance and effects of the weight of ice on the leading edge could cause you to overcontrol at low speeds (it depends on how closely balanced the flying tail is in the clean condition). This should be something to be considered if you still have ice on the airplane during the approach.

Ice on the wings and tail can be removed by pneumatic de-icer "boots," by heat, or by chemical fluid being continually "oozed out" through orifices in the leading edge. The de-icer boot is the most popular. It expands in sections, and the ice is broken up to be blown away in the airstream. The ice should be allowed to build up slightly before using the boots, but in heavy icing conditions, they may have to be operated continuously. The pneumatic system normally operates through the vacuum pump system, and "all-weather" twins usually have a pump on each engine, each capable of carrying the de-icer load *plus* that required for the vacuum-driven instruments. The term "all-weather" is not a good one, as some weather is too severe to be safely penetrated.

The boot system usually has a timer which inflates and deflates the boot sections at preset intervals. Some systems have a manual switch which the pilot (or copilot) can operate whenever he feels it to be necessary if the cycle is too short or too long for the existing conditions. The de-icer boots come in spanwise or chordwise tube installations, as shown by Figure 7-10.

types (but both types may be mixed at a particular time):

Rime Ice — This is a milky granular deposit of ice with a rough surface. It's formed by instantaneous freezing of *small* supercooled water droplets, as the airplane encounters them. Rime ice contains trapped air which contributes to its appearance and brittleness. Rime ice forms on leading edges and protrudes forward as a sharp nose. It is more easily removed than clear ice but spoils the airflow more because of its roughness. Rime ice is most often found in stratus type clouds but may also be present in cumulus buildups at temperatures below -10° C. Rime ice is somewhat similar in appearance to the thick frost in the ice compartment of an older type home refrigerator but is rougher.

Clear Ice — You've seen this type of ice in "ice storms" (freezing rain). It's clear, solid, and very hard to remove. Clear ice is the result of large droplets and comparatively slow freezing. It is normally smoother than rime ice, unless solid precipitation (snow, sleet, or small hail) is trapped in it — this results in an airflow spoiling, hard to remove combinations.

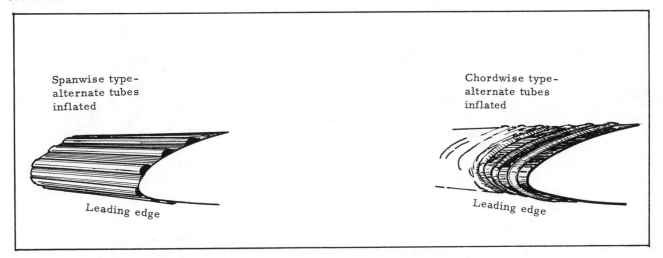

Fig. 7-10. Types of de-icer boots.

During your training, or later when you are flying that de-icer equipped airplane, climb to altitude VFR and check the stall characteristics of the airplane with the *de-icers working* to check the possibility of landing in that condition if necessary. Your airplane may have a placard or warning against landing with the de-icer boots working, and you must follow this. Other airplanes note that the stall speed is increased. It's also likely that ice remaining on the wing during the landing will be much more disturbing to the flight characteristics than the operating boots, but check any limitations for your airplane.

The other method of clearing ice from the wings is to circulate hot air from the exhaust inside the wing structure, so that it is warm enough to prevent freezing on the leading edge. The disadvantages of this system are that a considerable amount of power is used for the heat, and the thawing ice may slide rearward and refreeze on the unheated part of the structure. The heating of the boundary layer (the layer of air next to the wing surface) causes it to become more unstable with possible slight changes in stall characteristics, but this is considered to be a very minor problem.

There are commercial products available designed to decrease the holding power of ice. Some are used on the bare (or painted) wing, and others are used on the boots to decrease ice adhesion and aid in the breakaway process.

For fighting propeller icing, there are two main methods: (1) fluid *anti*-icing and (2) electric *de*-icing. The fluid (an alcohol mixture) is thrown out along the blades by centrifugal force and is most effective if the procedure is initiated *before* ice starts accumulating. The propeller blades may have rubber "feed shoes" to direct the fluid in the most effective direction. Fluid is used as necessary to keep the blades "wet."

If the other parts of the airplane are beginning to pick up ice, you can expect that the prop(s) is also

getting its share. The anti-icing (fluid) systems have recommendations as to the procedure to be used. The amount of fluid available is limited but will last long enough to cover most icing situations if conserved.

The electric de-icer for the propeller can be quite effective *after* ice has started to form. Figure 7-11 shows the two general methods of combating prop ice.

Icing can cause problems on antennas and in extreme conditions may cause them to be carried away. This is somewhat disconcerting particularly if, for instance, both VOR receivers are using that one antenna.

Windshield icing can sometimes be more of a menace than ice on other parts of the plane. The small storm pane in the side window has been a great aid for more than one pilot in landing the airplane with a load of windshield ice. Freezing rain is particularly bad in this regard because the ice film may be forming fast and thick and can get well ahead of the windshield defroster. Do whatever your *Pilot's Operating Handbook* says to get the maximum defrosting effect. You may have to deflect some of the cabin heat to get added defrost heat. It's better to be uncomfortable and be able to see out than vice versa.

Icing of the pitot tube and static vent can be a problem. If you expect icing, or if it's starting, use the pitot heat. Pitot heat is a severe drain on the electrical system, so its use should be tempered with judgment if communications and navigation equipment are already using large electrical loads. Check the volt-ammeter when the heat is turned on — its effect will be indicated.

The windshield may frost over on high-performance airplanes when a fast letdown is made from subfreezing temperatures to warm moist air. The surface of the windshield and airframe is still cold enough so that the moist air freezes on contact.

You may go to your airplane after it has been tied out for some time and find that it's covered with

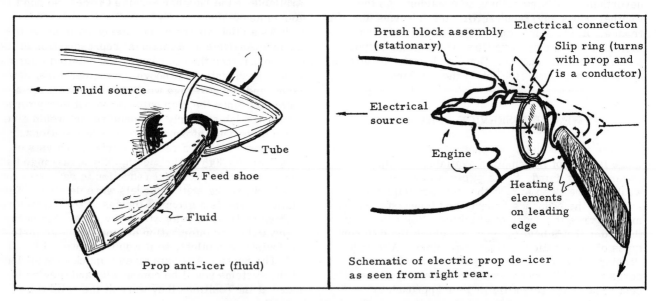

Fig. 7-11. A comparison of fluid anti-icing and electric de-icing systems for propellers.

ice. Obviously, hot water will freeze again, and you'd be back where you started, or worse, if you poured it on the plane. One thing sometimes forgotten by pilots who move an icy plane into a heated hangar is that the water from the melting ice can collect in control surface hinges, landing gear assemblies, and other vital spots. When the plane is moved back out into the freezing temperature, the water refreezes, and problems can result. Wipe the water out of such places and make sure they are dry before moving the plane out of the hangar. Along this same line, if you take off through puddles or slush and the temperature is near freezing, leave the gear down longer after take-off to allow the airflow to blow off most of the moisture. Otherwise, if a large amount of water is collected on the gear and this freezes, it might cause extension problems later. In some cases, you may want to cycle the gear a time or two before leaving it up. Your actions in this case will depend on whether other factors (ceiling, visibility, and obstructions) will allow cycling without risking possible loss of control.

We have mentioned that accretion of ice on a stabilator in flight could cause problems. There have been incidents of control flutter and crashes immediately after take-off — caused by ice that accumulated inside or on the surfaces while the airplane was sitting on the ramp (maybe in freezing rain) or had been moved out of a heated hangar with water on (or in) the surfaces. Flutter occurred and the control surfaces were destroyed with a resulting loss of control.

Frost — This phenomenon was mentioned earlier as occurring in flight, but an airplane left out overnight will be covered with frost on certain occasions. For some reason, pilots tend to ignore the effects of frost on take-off performance, and more than one accident has been caused by an almost paper-thin coating of frost on the airplane. Of course, frost of this type forms during clear cold nights, so you're most likely to encounter it on early morning VFR flights, nevertheless, it's something to consider. As the weather books say, frost sublimates (changes directly from a solid to a gas) quickly in warmer air and in motion, but you may have flown through the airport fence before this occurs. Don't underrate the effects of even a thin layer of frost on the airplane.

WEATHER SERVICE SERVICES

The final decision whether to go is up to you, but there is a lot of information available to help you make up your mind. Some pilots, when checking the weather, prefer to look at the weather maps first and *then* the sequence reports and forecasts. Others may reverse the order of checking. You can set up the order that's best for you, but check the different types of information against each other. Although flipping a coin or the twinges of rheumatism and corns may work pretty well for some endeavors, it's best to be more scientific in your approach to weather for instrument flying.

The following look at weather service information sources is based on what a pilot would check in planning an IFR flight from Memphis to Nashville on the 30th of the month with an estimated departure time of 2040Z. The charts, forecasts and weather reports are actual data as issued by the National Weather Service on the 30th of the month (December) for the time range indicated. Some of the charts (*which were used in an FSS*) had some wear and tear, plus added notes and markings, so have been recopied for clarity. Look at the actual weather reports as given in Figure 7-21 and compare them with what the various forecasts said would be happening at that time.

Weather Charts

Surface Analysis

The Surface Analysis is also called a surface weather map, and is transmitted every 3 hours in the 48 state portion of the U.S.

The valid time indicated on the map is that of the plotted observations. The date-time group (GMT) tells when conditions plotted on the map were occurring.

The solid line isobars are spaced at 4 millibar intervals; but if the pressure gradient is weak, dashed isobars are put at 2 millibar intervals to better define the system. A "24" is 1024.0 mb, and a "92" stands for 992.0 mb. The Highs and Lows have a two-digit underlined number ("32" means that the pressure at the center is 1032 millibars, etc.). In color presentations, Highs are blue and Lows are red. Figure 7-2 shows the frontal presentation for black and white and color.

The Weather Service has a 3 digit number on the map by the frontal systems to show type, intensities and character of the front. For instance "453" by a front means that it is a cold front at the surface (4), moderate, little or no change in intensity (5), with frontal area activity increasing (3). The codes are available in the Weather Service Offices, so don't try to memorize them.

As a pilot, you're more interested in using the Surface Analysis as a pictorial representation of the weather, rather than in reading the detailed information at each station (although the temperature, dewpoint, and sky coverage would be of interest for a pattern). Since the map only comes out every three hours, the latest hourly sequence report would give more up to date information on various stations.

Figure 7-12 is a redrawn Surface Analysis of 1800Z on the 30th of the month. The actual map had too much detail for useful reduction to page size for this book, so the station models were deleted. (The station model is a group of figures at the various reporting stations showing wind, sky cover, temperatures and other information of more value to meteorologists than pilots, so it won't be covered.)

The Surface Analysis gives a quick overall look at the systems you'll be coping with and they're located at the positions they were at the valid time (1800Z here).

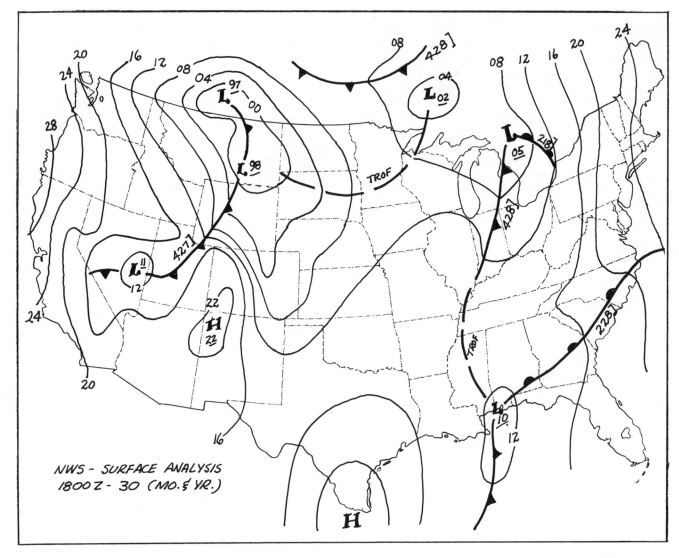

Fig. 7-12. A redrawn Surface Analysis for 1800Z on the 30th of the month for the same time span as the other data in this chapter.

Looking at Figure 7-12 you can get a couple of ideas for a Memphis to Nashville IFR trip: (1) The cold front coming out of the Low just north of Michigan turns into a trough through west Tennessee, losing its definite cold front characteristics and (2) the isobars are very far apart (the 1016 isobar is in western North Carolina and the 1012 isobar runs through Kansas and western Oklahoma). The pressure gradient is very shallow. These two items suggest a stagnant situation with widespread bad weather that could be around for days. This would be an input to your personal computer while planning (or forgetting) the trip. It could be an adverse factor for safety because you might decide that you'd better go on because you "might be stuck here for a week." However, being stuck someplace for a week might be better than flying in icing or other conditions beyond the scope of your abilities and the airplane's equipment.

Note that there is a High at the tip of Texas that might affect the movement of the systems. Look at several previous Surface Analysis charts to see if much movement is occurring.

By looking at the codes in the Weather Service Office you'd see that the warm front running from Alabama through North Carolina (228) is a warm front at the surface, weak with little or no change, and diffuse. This could give you another clue as to possible stagnation conditions.

Weather Depiction Chart

Figure 7-13 is an example of a Weather Depiction Chart. Because the weather (ceilings, visibilities, and other factors) as given for each station may be well out of date (or time), you should check the hourly reports for that information. The best use of this chart is to give you a *general* look at the areas of low ceilings and visibilities. This is an important factor should conditions go below minimums at the destination and alternate, or if you lost total electric power and have to fly to an area to let down in MVFR or VFR conditions. Admittedly the odds are long against such problems, particularly the last one, but still you should know that, if you are flying in the IFR area of Figure 7-13, turning north or south the conditions here would *not* be the quickest way to better weather.

Radar Summary Chart

Figure 7-14 is part of an actual Radar Summary Chart. At the bottom of the full chart the day of the week, the time (Zulu), day of the month and "RADAR SUMMARY" are indicated.

The shaded areas show radar echo areas and the contours outline intensities. In north central Alabama, the intensities are shown building toward

the center of the area with tops at 32,000 (*320*). Bases are noted by a bar *over* the height value in hundreds. Note also that there is a squall line in that area as indicated by the solid line. The direction and velocities of movement of areas and lines are shown by pennants (that Alabama group is moving east-southeast at 20 knots). Solid areas are indicated by a "SLD."

There is a rain shower area of increasing intensity (RW+) south of the Florida panhandle with tops at 15,000 feet ASL. Rain showers of decreasing intensity are indicated. There is an area of thunderstorms and rain showers (TRW) south of Pensacola moving south (arrow) at 5 knots (05).

NE means "no echoes" and OM indicates that a station is out for maintenance.

There is an area with rain and rain showers (tops at 23,000 feet) moving northeast at 25 knots

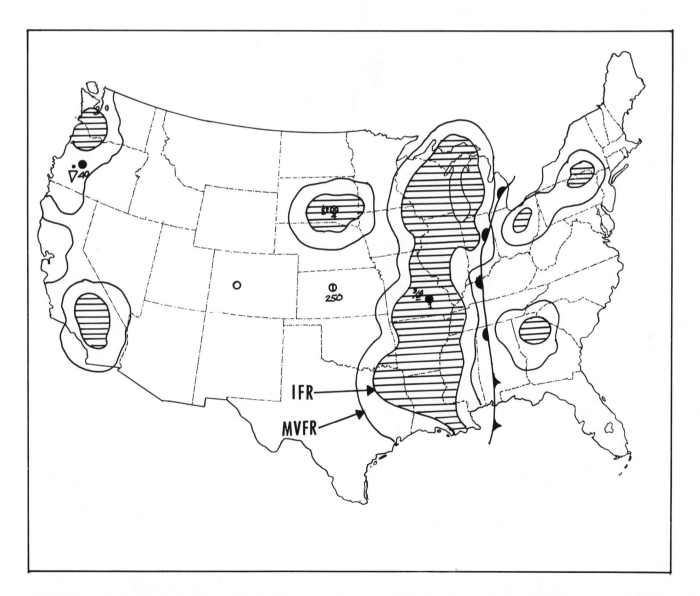

Fig. 7-13. A Weather Depiction Chart. Note that the areas of less than 1000 feet *and/or* 3 statute miles (IFR) are cross-hatched. Areas of marginal VFR (MVFR — ceiling 1000 to 3000 feet and/or visibility 3 to 5 statute miles) are contoured without shading. VFR areas have ceilings greater than 3000 feet and visibilities greater than 5 miles and are not contoured.

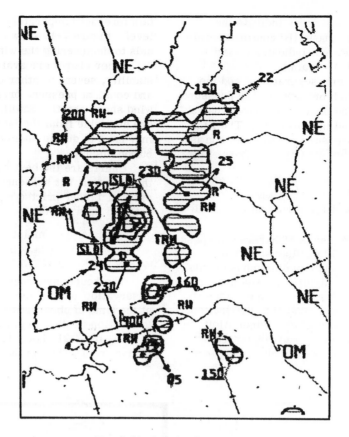

Fig. 7-14. A Radar Summary Chart.

(arrow) in the Atlanta area.

The Radar Summary Chart is valuable to you as an instrument pilot in showing intensities, movements and tops of areas of significant precipitation. Remember that these areas may change in size, position, and intensities by the time you get there.

Significant Weather Prognostics

The Low Level Prog is a four panel chart as shown by Figure 7-15. The top panels (A and B) are 12 and 24 hours Significant Weather Progs from the surface to 400 millibars (24,000 feet). The lower charts (C and D) are 12 and 24 hour Surface Progs. The charts show the conditions as they are forecast to be at the valid time of the chart. Note that the 12 and 24 hour Progs are valid at 1200Z on the 30th and 0000Z on the 31st, and "straddle" your intended departure.

Figure 7-16 is a larger and more detailed version of Chart B in Figure 7-15. It depicts ceiling, visibility, turbulence and freezing level at the valid time. Like the Weather Depiction Chart, MVFR areas are enclosed by scalloped lines and IFR areas are inclosed by smooth lines. "MVFR" and "IFR" have been printed in here to aid in picking out the areas, you won't see this on the real chart.

The dashed lines with accompanying numbers indicate the freezing levels (MSL) at 4000 foot intervals (40,80,120) and the surface freezing level is indicated by a dotted line and "32° F."

Turbulence areas are outlined by dashed lines, and symbols indicate the intensity of the turbulence (A). There's a turbulent area extending into the western U.S. from Canada as indicated by the dashed line (the "T's" were added by the writer). The symbol ($\_\wedge\_$) indicates that it is moderate and from the surface to 18,000 MSL. Severe turbulence would be indicated by another inverted "V" over the center of the symbol ($\_\wedge\!\!\wedge\_$). The northeastern U.S. is almost covered by a moderately turbulent area that extends from 10,000 to 24,000 feet MSL. The Southeastern U.S. has an area of moderate turbulence from the surface to 10,000 MSL.

When you first look at a Significant Weather Prog it will appear to be a jumble of solid and dashed lines and numbers, but as you spend a little more time with a particular chart things begin to take shape. Look for each feature (ceiling and visibility or turbulence or freezing) alone to see its significance.

This chart would be valid at slightly over 2 hours after your ETA at Nashville on the example trip.

Figure 7-17 gives the symbols and meanings used on the significant weather progs.

Figure 7-18 is a detailed version of the 24-hour surface prog (D in Fig. 7-15).

126

The shaded precipitation areas indicate the presence of (A) continuous snow, (B) continuous rain and (C) continuous drizzle. (D) indicates a rain shower area. (See Fig. 7-17 again.)

The isobars are in 4 mb. separations, and the two numbers with each High or Low are the pressure values. (The Low centered over South Dakota has a pressure of 1004.0 mb. and is stationary. The High over south Texas has a pressure of 1018 mb. and is moving in the direction indicated by the arrow at 10 knots.

Other Charts

High Level Progs — These are charts encompassing airspace from 400 to 150 millibars, including turbulence, and cirriform and cumulonimbus clouds.

International Flights — Significant Weather Prog Charts are available for international operations if you think you might overshoot your U.S. destination.

Winds and Temperatures Aloft — Winds Aloft Charts, forecast and observed, are transmitted by facsimile. The forecast winds aloft charts also contain forecast temperatures for that altitude or flight level. Figure 7-19 shows a few sample stations and aids to deciphering the wind direction and velocity.

Other charts are available for freezing levels, stability, severe weather outlook, constant pressure and constant pressure prognostics, tropopause and wind shear charts. Satellite pictures of U.S. weather are available from the facsimile machine, but you may have to dig for them because they aren't always out on display.

Charts are the best way to get an overall look at the situation, but sequence reports, terminal forecasts and other printed matter are the sources for detailed information.

Sequence Reports

The hourly sequence report is a surface weather observation and, as such, gives the actual conditions at the time of observation. Figure 7-20 shows the key to aviation weather reports and aviation forecasts.

Figure 7-21 shows 12 actual reports covering a period from 1000Z on the 30th (upper left hand cor-

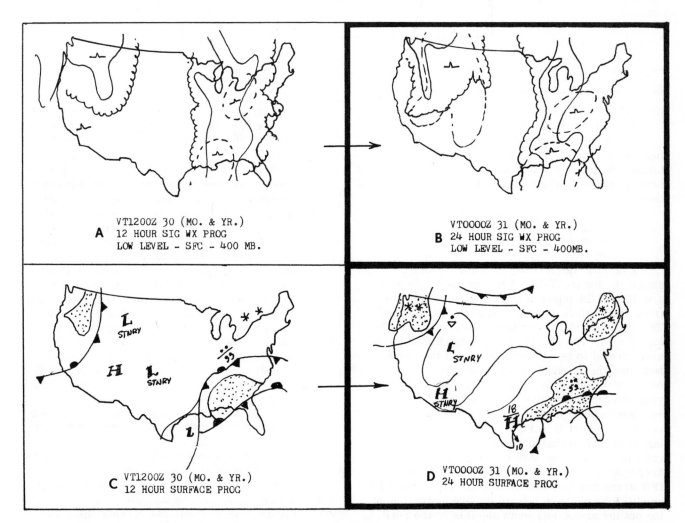

Fig. 7-15. The layouts of the Significant Weather Prognostics, familiarly known as "progs" if you want to sound like a pro.

127

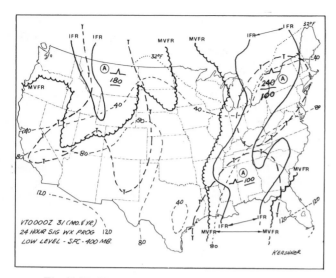

Fig. 7-16. Significant Weather Prog chart (24 hour).

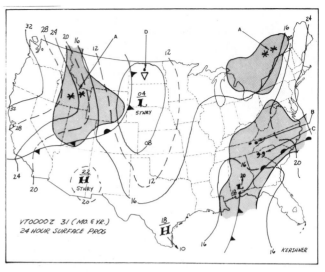

Fig. 7-18. A 24 hour surface prog chart, redrawn from the actual chart.

ner) and ending with the 2100Z report on the date. (You would be leaving on the flight at about 2040Z and wouldn't have a chance to check this one, but it's interesting to see what kind of weather would be occurring at some of the area stations while you're enroute.)

The three stations shown are Crossville (CSV), which is east of Nashville, but would give some idea

of conditions in that direction if you need such information; Nashville (BNA) and Jackson, Tennessee, and McKellar Field (MKL), which is between Memphis and Nashville.

The headings show that these are Aviation Sequences Circuit 14, (SA 14) for the 30th of the month, for the hours given in Zulu time.

Symbol	Meaning
∧	Moderate turbulence
⋀	Severe turbulence
⟆	Moderate icing
⟆	Severe icing
●	Rain
✳	Snow
❜	Drizzle
●▽	Rain shower
✳▽	Snow shower
⍐	Thunderstorm
∿	Freezing rain
⟲	Tropical storm
⟲	Hurricane (typhoon)
●●	Continuous rain
✳	Intermittent snow
❜❜	Continuous drizzle

Fig. 7-17. Symbols and meanings for the prog charts. Continuous precipitation is indicated by two symbols, intermittent by one symbol as shown at the bottom of the illustration.

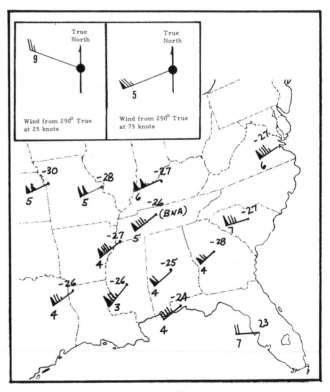

Fig. 7-19. Sample forecast winds aloft chart for a few stations at 28,000 feet. The temperature in C° is written in above each station. Each pennant is 50 knots, each full barb is 10 knots, and each half barb is 5 knots. The number by the barbs helps pin down the exact direction; for instance you see that the wind is from the west-southwest at Nashville (BNA) and the "5" indicates that it is from 250° true. The wind is 85 knots and the temperature is -26° at BNA.

KEY TO AVIATION WEATHER OBSERVATIONS . . . .

LOCATION IDENTIFIER TYPE AND TIME OF REPORT°	SKY AND CEILING	VISIBILITY WEATHER AND OBSTRUCTION TO VISION	SEA-LEVEL PRESSURE	TEMPERATURE AND DEW POINT	WIND	ALTIMETER SETTING	REMARKS AND CODED DATA
MKC SA 0758	15 SCT M25 OVC	1R-K	132	/58/56	/1807	/993/	R04LVR20V40

SKY AND CEILING

Sky cover contractions are in ascending order. Figures preceding contractions are heights in hundreds of feet above station. Sky cover contractions are:

CLR Clear: Less than 0.1 sky cover.
SCT Scattered: 0.1 to 0.5 sky cover.
BKN Broken: 0.6 to 0.9 sky cover.
OVC Overcast: More than 0.9 sky cover.

— Thin (When prefixed to SCT, BKN, OVC)

—X Partial obscuration: 0.9 or less of sky hidden by precipitation or obstructon to vision (bases at surface.)

X Obscuration: 1.0 sky hidden by precipitation or obstruction to vision (bases at surface.)

Letter preceding height of layer identifies ceiling layer and indicates how ceiling height was obtained. Thus:

E   Estimated height
M   Measured
W   Indefinite

V=Immediately following numerical value, indicates a variable ceiling.

VISIBILITY

Reported in statute miles and fractions.
(V=Variable)

WEATHER AND OBSTRUCTION TO VISION SYMBOLS

A	Hail	IC	Ice crystals	S	Snow
BD	Blowing dust	IF	Ice-fog	SG	Snow grains
BN	Blowing sand	IP	Ice pellets	SP	Snow pellets
BS	Blowing snow	IPW	Ice pellet showers	SW	Snow showers
D	Dust	K	Smoke	T	Thunderstorms
F	Fog	L	Drizzle	T+	Severe thunderstorm
GF	Ground fog	R	Rain	ZL	Freezing drizzle
H	Haze	RW	Rain showers	ZR	Freezing rain

Precipitation intensities are indicated thus: — Light; (no sign) Moderate: + Heavy

WIND

Direction in tens of degrees from true north, speed in knots. 0000 indicates calm. G indicates gusty. Peak speed of gusts follows G or Q if gusts or squall are reported. The contraction WSHFT followed by GMT time group in remarks indicates windshift and its time of occurrence. (Knots × 1.15=statute mi/hr.)

EXAMPLES: 3627=360 Degrees, 27 knots; 3627G40=360 Degrees, 27 knots, peak speed in gusts 40 knots.

ALTIMETER SETTING

The first figure of the actual altimeter setting is always omitted from the report.

RUNWAY VISUAL RANGE (RVR)

RVR is reported from some stations. Extreme values during 10 minutes prior to observation are given in hundreds of feet. Runway identification precedes RVR report.

PILOT REPORTS (PIREPS)

When available, PIREPS, in fixed-formats are appended to weather observations. The PIREP is desgnated by UA.

DECODED REPORT

Kansas City: Record observation taken at 0758 GMT 1500 feet scattered clouds, measured ceiling 2500 feet overcast, visibility 1 mile, light rain, smoke, sea-level pressure 1013.2 millibars, temperature 58°F, dewpoint 56°F, wind 180°, 7 knots, altimeter setting 29.93 inches. Runway 04 left, visual range 2000 feet variable to 4000 feet.

°TYPE OF REPORT

SA—a scheduled record observation

SP—an unscheduled special observation indicating a significant change in one or more elements

RS—a scheduled record observation that also qualifies as a special observation.

All three types of observations (SA, SP, RS) are followed by a 24 hour-clock-time-group in GMT.

KEY TO AVIATION WEATHER FORECASTS . . . .

TERMINAL FORECASTS contain information for specific airports on expected ceiling, cloud heights, cloud amounts, visibility, weather, and obstructions to vision and surface wind. They are issued 3 times/day and are valid for 24 hours. The last six hours of each forecast are covered by a categorical statement indicating whether VFR, MVFR, IFR or LIFR conditions are expected (L in LIFR and M in MVFR indicate "low" and "marginal"). Terminal forecasts will be written in the following form:

CEILING: Identified by the letter "C"
CLOUD HEIGHTS: In hundreds of feet above the station (ground)
CLOUD LAYERS: Stated in ascending order of height
VISIBILITY: In statute miles but omitted if over 6 miles
WEATHER AND OBSTRUCTION TO VISION: Standard weather and obstruction to vision symbols are used
SURFACE WIND: In tens of degrees and knots; omitted when less than 10

EXAMPLE OF TERMINAL FORECAST

DCA 221010: DCA Forecast 22nd day of month—valid time 10Z-10Z.
10 SCT C18 BKN 5SW—3415G25 OCNL C8 X 1/2SW: Scattered clouds at 1000 feet, ceiling 1800 feet broken, visibility 5 miles, light snow showers, surface wind 340 degrees 15 knots Gusts to 25 knots, occasional ceiling 8 hundred feet sky obscured, visibility ½ mile in moderate snow showers.
12Z C50 BKN 3312G22: At 12Z becoming ceiling 5000 feet broken, surface wind 330 degrees 12 knots Gusts to 22.
04Z MVFR CIG: Last 6 hours of FT after 04Z marginal VFR due to ceiling.

AREA FORECASTS are 18-hour aviation forecasts plus a 12-hour categorical outlook prepared 2 times/day giving general descriptions of cloud cover, weather and frontal conditions for an area the size of several states. Heights of cloud tops, and icing are referenced ABOVE SEA LEVEL (ASL); ceiling heights, ABOVE GROUND LEVEL (AGL); bases of cloud layers are ASL unless indicated. Each SIGMET or AIRMET affecting an FA area will also serve to amend the Area Forecast.

SIGMET or AIRMET messages broadcast by FAA on NAVAID voice channels warn pilots of potentially hazardous weather, SIGMET concerns severe and extreme conditions of importance to all aircraft. (i.e. icing, turbulence, and duststorms/sandstorms). Convective SIGMETS are issued for thunderstorms by the Severe Storms Forecast Center at Kansas City for the conterminous U.S. AIRMETS concern less severe conditions which may be hazardous to some aircraft or to relatively inexperienced pilots.

WINDS AND TEMPERATURES ALOFT (FD) FORECASTS are 12-hour forecasts of wind direction (nearest 10° true N) and speed (knots) for selected flight levels. Temperatures aloft (°C) are included for all but the 3000-foot level.

EXAMPLES OF WINDS AND TEMPERATURES ALOFT (FD) FORECASTS:
FD WBC 121745
BASED ON 121200Z DATA
VALID 130000Z FOR USE 1800-0300Z. TEMPS NEG ABV 24000

FT	3000	6000	9000	12000	18000	24000	30000	34000	39000
BOS	3127	3425-07	3420-11	3421-16	3516-27	3512-38	311649	292451	283451
JFK	3026	3327-08	3324-12	3322-16	3120-27	2923-38	284248	285150	285749

At 6000 feet ASL over JFK wind from 330° at 27 knots and temperature minus 8°C.

TWEB (CONTINUOUS TRANSCRIBED WEATHER BROADCAST)—Individual route forecasts covering a 25-nautical-mile zone either side of the route. By requesting a specific route number, detailed en route weather for a 12- or 18-hour period (depending on forecast issuance) plus a synopsis can be obtained.

PILOTS . . . report inflight weather to nearest FSS. The latest surface weather reports are available by phone at the nearest pilot weather briefing office by calling at H+10.

U.S. DEPARTMENT OF COMMERCE — NATIONAL OCEANIC AND ATMOSPHERIC ADMINISTRATION — NATIONAL WEATHER SERVICE

Fig. 7-20. Key to Aviation Weather Reports and Forecasts.

When you walk into the Weather Service Office you won't just check the latest weather, but will also look back at past observations to check for a trend. Look at Figure 7-21, *starting* at the 1000Z observations and you'll see that:

1. Crossville is up and down throughout the period with little improvement. The worst weather for the day was at 1900Z, although it has been hovering just above ILS minimums throughout the report times.

2. Nashville is holding above instrument minimums throughout the period and has generally had the best weather of the three stations. Something that might alert you to possible problems on your flight is that at 2000Z the ceiling at Nashville is the lowest it has been all day. Is this a temporary drop or does it mark a trend that may end with weather below IFR minimums? Check the temperature dewpoint spread for the preceding reports. Note that it

started out as a 1° F spread, moved to a 2° F spread at 1300Z and a "no-spread" at 1600Z, but has been hovering generally around a 1° F spread.

3. MKL has basically been deteriorating in ceiling and visibility in the period from 1000Z to 1500Z with the spread of 3° F decreasing to 1° F and hanging at 2° F until the 2100Z observation. (Note that at 2100Z there is no spread.) Visibility is back up to 5 miles at 2000Z.

A brief review of some pertinent points for each time period:

301000 — BNA has 700 broken with 1500 overcast, visibility is 4 statute miles in light rain (R-) and fog (F). The barometric pressure is 1015.6 millibars (not much use to you as a pilot) the temperature and dewpoint are 48° F and 47° F respectively. The wind is from 170° true at 10 knots. The altimeter setting is 29.99 inches of mercury. Remember that broken clouds constitute a ceiling and the ceiling is always

```
        SA 301000                                    SA 301700
CSV SA 0954 M3 OVC 3R-F 173/42/42/1409/001      CSV RS 1654 E3 OVC 21/2R-F 154/47/47/1708/997/RB21
BNA SA 0953 M7 BKN 15 OVC 4R-F 156/48/47/1710/999  BNA SA 1653 M8 OVC 3R-F 149/51/50/1607/997
MKL SA 0956 M8 BKN 12 OVC 5RW-F 147/47/44/1709/996 MKL SA 1656 E5 OVC 4L-F 143/50/48/1905/995 LB40

        SA 301100
CSV SA 1054 M3 OVC 3R-F 173/43/42/1306/001/ CIG RGD
BNA SA 1053 M9 OVC 4F 150/49/47/1608/997/INTMT R-
MKL SA 1056 M8 BKN 14 OVC 5RW-F 143/47/45/1807/995

        SA 301200                                    SA 301800
CSV SA 1154 M4 OVC 4F 168/44/43/1412/000/ RE40  CSV SA 1754 E4 OVC 4L-F 147/48/48/1610/994
BNA SA 1153 M9 OVC 4F 148/49/48/1707/997/INTMT R-   BNA SA 1753 M8 OVC 3F 135/52/51/1707/993
MKL SA 1156 M8 BKN 14 OVC 5RW-F 138/48/46/1607/994  UA 1742 50S BNA TOPS LYRD TO 26.5.
                                                    UA 1742 3E BNA TOPS 45 55Φ65
        SA 301300                                    UA 14N BNA OVC40 CLDS ABV
CSV SA 1254 E4 OVC 4R-F 162/45/44/1811/999/ RB03    MKL E5 OVC 4R-F 128/50/48/2105/991
BNA SA 1253 M9 4R-F 150/50/48/1204/997
MKL RS 1256 E8 BKN 12 OVC 21/2RW-F 146/48/46/1908/996

        SA 301400                                    SA 301900
CSV SA 1354 E4 OVC 4R-F 161/46/45/2108/999/ CIG RGD CSV RS 1854 W2X3/4RF 138/48/48/2205/992
BNA SA 1353 M9 OVC 2R-F 150/49/48/1706/997          BNA SA 1853 M7 OVC 3R-F 125/52/51/1808/990/RB25
MKL SA 1356 E6 OVC 21/2RW-F 146/48/46/1706/996      MKL SA 1856 E5 OVC 4R-F 122/50/48/1905/989

        SA 301500                                    SA 302000
CSV SA 1454 E4 OVC 3L-F 168/46/46/2009/000      CSV RS 1954 W5X1R-F 130/48/48/2003/990
BNA SA 1454 M7 OVC 4F 159/48/47/1806/000/INTMT R-   BNA SA 1953 M6 OVC 3R-F 122/53/52/1706/989
MKL RS 1456 E3 OVC 2L-F 145/49/48/1706/996          MKL SA 1956 E5 OVC 5R-F 115/50/49/0000/987

        SA 301600                                    SA 302100
CSV RS 1554 3 OVC 3L-F 161/47/46/1709/999       CSV RS 2054 W3X 2L-F 134/48/48/1704/991
BNA RS 1553 M7 OVC 11/2R-F 152/50/50/1802/998       BNA RS 2053 M7 OVC 21/2R-F 125/53/52/1705/990
MKL SA 1556 E5 OVC 4F 142/50/48/1806/995/LE35       MKL RS 2056 E8 OVC 5R-F 107/50/50/000/985
```

Fig. 7-21. Sequence reports for Crossville, Nashville, and Jackson (McKellar Field), Tennessee from 1000Z through 2100Z on the 30th of the month.

preceded by a letter showing how it was obtained (see Fig. 7-20). Thin broken (BKN) or scattered (SCT) do not constitute a ceiling.

301100 — Note that the ceiling is ragged (CIG RGD) at Crossville at 1100Z and Nashville has intermittent light rain.

301200 — The rain at CSV ended at 40 minutes past the last hour and BNA still has intermittent light rain.

301300 — The rain began at 3 minutes past the hour at CSV; MKL visibility has dropped to 2 1/2 (statute) miles in light rain showers and fog (RW-F).

301400 — The ceiling is ragged at CSV and the ceiling at MKL has dropped from 800 broken (last hour) to 600 overcast.

301500 — BNA is still suffering intermittent light rain and MKL has a record special report (RS) noting that significant weather changes have occurred since the last hourly observation. At MKL the ceiling is 300 feet overcast with a visibility of 2 miles in light drizzle (L-) and fog (F). Note the 1° F spread now.

301600 — Crossville and BNA have an RS. (Compare the visibility at BNA with the last hour.) The drizzle ended at MKL at 35 minutes past the last hour.

301700 — CSV visibility has deteriorated slightly and noted by a record special; also the rain began at 21 minutes past the last hour. Figure 7-22 indicates

how BNA (Nashville) NOTAMS (Notices to Airmen) would be listed under the hourly report. (This is from a different hourly report but is included to show the idea.) The NOTAM is preceded by an exclamation mark (!) to get your attention. The first NOTAM is the 9th one for the 9th month (September) and indicates that the Tullahoma (Tennessee) (THA) Simplified Directional Facility (SDF) is out of service (OTS).

The 37th NOTAM issued in the 11th month shows that the Clarksville (Tennessee) Remote Communications Outlet (RCO) is out of service. The 17th NOTAM for February indicates that the Smyrna (MQY) pilot controlled lighting (PCL) is out of service.

There is also a pilot report (PIREP) indicating that over MSL (Muscle Shoals) at 1530, present flight level unknown (FLUKN), a PA-31 had encountered moderate to severe turbulence at 5000 MSL.

```
BNA SA 1647 E29 BKN 80 OVC 8R- 134/52/47/0807/993

!BNA 09/008 THA SDF OTS
!BNA 11/037 CKV RCO OTS
!BNA 02/017 MQY PCL OTS
BNA UA /OV MSL 1530 FLUKN /TP PA31 /TB MDT-SVR  AT 50
BNA 221515 C50 OVC. 16Z C30 OVC 5F 0712 CHC 3RW/TRW.
09Z MVFR CIG RW..
```

Fig. 7-22. NOTAMS accompanying an hourly weather report.

```
MSL SA 1650 E40 BKN 10 087/56/54/0712/979/ RE36
!MSL 09/009 HUA ARPT PPR EXC MIL
!MSL 12/010 HUA ATCT/GCA 1230-0030 WKDAY EXCP HOL
!MSL 02/004 HUA NDB UNMON
!MSL 02/005 1M4 18-36 CLSD
!MSL 02/008 MSL 18-36 CLSD
```

Fig. 7-23. An actual weather report and NOTAMS for Muscle Shoals, Alabama, (MSL). You will note that there is a NOTAM left over from the last September (Ø9) and December (12), but the rest were issued in February (Ø2). (1) HUA (Huntsville Army Air Field — Redstone) notes that Prior Permission is Required (PPR) except for military traffic. (2) The Redstone air traffic control tower (ATCT) and ground control approach (GCA) are operative from 1230Z to 0030Z on weekdays, except holidays. (3) The Redstone nondirectional beacon (NDB) is unmonitored. (4) The 1M4 airport (Haleyville, Alabama) runway 18-36 is closed. (5) Runway 18-36 is closed at Muscle Shoals.

Check Figure 7-30 for more about decoding pilot reports.

Note that this sample (Fig. 7-22) includes the pilot report (UA) just mentioned *and* a terminal forecast (FT). Terminal forecasts will be covered shortly and you might come back to see that after 1600Z on the 15th of the month the Nashville weather is forecast to be ceiling 3000 overcast, visibility 5 miles in F (fog) with winds from 070° true at 12 knots, with a chance of 3 miles visibility in rain showers and thunderstorms.

301800 — Some pilot reports (UA) are issued at BNA. Taking them in order:

1. At 1742Z a pilot reports at 50 nautical miles south of BNA that the tops are layered to 26,500 MSL.

2. At 1742Z, 3 nautical miles east of Nashville the first top is at 4500 with an overcast layer from 5500 to 6500 (all MSL values).

3. A pilot 14 nautical miles north of BNA reports the top of the overcast at 4000 MSL with clouds above.

The following sequence reports (1900Z, 2000Z, and 2100Z) show that Crossville has ceiling and visibility problems. Nashville is still above IFR minimums.

Figure 7-23 gives another look at the combination of hourly reports and NOTAMS.

As noted earlier, the issuance of special weather reports indicates significant changes and Figure 7-24 is a series of specials issued by Memphis during the period of from 1507Z to 2237Z on the 30th. The last one you would see before leaving would be the 2009Z report, but the last two have been included to show "what had happened" at Memphis after you had departed there. Note that Runway 35L Visual Range is varying considerably. Looking at the 1704Z report you see that there is an indefinite ceiling 200, sky obscured, visibility 1/4 mile in light drizzle and fog. The wind is from 290° at 8 knots; the altimeter setting is 29.96 inches of mercury. Runway 35L Visual Range 1800 variable to 2800. Visibility from the tower is 1/16 of a mile.

Note at 1822Z the pressure is falling rapidly. At 2237Z the weather is "not very good."

Forecasts

Terminal Forecast

Terminal forecasts are for specific airports, and, in the U.S., for an area within a 5 mile radius of the runway complex. They are issued three times daily for the East and Central portion of the U.S. at 0940Z, 1440Z, and 2140Z. The Mountain and Pacific areas have forecasts at 0940Z, 1540Z, and 2240Z. Each terminal forecast issued is for a 24 hour period with the valid time starting on the hour following the issuance time. Incidentally, the issuance times may be changed by the time you read this, so check with the Weather Station specialists when you go in to check the weather.

Figure 7-25 are actual terminal forecasts for the stations and periods covered by the charts and sequence reports.

```
MEM SP 1507 M3 OVC 21/2F 2907/997/R35LVR 60+

MEM SP 1704 W2X1/4L-F 2908/996/R35LVR18V28/TWR VSBY 1/16

MEM SP 1834 M5 OVC 11/2F 2708/994

MEM SP 1822 M4 OVC 2F 2404/991/PRESFR

MEM SP 2009 M2 OVC 1/4F 2907/987/ R35LVR 40V60+

MEM SP 2118 W2X1/2F 2905/988/ R35LVR26V50

MEM SP 2237 W1X1/2F 2909/989/R35LVR26V35
```

Fig. 7-24. Special weather reports from Memphis.

```
    FT 300940
BNA 301010 C4 OVC 2R-F 1710. 19Z C8 OVC 5F OCNL L-. 01Z C10 OVC.
    04Z MVFR-CIG..
CSV 301010 C3 OVC 1R-F 1510. 21Z C8 OVC 4F OCNL L-. 04Z IFR-CIG..
CHA 301010 2 SCT C5 OVC 2R-F SCT V BKN. 19Z C4 OVC 1F OCNL R-L-.
    04Z IFR-CIG-L-..
MKL 301010 C4 OVC 2R-L-F 1610. 14Z C4 OVC 3F OCNLY 1L-F. 17Z
    C8 OVC 6F 3010. 21Z C12 OVC. 04Z MVFR-CIG..
```

```
    FT 301440
BNA 301515 2 SCT C4 OVC 2R-F VRBL C2 OVC 1R-L-F. 19Z C8 OVC 4F OCNL L-.
    01Z C10 OVC. 09Z MVFR CIG..
CSV 301515 C2 OVC 1R-L-F. 23Z C6 OVC 3F OCNL L-. 09Z IFR CIG BCMG MVFR..
CHA 301515 4 SCT C8 OVC 1R-F SCT V BKN. 17Z C4 OVC 1R-L-F. 00Z C8 OVC 5F
OCNL L-. 09Z IFR CIG..
MKL 301515 C2 OVC 1R-L-F 1910. 17Z C4 OVC 3F 3010 OCNL L-. 19Z C8 OVC
    5F 3010. 21Z C12 OVC 3010. 09Z MVFR CIG..
```

Fig. 7-25. Terminal forecasts issued at 0940Z and 1440Z for 4 stations.

Looking at the *top forecast group* you'll note that it is a "Forecast, Terminal" (FT) issued on the 30th of the month at 0940Z. Each terminal gives the day of the month and valid time (301010), the forecast is valid from the 30th at 1000Z to 1000Z the next day.

Nashville (BNA) is forecast to have a ceiling of 400 overcast with 2 statute miles visibility in light rain (R-) and fog (F). The wind is forecast to be from 170° true at 10 knots. Remember from Figure 7-20 that the *visibility* is included if it's *6 statute miles or less* and the *wind* is included if *10 knots or more*. Both items were required to be included here. After 1900Z the ceiling is forecast to be 800 overcast, visibility 5 miles in fog with occasional light drizzle (L-). The wind was omitted, so it's expected to be less than 10 knots during that forecast period.

After 0100Z (on the 31st) the ceiling is forecast to be 1000 feet overcast. Both visibility and wind have been omitted in this period, so the visibility is expected to be more than 6 statute miles and the wind to be less than 10 knots.

The last 6 hours of a terminal forecast are given in general terms. So after 0400Z on the 31st, conditions at BNA will be marginal VFR because of ceiling (CIG). Note that Chattanooga (CHA) during that last 6 hours is expected to be IFR because of the ceiling, plus light drizzle as a restriction to visibility.

The period starting at 1900Z is of most interest and you should compare Figure 7-21, the actual weather at 1900Z, with the forecast for that time. Figure 7-21 BNA had 700 overcast with 3 miles in light rain and fog, as compared to a forecast of 800 overcast and 5 miles, and at 2000Z the ceiling was 600 overcast with 3 miles.

You would naturally want to look at the latest terminal forecast before the departure, and the 1440Z forecast at the bottom of Figure 7-25 would be the most recent issued. (It's valid from 1500Z on the 30th to 1500Z on the next day.) The two forecasts were included so that you could compare predictions. If the forecasts for the 1900Z-0100Z period had been radically different you would know that the weather is unstable (particularly if the most recent showed a worse condition). The only difference is that in the 1440Z forecast the visibility is noted as 4 miles instead of 5. Note that the actual weather at BNA at 1900Z and 2000Z (Fig. 7-21) is close to that forecast (slightly worse) so it's likely that you could trust later predictions in that forecast. You should compare forecasts and actual weather for other stations in the area as well.

Figure 7-26 shows a forecast for Huntsville, Alabama (HSV) for the 30th at 1500Z to 1500Z the next day and an amendment issued on the 30th for 1600Z to 1500Z the next day. (A quick change.) Note that the weather is expected to deteriorate sharply as indicated by the amended forecast.

```
AL 301442
HSV 301515 5 SCT C10 OVC 5L-F SCT V BKN. 09Z LIFR CIG R F..


HSV FT AMD 1 301515 1610Z C3 OVC 2L-F VRBL C2 OVC 1/2RF. 09Z LIFR
    CIG R F..
```

Fig. 7-26. Amended forecasts for Huntsville.

Area Forecast (FA)

Figure 7-27 is an actual area forecast issued by New Orleans (MSY) on the 30th of the month at

132

1240Z, and is valid for an 18 hour period (from 1300Z on the 30th to 0700Z on the 31st) with a 12 hour outlook (0700Z-1900Z on the 31st). Note that it covers the time period of the proposed flight.

The FA is for Tennessee, Arkansas, Louisiana, Mississippi, Alabama, Florida west of 85° west longitude and the coastal waters adjoining of the Gulf states.

The heights are above sea level (ASL) unless noted . . . .

SYNOPSIS — A frontal wave developing off the southeast Louisiana coast will move into southeastern Alabama by 0700Z Wednesday and continue northeastward.

SIGNIFICANT CLOUDS AND WEATHER (SIGCLD AND WX) —

(A) (For) Tennessee east of Nashville (BNA), Alabama north of Montgomery (MGM) . . . .

Ceilings are forecast to be at or below (AOB) 1000 overcast, layered to above 16,000 feet, with visibilities at or below 3 statute miles in light rain, light drizzle and fog. Outlook . . . IFR due to ceilings, light rain, light drizzle and fog.

(B) (For) Tennessee west of Nashville, Arkansas, Mississippi north of Jackson (JAN), Louisiana west of a Natchez-Lafayette (HEZ-LFT) line and adjacent coastal waters . . . .

Ceilings are forecast to be at or below 1000 overcast layered to 10,000 feet (ASL), visibilities at or below 3 miles in light rain, light drizzle and fog, improving from the west, becoming at 1900Z, 2000 broken to overcast, tops 10,000 feet with visibility unrestricted. Outlook . . . . Marginal VFR because of ceilings.

(C) Louisiana east of a Natchez-Lafayette line, Mississippi south of Jackson and adjacent coastal waters . . . . Ceilings will be at or below 1000 overcast (with tops to) 15,000 feet, visibility below 3 miles in light rain, light drizzle and fog with (a) few thunderstorms with tops near 35,000 feet. Improving after 1700Z becoming 2000 broken, tops to 10,000, visibility near 5 miles in smoke. Outlook . . . . Marginal VFR because of ceilings.

(D) (For) Alabama south of Montgomery, northwest Florida and adjacent coastal waters . . . .

Ceilings are forecast to be below 1000 overcast (with) tops 15,000 ASL, visibilities at or below 3 miles in light rain, light drizzle and fog with widely scattered thunderstorms (with) tops near 35,000 ASL. Outlook low IFR because of ceilings and light rain, light drizzle and Fog.

ICG (icing) — Moderate, mostly rime icing (ICG)

MSY FA 301240.
13Z TUE-07Z WED.
OTLK 07Z WED-19Z WED.

TENN ARK LA MISS ALA FLA W OF 85 DEG CSTL WTRS...

HGTS ASL UNLESS NOTED...

SYNS... FRONTAL WV DVLPG OFF SE LA CST WL MOV INTO SERN ALA BY 07Z WED AND CONT NEWD.

SIGCLD AND WX...
A TENN E OF BNA ALA N OF MGM...
CIGS AOB 10 OVC LYRD TO ABV 160 VSBY AOB 3 MI R-L-F. OTLK... IFR CIG R-L-F.

B TENN W OF BNA ARK MISS N OF JAN LA W OF HEZ-LFT LN ADJ CSTL WTR...
CIGS AOB 10 OVC LYRD TO 100 VSBY AOB 3 MI R-L-F IPVG FM W BCMG 19Z 20 BKN TO OVC 100 VSBY UNRSTD. OTLK... MVFR CIG.

C LA E OF HEZ-LFT LN MISS S OF JAN ADJ CSTL WTRS...
CIGS AOB 10 OVC 150 VSBY BLO 3 MI R-L-F WITH FEW TSTMS TOPS NR350 IPVG AFT 17Z BCMG 20 BKN 100 VSBY NR 5 MI K. OTLK... MVFR CIG.

D ALA S OF MGM NW FLA ADJ CSTL WTRS...
CIGS BLO 10 OVC 150 VSBY AOB 3 MI R-L-F WITH WDLY SCTD TSTMS TOPS NR 350. OTLK L IFR CIG R-L-F.

ICG... MDT MSTLY RIME ICGICIP ABV FRZLVL NR 50 NW ARK TO NR 120 SE CSTL WTRS.

Fig. 7-27. Area forecast.

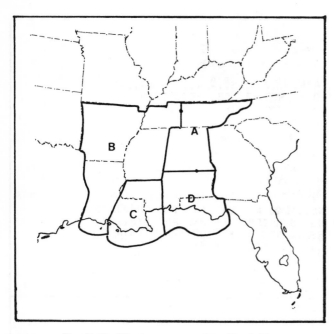

Fig. 7-28. The areas covered by Fig. 7-27.

in clouds (IC) and in precipitation (IP) above the freezing level (which is) near 5000 ASL in northwest Arkansas to near 12,000 in the southeastern coastal waters.

Figure 7-28 shows the areas (A, B, C, and D) just discussed.

Advisories

Figure 7-29 shows actual AIRMETS and SIG-METS issued by New Orleans which would be valid for the period including the flight.

AIRMETS

Airmets are for weather that might be hazardous to single engine and light aircraft and in some cases to other aircraft as well. An AIRMET includes information on the following:
1. Moderate icing
2. Moderate turbulence
3. Sustained winds of 30 knots or greater at or within 2000 feet of the surface
4. Onset of extensive areas of visibility below 3 miles and/or ceilings of less than 1000 feet, including mountain ridges and passes

An AIRMET continued (WAC) is issued for continued moderate turbulence over mountain terrain and continued ceilings below 1000 feet and/or visibility less than 3 miles over an extensive area. Unlike the regular AIRMETS which have a valid time period the WAC's remain valid until cancelled.

Looking at the first AIRMET in Figure 7-29 you'll note that it was issued by MSY on the 30th at 1500Z and is valid from 1500Z to 2100Z on that date.

The "AIRMET DELTA 3" indicates that it was the 3rd one issued for the New Orleans Delta area for that date. *Translated:*

Flight Precautions for Tennessee, Arkansas, Louisiana, Mississippi, Alabama, Northwest Florida and adjacent coastal waters for IFR conditions. Ceilings below 1000 feet and visibilities less than 3 miles in clouds and precipitation. Continuing beyond 2100Z. (Note that this is the area covered by the area forecast — Fig. 7-27.)

The second AIRMET warns of turbulence and strong low level winds over the area noted.

SIGMETS

A SIGMET advises of weather potentially hazardous to all categories of aircraft specifically:
1. Tornadoes
2. Lines of thunderstorms (squall lines)
3. Embedded thunderstorms
4. Hail or 3/4" or greater in diameter
5. Severe or extreme turbulence
6. Severe icing
7. Widespread sandstorms/duststorms lowering visibilities to below 3 miles.

The SIGMET in Figure 7-29 was issued by New Orleans on the 30th at 1845Z and is valid from 1845Z to 2300Z. It's the first one for the Foxtrot area and says:

Flight Precaution. Arkansas and western Tennessee for turbulence. Over the area northwest of a line through Shreveport-Jackson (McKellar Field) Tennessee, moderate or greater (MOGR) turbulence likely from 15,000 to 35,000 feet due to strong wind shear. (Conditions will continue beyond 2300Z.)

The altitudes affected in this SIGMET are well above the altitude you'd use for the trip, but any unusual conditions over the area should be noted and stored away in your mind.

The first four items listed above plus thunderstorm areas of 40 per cent coverage or more are issued as CONVECTIVE SIGMETS.

PILOT AND RADAR REPORTS

Figure 7-30 is an example of a pilot report.

The format of the PIREPS is in the following order of presentation to the reader: Location (DV), flight level (FL), type of airplane (TP), sky cover (SK), temperature (TA),

```
              MSY WA 301500
              301500-302100

              AIRMET DELTA 3. FLT PRCTN. TENN..ARK..LA..MISS..ALA..NW FLA..ADJ
              CSTL WTRS FOR IFR CONDS. CIGS BLO 1 THSD FT VSBY LESS THAN 3 MI
              CLDS AND PCPN. CONT BYD 21Z.
```

AIRMETS

```
              MSY WA 301600
              301600-302200

              AIRMET BRAVO 1. FLT PRCTN. MDL TENN..ALA..NW FLA..AND THE ADJ CSTL
              WTRS FOR TURBC AND STG LO LVL WNDS. OVR TENN BTW MKL AND TYS..AND
              OVR ALA..NW FLA..AND ADJ CSTL WTRS...WNDS 30 KTS OR GTR WITHIN 2
              THSD FT OF SFC AND MDT TURBC BLO 7 THSD FT. CONT BYD 22Z.
```

SIGMET

```
              MSY WS 301845
              301845-302300

              SIGMET FOXTROT 1. FLT PRCTN.ARK AND WRN TENN FOR TURBC. OVR THE
              AREA NW OF A LN THRU SHV-MKL MOGR TURBC LKLY 150-350 DUE TO STG
              WND SHEAR. CONT BYD 23Z.
```

Fig. 7-29. AIRMETS and SIGMETS.

ENCODE

```
UA /OV MRB-PIT 1600 FL080 /TP BE55 /SK 004 BKN 012/022 BKN-OVC
/TA 01 /IC LGT-MDT RIME 035-060 /RM WIND COMP HEAD 020 MH310
TAS 180
```

DECODE

Pilot report, Martinsburg to Pittsburg at 1600Z at 8000 feet; a Beechcraft Baron; cloud base 400 broken tops 1200, second layer 2200 broken variable overcast; air temperature 1 degree Celsius; light to moderate rime icing 3500-6000 feet; headwind component 20 knots, magnetic heading 310, true air speed 180 knots.

Fig. 7-30. Pilot reports.

```
SDUS KNKA 301928
 BNA 1831 AREA3R-/NC 100/125 235/125 140W C2325 MT 220 AT 205/75
  0110 1111 1211 11①
```

Fig. 7-31. Radar report.

winds (WV), turbulence (TB), icing (IC), and remarks (RM).

Figure 7-31 shows a radar report (RAREP) for Nashville issued on the 30th. Looking at the RAREP (issued at 1928Z) you might note the following: The radar at Nashville at 1831Z plotted an area 3/10th coverage containing light rain, with no change (NC) in intensity and defined by the following area from Nashville radar: 100°-125 nautical miles, 235°-125 N.Mi, 140 miles wide. What it means is that the weather is defined by a line between the two points (100°-125 N.Mi and 235°-125 N.Mi) and extends 70 miles on either side of that line, or the area is 140 miles total in width. Other RAREPS may give a number of reference points from the radar, and the area is found by joining these points.

C2325 — There are cells (TRW) moving from 230° (true) at 25 knots.

MT 220 AT 205/75 — The maximum tops are at 22,000 feet at 205°-75 nautical miles from BNA.

The bottom row of numbers (0110, etc.) are codes for the weather specialist and are not for the pilot.

WINDS AND TEMPERATURES
ALOFT FORECAST

The winds are covered last because there's not much point in finding out the winds until you've looked at the weather and decided that you can go. Figure 7-32 is an actual winds aloft forecast (FD)

issued on the 30th at 0545Z, and the data is based on 0000Z data on the 30th. It's valid at 201800Z for use from 1500-2100Z. All temperatures are negative above 24,000 feet MSL. (Since air temperatures are below 0° C above 24,000 feet, this saves putting in all those "minuses.")

Through 12,000 feet the levels are true altitude (ASL), 18,000 feet and above are pressure altitudes. The locations are listed in alphabetical order. No winds are forecast within 1500 feet of the station elevation. No temperatures are forecast for the 3000 foot level or within 2500 feet of the station elevation. So for some of the stations of higher elevations, both 3000 and 6000 foot winds may be left off.

Looking at the Nashville forecast, at 3000 feet the wind is forecast to be from 240° *true* at 20 knots.

6000 — Wind from 250° at 22 knots with an outside temperature of +4° C.

9000 — From 240° true at 36 knots, temperature -1° C.

At 30,000 in the Nashville area the wind is from 230° true at 93 knots and the temperature is a slightly chilled -43° C.

Take a look at Allegheny County (AGC) at Pittsburgh at 30,000 feet. There the wind is from 750° at 10 knots. The Weather Service wants to hold things down to 6 digit figures, so when the wind is over 100 knots they add 50 to the direction and subtract 100 from the speed. You subtract 50 from the two digits for the direction, 75 - 50 = 25 or 250° true at 10 + 100 = 110 knots. The temperature is -46° C.

As another example, Louisville is forecast for a

```
        FD US2 KWBC 300545
    DATA BASED ON 300000Z

    VALID 301800Z    FOR USE 1500-2100Z.  TEMPS NEG ABV 24000

    FT   3000    6000    9000    12000   18000    24000   30000   34000   39000


    AGC  1927  2346+02  2350-03  2349-08  2461-21  2480-32  751046  752354  751259
    ATL  2218  2433+07  2438+03  2340-03  2350-16  2363-28  237842  248352  248663
    BHM  2426  2429+07  2433+02  2337-04  2352-16  2366-28  238142  238452  238662
    BNA  2420  2522+04  2436-01  2450-07  2365-19  2379-30  239343  239452  248559
    CAE  2021  2233+08  2337+03  2338-02  2445-15  2458-27  257442  258351  259562
    CLF  2122  2439+01  2344-05  2348-11  2461-23  2477-34  740547  751354  259756
    CRW  2028  2447+04  2452-01  2452-07  2464-20  2584-31  751144  752353  751861
    DAL  2709  2616+00  2616-05  2616-10  2423-21  2331-31  233944  234150  243353
    DSM  2611  2818-01  3121-05  3226-09  3334-22  3443-33  355746  355051  343353
    EMI  1925  2341+02  2345-03  2446-08  2557-20  2577-31  760646  762155  761762
    ILM  2021  2233+05  2339+01  2441-04  2550-16  2566-28  268542  269451  760463
    IND  2421  2526+00  2536-05  2545-11  2458-22  2472-33  249245  249351  25825A
    JAN  3113  2518+05  2429+00  2348-05  2358-17  2373-28  239042  239651  239259
    JAX  2228  2228+10  2128+05  2128+00  2336-13  2448-25  256540  257650  258961
    JFK  2219  2326-01  2430-05  2533-09  2649-21  2771-33  770647  772556  771760
    JOT  2714  2717-03  2724-08  2832-13  2636-24  2642-35  265047  265050  264650
    LIT  2712  2421+01  2425-04  2430-09  2345-20  2360-31  237544  237550  236055
    LOU  2323  2532+02  2542-03  2450-09  2465-21  2482-31  740244  740552  249557
    MEM  2619  2524+03  2432-02  2342-07  2360-19  2375-30  239043  239051  237657
```

Fig. 7-32. Winds and temperatures aloft forecast.

wind at 30,000 feet from 240° at 102 knots and a temperature of -44° C.

AUTOMATION OF FIELD OPERATIONS AND SERVICES (AFOS)

The National Weather Service is setting up the AFOS program to make information available to the pilot in a more efficient and less wasteful manner. The present system teletypes each hour yards of information, some of which is not used by the pilots. (For instance, a particular Weather Service Office may have received all sorts of weather information on Gitchygoomy International Airport for years. It may turn out that no one from this station has had any inclination to go there. However, the teletype has been dutifully printing reams of paper containing all sorts of information about Gitchygoomy and other stations. The result being enough extra paper to cover the Pentagon walls.)

The AFOS will use a computer, and instead of typing *all* of the information, it will be stored. There will be a console with three scopes in the Weather Service Office. The specialist will punch in your route and other data, and you'll see the desired information (teletype material and graphic — charts — displays) as needed.

It may be some time before AFOS are widespread, but you might get a chance to use it on one of your trips into an airport having the facility.

ENROUTE FLIGHT ADVISORY SERVICE

This is a weather service available to you from selected Flight Service Stations along heavily traveled airways at a service criterion of 5000 feet above ground level at 80 miles from an EFAS outlet. All communications will be conducted on a designated frequency 122.0 MHz using the radio call (name of station) FLIGHT WATCH.

Routine weather information plus current reports on the location of thunderstorms and other hazardous weather as reported by ground observers or pilots (or radar) may be obtained from the nearest flight watch facility.

This service is not intended to be used for flight plan filing, routine position reporting or to obtain a complete preflight briefing in lieu of contacting an FSS or Weather Service Office.

SUMMARY

By the time you read this, some of the services covered may be changed or eliminated, or new ones added. The main point is, to check the various inputs against each other. How have the earlier forecasts been comparing with the earlier weather? Look at PIREPS and other reports that may be more up to date than the last sequence or forecast. Remember that some of the charts may be several hours out of date, and wind and temperature aloft

Fig. 7-33. "Jonesville Flight Watch, would you drive that part about 'sunny and mild here' by me one more time?"

forecasts are just that — *forecasts*.

These are actual reports or copies thereof, and later chapters are based on a flight from Memphis to Nashville. (The Chattanooga weather will be found to be good enough for it to be the alternate — it's easy to manipulate things on paper.)

It's going to be up to you to ask the proper questions, because it's your neck. You may want to set up a weather checklist for your use in the WSO or FSS so that you don't forget a vital check.

Remember that the people in the Weather Service Offices are there to help you. If you have questions, ask them, because there may have been changes in weather presentation or release times since that last flight a few days ago.

Chapter 8

THE CHARTS AND OTHER PRINTED AIDS

THE ENROUTE LOW ALTITUDE CHARTS

THE Enroute Chart System is broken down into two main segments:

    1. *Enroute High Altitude Charts* which cover the conterminous United States in four charts and concern airspace 18,000 feet MSL and above. These won't be covered in this book.

    2. *Enroute Low Altitude Charts* which cover altitudes up to but not including 18,000 MSL and which cover the conterminous United States in 28 charts. (There are also charts for Alaska and Hawaii and one for the Caribbean (Miami-Nassau-Puerto Rico). The U.S. charts are printed back to back; for instance, L-13 and L-14 go together as shown by Figure 8-1.

    In the back of this book there is a part of the L-14 Enroute Low Altitude Chart with a legend block. Study this chart and legend until the symbols become familiar to you. (You'll still be allowed to use the legend later if you need it, though.) Note the boundary between Memphis and Atlanta Centers just west of Chattanooga.

    The arrow and enclosed "38" west of Chattanooga indicates that the COWAN intersection is 38 DME (nautical) miles from the Chattanooga VOR.

    Low frequency facilities are, as on the sectional charts, in red or magenta on the actual chart (we had to stick to black and white, here). North Chattanooga radio beacon and WDOD, a commercial broadcast station, are indicated in red.

    The tower (local, ground, departure and approach controls) communications frequencies of each Enroute Low Altitude Chart are placed on the particular

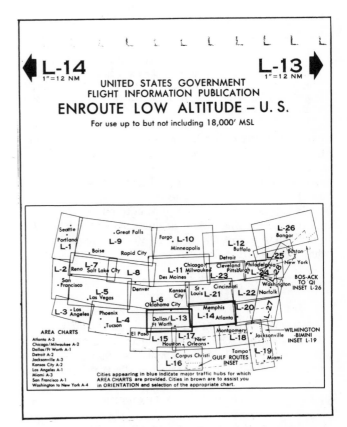

Fig. 8-1. The Enroute Low Altitude Charts are published "back to back" and are issued every 56 days.

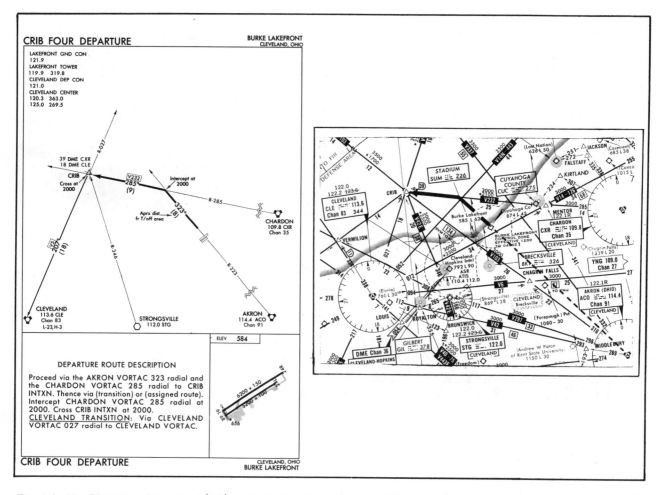

Fig. 8-2. The "Crib Four Departure" (SID) for Burke Lakefront Airport at Cleveland. A comparison is shown between the SID Chart and the departure as would be seen on an Enroute Low Altitude Chart. (The heavy and dashed lines on the Enroute Chart were drawn in by the writer.) Note in the departure route description that if you want to make a transition to Cleveland you use the 027R (course 207°). Note on the SID that information by the Cleveland VOR notes that this area is on the L-23 (Low Altitude) Chart and H-3 (High Altitude Chart). Like all charts in this book, this is an example only.

chart. In a box in the legend section of the sample chart are some pertinent airport tower frequencies which could be used in a later example trip. The towers are always listed alphabetically by city (these have been reshuffled a little).

If you are a refresher pilot note on the Chart that the controlling Flight Service Station is now listed under each VOR box. Review the legend for other presentations new to you.

AREA CHARTS

Certain congested terminal areas (Atlanta, Miami, etc.) have larger scale charts for close-in work. Several of these are published on one sheet and are included with the Enroute Chart subscription.

STANDARD INSTRUMENT DEPARTURES (SIDS)

A standard Instrument Departure (SID) is an air traffic control coded departure routing which has been established at certain airports to simplify clearance delivery procedures.

Pilots of civil aircraft operating from locations where SID procedures are effective may expect ATC clearances containing a SID. Use of a SID requires pilot possession of at least the textual description of the approved effective SID. If the pilot does not possess a preprinted SID description or for any other reason does not wish to use a SID, he is expected to advise ATC. Notification may be accomplished by filing "NO SID" in the remarks section of the filed flight plan or by the less desirable method of verbally advising ATC.

All effective SIDs are published in textual and graphic form by the National Ocean Survey in East and West SID booklets.

Figure 8-2 is sample (and simple) SID information. These charts are published every eight weeks, or with issues of Low Altitude Enroute Charts and Area Charts.

STANDARD TERMINAL ARRIVAL ROUTES (STARS)

A standard terminal arrival route (STAR) is an air traffic control coded instrument flight rules (IFR)

140

arrival route established for application to arriving IFR aircraft destined for certain airports. Its purpose is to simplify clearance delivery procedures.

Pilots of IFR civil aircraft destined to locations for which STARs have been published may be issued a clearance containing a STAR whenever ATC deems it appropriate. Until military STAR publications and distribution is accomplished, STARs will be issued to military pilots only when requested in the flight plan or verbally by the pilot.

Use of STARs requires pilot possession of at least the approved textual description. As with any ATC clearance or portion thereof, it is the responsibility of each pilot to accept or refuse an issued STAR. A pilot should notify ATC if he does not wish to use a STAR by placing "NO STAR" in the remark section of the flight plan or by the less desirable method of verbally stating the same to ATC.

A bound booklet containing all STAR charts is available on subscription from the National Ocean Survey. STAR's implemented on an urgent/emergency basis will be published in textual form in NOTAMS (Class II) until charted.

INSTRUMENT APPROACH PROCEDURE CHARTS

The Instrument Approach Charts will be of the greatest importance for the successful completion of an instrument flight. If you're taking an instrument refresher you'll find the chart layouts have changed. The new term you'll hear for approach charts is "TERPS" (Terminal Instrument Procedures).

Going first into some definitions. (Most of the following material on Approach Charts is taken from the FAA's *Advisory Circular* 90-1A.)

Definitions

MDA — "Minimum descent altitude" means the lowest altitude, expressed in feet above mean sea level, to which descent is authorized on final approach, where no electronic glide slope is provided, or during circle-to-land maneuvering in execution of a standard instrument approach procedure.

DH — "Decision height," with respect to the operation of aircraft, means the height at which a

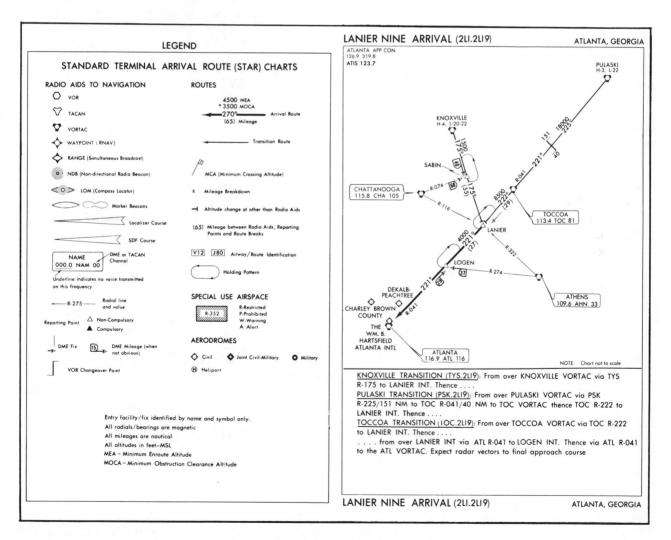

Fig. 8-3. The Lanier Nine Arrival (STAR) and legend.

decision must be made, during an ILS or PAR instrument approach, to either continue the approach or to execute a missed approach. This height is expressed in feet above mean sea level (MSL), and for Category II ILS operation the decision height is additionally expressed as a radio altimeter setting. (Category II has special lower minima and depends on equipment at the airport *and* in the airplane.)

HAA — "Height above airport" indicates the height of the MDA above the published airport elevation. HAA is published in conjunction with circling minimums for all types of approaches.

HAT — "Height above touchdown" indicates the heights of the DH or MDA above the highest runway elevation in the touchdown zone (first 3000 feet of the runway). HAT is published in conjunction with straight-in minimums.

NoPT — Means No Procedure Turn required.

"Precision approach procedure" means a standard instrument approach in which an electronic glide slope is provided (ILS or PAR).

"Non-precision approach procedure" means a standard instrument approach in which no electronic glide slope is provided.

IFR Landing Minimums

Earlier, the ceiling at the airport, as well as the visibility, were used as landing limits. *This is no longer the case.* The published visibility is the required weather condition for landing as cited in FAR 91. FAR 91 now allows approach down to the minimum descent altitude (MDA) or decision height (DH) as appropriate to the procedure being executed, without regard to the reported ceiling.

FAR 91 states the situation as follows:

Landing minimums. Unless otherwise authorized by the Administrator, no person operating an aircraft (except a military aircraft of the United States) may land that aircraft using a standard instrument approach procedure described in Part 97 of this chapter unless the visibility is at or above the landing minimum prescribed in that Part for the procedure used. If the landing minimum in a standard instrument approach procedure prescribed in Part 97 is stated in terms of ceiling and visibility, the visibility minimum applies. However, the ceiling minimum shall be added to the field elevation and that value observed as the MDA or DH, as appropriate to the procedure being executed.

You can't go below MDA or DH unless:

(a) The aircraft is in a position from which a normal approach can be made to the runway of intended landing.

(b) The approach threshold of that runway, or approach lights or other markings identifiable with the approach end of that runway, is clearly visible to the pilot.

(c) If, upon arrival at the missed approach point, or at any time thereafter, either of the above requirements are not met, the pilot shall immediately execute the appropriate missed approach procedure.

The Government-produced charts always contain the following information listed in this order: MDA or DH, visibility, HAA or HAT, and military minimums (ceiling and visibility) for each aircraft approach category.

Figure 8-4 is the legend sheets for the U.S. Coast and Geodetic Survey Instrument Approach Charts.

The chart itself is broken down into five main sections as shown in Figure 8-5.

(1) Top (and Bottom) Margin Identification

The procedure number (ILS RWY 14) states the type of facility and the runway involved. When the approach course is within 30° of the runway centerline (an ILS *always* is), the runway number is given such as NDB RWY 17, VOR RWY 12, etc.

When the final approach course is more than 30° from the runway centerline the procedures are listed as VOR-A, VOR/DME-A, NDB-A, etc.

A VOR/DME procedure number means that *both* VOR and DME receivers and ground equipment must be operating to use the procedure.

When DME arcs and DME fixes are authorized in a procedure and the procedure number does not include the three letters DME in the margin, the procedure may be used without DME equipment.

A VORTAC type procedure is a VOR/DME procedure that is authorized for an airplane equipped with either VOR/DME or TACAN receiver. (TACAN is military-type equipment.)

(2) Plan View

"NDB" (non-directional beacon) procedure number replaces the older ADF type procedure. "LOC" procedure number indicates that the course guidance is furnished by the localizer and there is no ground equipment available for glide slope. Back course ILS procedures are included under this listing.

"LDA" (localizer type directional aid) has the same procedure number but is not aligned with the runway centerline. The approach chart will give the word as to the direction and degrees of alignment away from the runway centerline.

Figure 8-6 is the plan view with explanations of the fictional ILS at "Lattiville Airport."

The plan view is normally shown to scale. *The data shown within the 10 nautical mile distance circle is always shown to scale.*

The plan view gives such information pertaining to the initial approach segment, procedure turn, obstructions, navigation and communications frequencies. The minimum safe altitude for each sector is also given. Take a look at Figure 8-6 and note that the minimum sector altitude (MSA) is 2700 MSL for an inbound course of between 180 and 270° magnetic. From 270° clockwise to 180° magnetic (270° of azimuth) the minimum sector altitude is 3600 feet. These are safe altitudes (1000 foot obstruction clearance) within 25 miles of the facility being used for the approach (VOR, ADF, and LOM for the ILS).

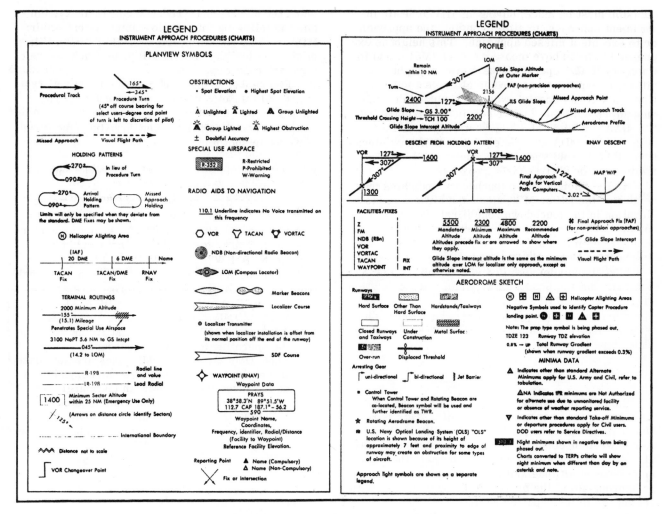

Fig. 8-4. Legend for Instrument Approach Charts.

These sectors cannot be less than 90° in spread.

Figure 8-7 shows the plan view section of a VOR approach with some additional information.

Note in Figure 8-7 the *Enroute Facilities Ring*. Radio aids to navigation, fixes and intersections that are part of the Enroute Low Altitude airway structure and used in the approach procedure are shown in their relative position(s) on this ring.

The *Feeder Facilities Ring* shows the relative positions of radio aids to navigation, fixes and intersections used by the air traffic controller to direct aircraft to intervening facilities and/or fixes between the enroute structure and the initial approach fix.

Since the subject of initial approach fixes has been mentioned, it might be well to see what the official idea of the initial approach is:

a. In the initial approach, the aircraft has departed the enroute phase of flight, and is maneuvering to enter an intermediate or final segment of the instrument approach.

b. An initial approach may be made along prescribed routes within the terminal area which may be along an arc, radial, course, heading, radar vector, or a combination of these. Procedure turns and high altitude teardrop penetrations are initial approach segments.

c. Initial approach information is portrayed in the plan view of instrument approach charts by course lines, with an arrow indicating the direction. Minimum altitude and distance between fixes is also shown with the magnetic course.

d. When the term "NoPT" appears, an intermediate approach is provided. These altitudes shown with the term "NoPT" cannot be used as an initial approach altitude for the purpose of determining alternate airport requirements under FAR 91.23(c) and 91.83(b).

If the facility has RADAR, it will be noted below the communications information and would be identified by the appropriate letters "ASR", "PAR", "ASR/PAR" or "RADAR VECTORING."

1. ASR — means Airport Surveillance Radar instrument approach procedures are available at the airport, and also radar vectoring is available for the procedure.

2. PAR — means that Precision Approach Radar instrument approach procedures are available.

3. RADAR VECTORING — means radar vectoring is available but radar instrument approach procedures are not available.

There will be more about ASR and PAR later.

Notice that a procedure turn barb is shown on Figure 8-6, but a one-minute holding pattern is shown on Figure 8-7. What about the procedure turn and when do you have to use it?

A *procedure turn* is the maneuver prescribed when it is necessary to reverse direction to establish the aircraft inbound on an intermediate or final approach course. It is a required maneuver *except* when the symbol "NoPT" is shown, when RADAR VECTORING is provided, when a one-minute holding pattern is published in lieu of a procedure turn, or when the procedure turn is not authorized. The altitude prescribed for the procedure turn is a *minimum* altitude until the aircraft is established on the inbound course. The maneuver must be completed within the distance specified in the profile view.

1. A barb indicates the direction or side of the outbound course on which the procedure turn is made. Headings are provided for course reversal

using the 45°-type procedure turn. However, the point at which the turn may be commenced and the type and rate of turn is left to the discretion of the pilot. Some of the options are the 45° procedure turn, the racetrack pattern, the tear-drop procedure turn, or the 80°-260° (or 90°-270°) course reversal.

2. Limitations on procedure turns:

a. In the case of a radar initial approach to a final approach fix or position, or a timed approach from a holding fix, or where the procedure specifies "No PT", no pilot may make a procedure turn unless, when he receives his final approach clearance, he so advises ATC and a clearance is received.

b. When a tear-drop procedure turn is depicted and a course reversal is required, this type turn must be executed.

c. When a one-minute holding pattern replaces the procedure turn, the standard entry and the holding pattern must be followed except when RADAR

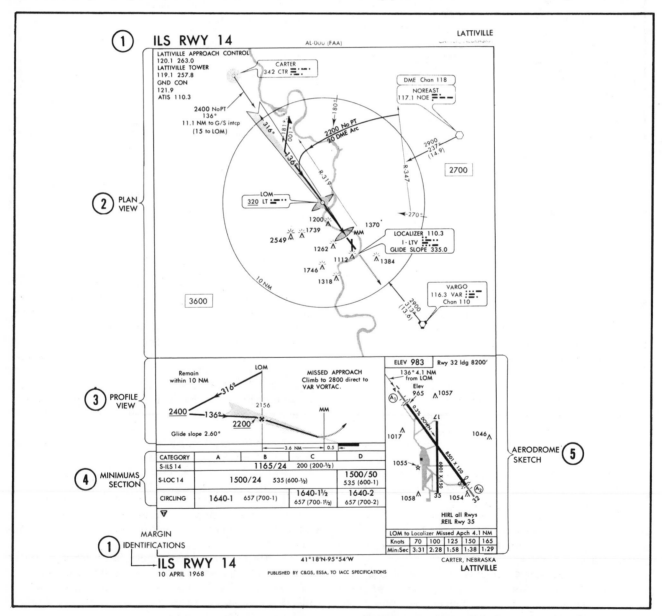

Fig. 8-5. A sample ILS Chart for "Lattiville Airport." The chart is made up of 5 main parts. (AC 90-1A).

144

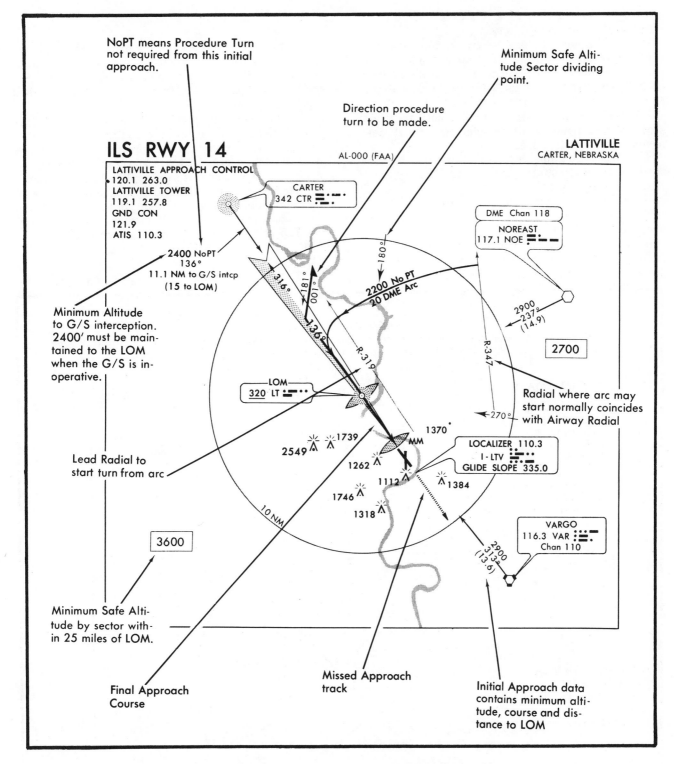

Fig. 8-6. Plan view of the "Lattiville ILS RWY 14" (AC 90-1A).

VECTORING is provided or when "NoPT" is shown on the approach course. As in the procedure turn, the descent from the minimum holding pattern altitude to the final approach fix altitude (when lower) may not commence until the aircraft is established on the inbound course.

d. The absence of the procedure turn barb in the Plan View indicates that a procedure turn is not authorized for that procedure.

3. A procedure turn is not required when the symbol "NoPT" appears on an approach course shown on the Plan View. If a procedure turn is desired, descent below the procedure turn altitude should not be made since some NoPT altitudes may be lower than the procedure turn altitude.

When an approach course is published on an ILS

145

procedure that does not require a procedure turn ("NoPT"), the following applies:

a. In the case of a dog-leg track, and no fix is depicted at the point of interception on the localizer course, the total distance is shown from the facility or fix to the LOM, or to an NDB associated with the ILS.

b. The minimum altitude applies until the glide slope is intercepted, at which point the aircraft descends on the glide slope.

c. When the glide slope is not utilized, this minimum altitude is maintained to the LOM (or to the NDB if appropriate).

d. In isolated instances, when proceeding NoPT to the LOM and the glide slope cannot be utilized, a procedure turn will be required to descend for a straight-in approach and landing. In these cases, the requirement for a procedure turn will be annotated on the Plan View of the procedure chart.

(3) Profile Views Section

These are side views of the procedures. These views include the *minimum* altitude and maximum distance for the procedure turn, altitudes over prescribed fixes, distance between fixes, and the missed approach procedure.

1. Precision approach glide slope intercept altitude. This is a minimum altitude for glide slope interception after completion of procedure turn. It applies to precision approaches and, except where otherwise prescribed, it also applies as a minimum altitude for crossing the final approach fix in case the glide slope is inoperative or not used.

Figure 8-8 is the profile view section of Figure 8-5 (ILS RWY 14 for Lattiville Airport) with comments added. ILS (with glide slope operative) and PAR are "precision" approaches.

2. Step down fixes in non-precision procedures.

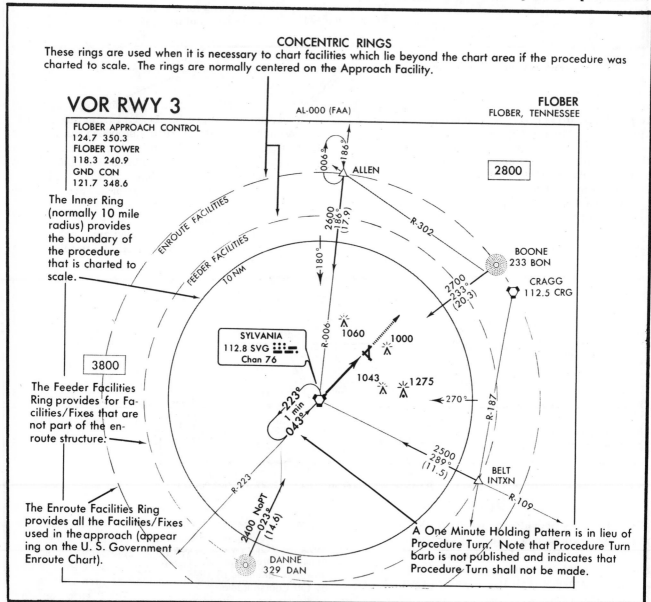

Fig. 8-7. Plan view of a VOR approach (AC 90-1A).

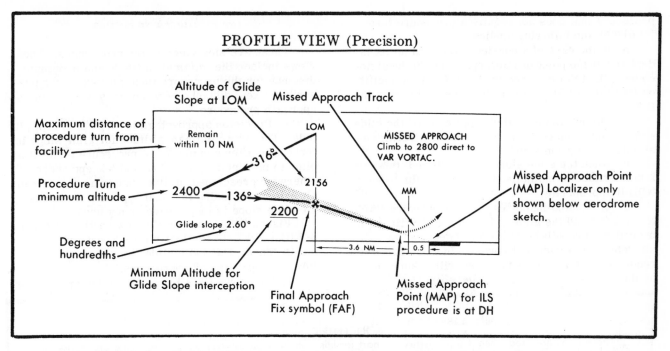

Fig. 8-8. An enlarged profile view of a precision approach, as taken from Figure 8-5 (AC 90-1A).

A stepdown fix may be provided on the final, i.e., between the final approach fix and the airport, for the purpose of authorizing a lower MDA after passing an obstruction. This stepdown fix may be made by an NDB bearing, fan marker, radar fix, radial from another VOR, or by a DME, when provided for as shown in Figure 8-9.

Figure 8-9 is a sample profile view of a non-precision (no glide slope information) approach. This particular one is for a VORTAC approach with a stepdown fix based on DME information.

3. Normally, there is only one stepdown fix between the final approach fix (FAF) and the missed approach point (MAP). If the stepdown fix cannot be identified for any reason, the altitude at the stepdown fix becomes the MDA for a straight-in landing. However, when circling under this condition, you must refer to the Mimimums Section of the procedure for the applicable circling minimum.

4. Missed approach point (MAP). It should be specifically noted that the missed approach points are different for the complete ILS (with glide slope)

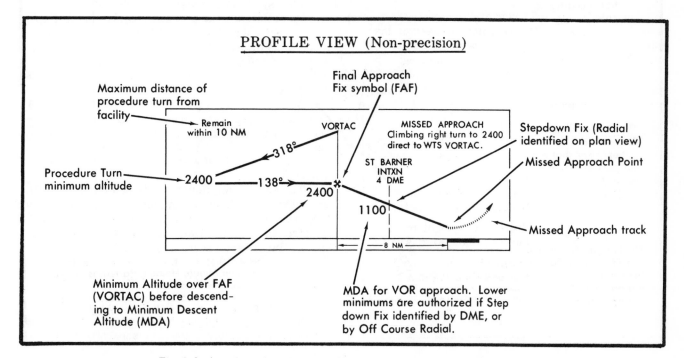

Fig. 8-9. An enlarged profile view of a non-precision approach (AC 90-1A).

and for the localizer only approach. The MAP for the ILS is at the decision height (DH) while the "localizer only" MAP is usually over the (straight-in) runway threshold. In some non-precision procedures, the MAP may be prior to reaching the runway threshold in order to clear obstructions in the missed approach climbout area. In non-precision procedures, the pilot determines when he is at the missed approach point (MAP) by timing from the final approach fix. The FAF has been clearly identified by use of the maltese cross symbol in the profile section. The distance from FAF to MAP and time and speed table, for easy calculation, are found below the aerodrome sketches. This does not apply to VOR/DME procedures, or when the facility is on the airport and the facility is the MAP.

When the missed approach procedure specifies holding at a facility or fix, holding shall be in accordance with the holding pattern depicted on the plan view, and at the minimum altitude in the missed approach instructions, unless a higher altitude is specified by ATC. An alternate missed approach procedure may also be given by ATC.

There are various terms in the missed approach procedure which have specific meanings with respect to climbing to altitude, to execute a turn for obstructions avoidance, or for other reasons. Examples:

"Climb to" means a normal climb along the prescribed course.

"Climbing right turn" means climbing right turn as soon as safety permits, normally to avoid obstructions straight ahead.

"Climb to 2400 turn right" means climb to 2400 prior to making the right turn, normally to clear obstructions.

(4) Minimums Section

The minimums are based on (among other things) the airplane category which is based on the approach speed. The approach speeds to be used are 1.3 times the stalling speed in the landing

Approach Category	Speed
A	Approach speed less than 91 knots.
B	Speed 91 knots or more but less than 121 knots.
C	Speed 121 knots or more but less than 141 knots.
D	Speed 141 knots or more but less than 166 knots.
E	Speed 166 knots or more.

configuration at maximum certificated landing weight.

If it is necessary to maneuver at speeds in excess of the upper limit of the speed range for each category, the minimum for the next higher category should be used. A B-727-100 falls in Category C but when circling to land in excess of 140 knots should use the approach category D minimum. It is quite likely that the airplane you are using in working on your instrument rating is in Category A (or if you're really up there, Category B). It may be some time before you have dealings with Categories C, D and E.

Figure 8-10 shows sample minimums sections for the ILS RWY 14 (Fig. 8-5), and a sample VOR approach.

In Figure 8-10(A), the ILS minimums section indicates that the Decision Height (DH) is 1165 MSL and the minimum visibility is 2400 feet Runway Visual Range. The RVR is separated from the DH by a slash (/). Visibility in miles is separated from the DH or Minimum Descent Altitude (MDA) by a dash (-). The minimums for a circling approach are an MDA of 1640 feet MSL and a visibility of 1 mile. Notice in Figure 8-10(A) that the Decision Height is given as HAT, or Height Above Touchdown (the airport elevation is 983 feet but the elevation of the field at the TDZ — Touchdown Zone is 965 feet). The circling section gives the HAA — Height Above Airport — since this is of most importance in that case.

Note in Figure 8-10(A) that special take-off minimums or departure procedures exist for that airport.

The same minimums apply to both day and night operations unless different minimums are specified at the bottom of the minimum box in the space provided for symbols or notes.

The minimums for straight-in and circling appear directly under each aircraft category. When there is no division line between minimums for each category on the straight-in or circling lines, the minimums apply to two or more categories under the A, B, C or D.

In Figure 8-10(A), the S-ILS 14 minimums apply to all four categories. The S-localizer 14 minimums are the same for Categories A, B, and C, and different for Category D. The circling minimums are the same for A and B and individually different for C and D.

The sample minimums section, Figure 8-10(B), procedure authorizes minimums for aircraft with one VOR receiver. Lower Minimums are authorized if the aircraft also has DME or dual VOR receivers and St. Barner Intersection is identified.

In Figure 8-10(B) is the minimums section for a VOR approach or Dual VOR or VOR/DME approach as indicated.

Side-Step Maneuver — Air Traffic Control may authorize an approach procedure which serves either one of parallel runways that are separated by 1,200

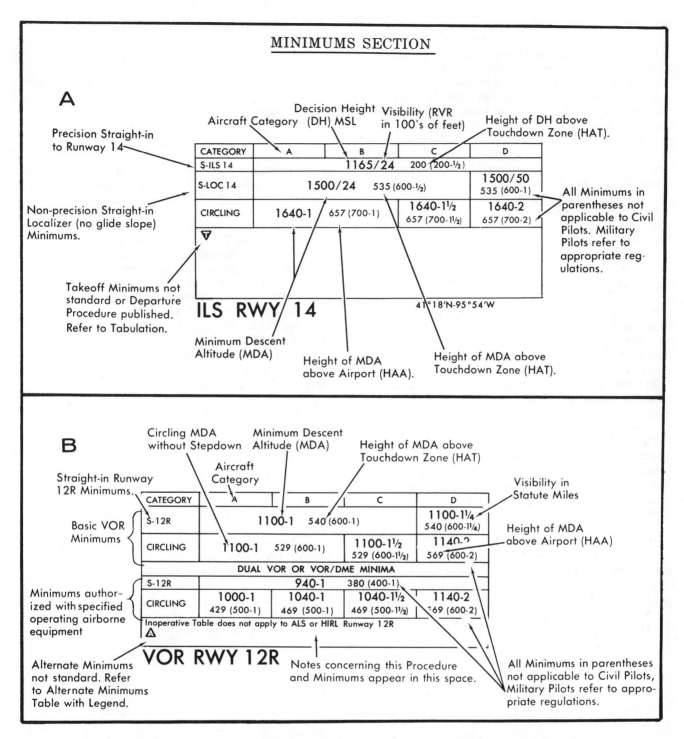

MINIMUMS SECTION

A

Precision Straight-in to Runway 14

Aircraft Category

Decision Height (DH) MSL

Visibility (RVR in 100's of feet)

Height of DH above Touchdown Zone (HAT).

CATEGORY	A	B	C	D
S-ILS 14		1165/24	200 (200-½)	
S-LOC 14	1500/24	535 (600-½)		1500/50 535 (600-1)
CIRCLING	1640-1	657 (700-1)	1640-1½ 657 (700-1½)	1640-2 657 (700-2)

Non-precision Straight-in Localizer (no glide slope) Minimums.

All Minimums in parentheses not applicable to Civil Pilots. Military Pilots refer to appropriate regulations.

Takeoff Minimums not standard or Departure Procedure published. Refer to Tabulation.

ILS RWY 14

41°18'N-95°54'W

Minimum Descent Altitude (MDA)

Height of MDA above Airport (HAA).

Height of MDA above Touchdown Zone (HAT).

B

Circling MDA without Stepdown

Minimum Descent Altitude (MDA)

Height of MDA above Touchdown Zone (HAT)

Aircraft Category

Straight-in Runway 12R Minimums.

Visibility in Statute Miles

CATEGORY	A	B	C	D
S-12R		1100-1	540 (600-1)	1100-1¼ 540 (600-1¼)
CIRCLING	1100-1	529 (600-1)	1100-1½ 529 (600-1½)	1140-2 569 (600-2)
DUAL VOR OR VOR/DME MINIMA				
S-12R		940-1	380 (400-1)	
CIRCLING	1000-1 429 (500-1)	1040-1 469 (500-1)	1040-1½ 469 (500-1½)	1140-2 569 (600-2)

Basic VOR Minimums

Height of MDA above Airport (HAA)

Minimums authorized with specified operating airborne equipment

Inoperative Table does not apply to ALS or HIRL Runway 12R

VOR RWY 12R

Alternate Minimums not standard. Refer to Alternate Minimums Table with Legend.

Notes concerning this Procedure and Minimums appear in this space.

All Minimums in parentheses not applicable to Civil Pilots, Military Pilots refer to appropriate regulations.

Fig. 8-10. Minimums sections for (A) and non-precision (B) approaches (AC 90-1A).

feet or less followed by a straight-in landing on the adjacent runway. Aircraft that will execute a side-step maneuver will be cleared for a specified approach and landing on the adjacent parallel runway. Example "cleared ILS runway 5 right approach, land on runway 5 left." Pilots are expected to commence the side-step maneuver as soon as possible after the runway or runway environment is in sight. Landing minima to the adjacent runway will be higher than the minima to the primary runway, but will normally be lower than the published circling

minima. ATC will not clear aircraft for landing on an adjacent runway unless weather conditions will permit successful completion of the side-step maneuver.

To repeat, landing minima for a side-step maneuver to the adjacent runway will be higher than the minima to the primary runway but will normally be lower than the published circling minima.

Circling Minimums. The circling minimums published on the instrument approach chart provide adequate obstruction clearance. The pilot should not

149

descend below the circling altitude until the airplane is in a position to make final descent for landing. Sound judgment and knowledge of his and the aircraft's capabilities are the criteria a pilot needs to determine the exact maneuver in each instance – since the airport design, aircraft position, altitude and airspeed must all be considered. The following basic rules apply:

1. Maneuver the shortest path to the base or downwind leg as appropriate under minimum weather conditions. There is no restriction from passing over the airport or other runways.

2. Many circling maneuvers may be made while VFR flying is in progress at the airport. Standard left turns or specific instruction from the controller for maneuvering must be considered when circling to land.

3. At airports without a control tower, it may be desirable to fly over the airport to determine wind and turn indicators, and to observe other traffic which may be on the runway or flying in the vicinity of the airport.

Notice the "A" symbol in Figure 8-10(B); this concerns alternate minimums.

If the weather is forecast to be above certain minimums at the destination airport an alternate won't be required. See Chapter 9, the section on the flight plan.

An example of one confusion that used to prevail: the alternate airport had a fine operating ILS (and required 600 and 2 minimum as an alternate). The pilot missed his approach at the destination and had to go to the alternate – what were his minimums then? Suppose the weather at the alternate is now down to 400 and 1, can he go? *Yes, in this case the alternate becomes the new destination and the normal ILS minimums apply*.

Figure 8-11 shows how the non-standard take-off and alternate minimums look in the separate

listings. These listings are issued (or reissued) with the Approach Charts subscription.

(5) Aerodrome Sketch

Figure 8-12 shows the Aerodrome Sketch part of the Approach Chart (A) for the ILS of Figure 8-5 and a sample VOR approach (B).

Lighting

One part of the approach procedures often overlooked in planning by pilots is the airport lighting available. It can make a difference in minimums. For instance, to authorize Runway Visual Range (RVR) minimums, the following components and visual aids *must be available in addition* to the basic components of the approach procedure:

Precision Approach Procedures – (a) RVR reported for the runway (b) HIRL (High Intensity Runway Lights) and (c) All weather runway markings.

Non-precision Approach Procedures – (a) RVR reported for the runway (b) HIRL and (c) Instrument runway markings.

Figure 8-13 shows all weather runway markings and instrument runway markings as well as basic runway markings.

Figure 8-14 shows various approach lighting systems in the United States. Most pilots take the lighting systems for granted, but each has a purpose. For example, the roll guidance bars are a great help in marginal visibility (particularly at night) in keeping you from dropping a wing or getting into other lateral/directional problems.

Inoperative Components

Inoperative components or visual aids tables are published by the National Ocean Survey on a separate sheet for insertion in the Approach Chart binders.

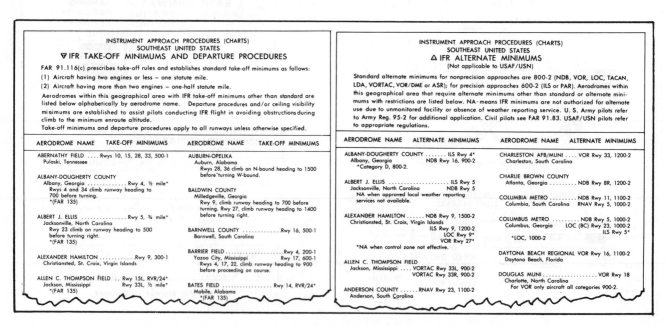

Fig. 8-11. Non-standard take-off and alternate minimums (AC 90-1A).

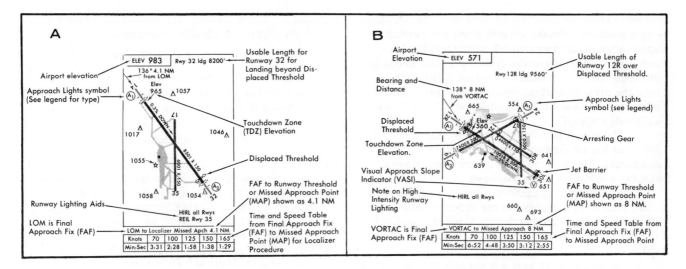

Fig. 8-12. Aerodrome sketch of a precision (A) and a non-precision approach (B), with notes (AC 90-1A).

They are to be available to the pilot to show the new minimums (which can be the same in some cases, but are generally higher).

Figure 8-15 shows a sample list of inoperative components and their effects on minimums. (Check the current National Ocean Survey publications for the latest tables.)

1. Operative runway lights are required for night operation. (That makes sense.)

2. When the facility providing course guidance is inoperative, the procedure is not authorized. On VOR/DME procedures: when either VOR or DME is inoperative, the procedure is not authorized.

3. When the ILS glide slope is inoperative or not utilized, the published straight-in localizer minimum applies.

4. Compass locator or precision radar may be substituted for the ILS outer or middle marker.

5. Surveillance radar may be a substitute for the ILS outer marker. DME, at the glide slope site, may be substituted for the outer marker when published on the ILS procedure.

6. Facilities that establish a stepdown fix, etc., 75 MHz FM, off course VOR radial, etc. are not components of the basic approach procedure, and applicable minimums for use, both with or without identifying the stepdown fix, are published in the minimums section.

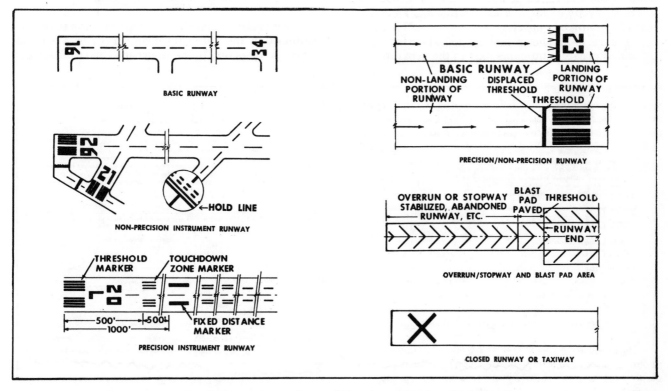

Fig. 8-13. Runway markings. (*Airman's Information Manual.*)

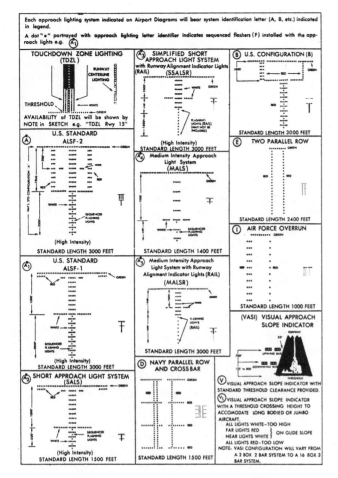

Fig. 8-14. Approach lighting systems.

7. Additional methods of identifying a fix may be used on the procedure, when authorized.

Where RVR visibility minimums are published and the runway markings become unusable, the necessary adjustment will be accomplished by NOTAM and by air traffic advisory. If RVR minimums for take-off or landing are published in an instrument approach procedure, but RVR is inoperative and cannot be reported for the runway at that time, it is necessary that the RVR minimums which are specified in the procedure be converted and applied as ground visibility in accordance with the table below.

RVR	Visibility (statute miles)
1600 feet	1/4 mile
2400 feet	1/2 mile
3200 feet	5/8 mile
4000 feet	3/4 mile
4500 feet	7/8 mile
5000 feet	1 mile
6000 feet	1 1/4 mile

FAA Radar and Visibility Minimums

Included with the Coast and Geodetic Survey Approach Charts subscription is a section listing FAA radar ceiling and visibility minimums for various airports. This data for Nashville (Tennessee) and

Charleston (South Carolina) is given as a part of a sample trip in Chapter 13.

Summary of the Instrument Approach Procedure Charts

You should be well familiar with all of the publications you'll use as an instrument pilot, but the Approach Charts will be the most important of all. Flying into something solid and expensive on an instrument approach just because you didn't bother to check an altitude (or obstruction problem) on the chart, is a spectacular but unrewarding way to make an arrival.

Figure 8-16 is a summary of some of the abbreviations you'll see on the approach chart as published with the NOS booklet of charts.

THE AIRMAN'S INFORMATION MANUAL AND AIRPORT/FACILITY DIRECTORY

The *Airman's Information Manual* is a FAA publication containing information of great value to pilots. You should have access to a copy when you get into the cross-country phase of your instrument training.

INOPERATIVE COMPONENTS OR VISUAL AIDS TABLE

ILS and PAR with visibility of 1/2 mile (RVR 2400) or greater.

Inoperative Component or Aid	Increase DH	Increase Visibility	Approach Category
OM*, MM*	50 feet	By None	ABC
OM*, MM*	50 feet	By 1/4 mile	D
ALS	50 feet	By 1/4 mile	ABCD
SALS	50 feet	By 1/4 mile	ABC

*Not applicable to PAR

ILS and PAR with visibility minimum of 1,800 or 2,000 feet RVR.

Inoperative Component or Aid	Increase DH	Increase Visibility	Approach Category
OM*, MM*	50 feet	To 1/2 mile	ABC
OM*, MM*	50 feet	To 3/4 mile	D
ALS	50 feet	To 3/4 mile	ABCD
HIRL, TDZL, RCLS	None	To 1/2 mile	ABCD
RVR	None	To 1/2 mile	ABCD

*Not applicable to PAR

VOR, VOR/DME, LOC, LDA, and ASR.

Inoperative Visual Aid	Increase MDA	Increase Visibility	Approach Category
ALS, SALS	None	By 1/2 mile	ABC
HIRL, MALS, REILS	None	By 1/2 mile	ABC

NDB (ADF) and RNG.

Inoperative Visual Aid	Increase MDA	Increase Visibility	Approach Category
ALS	None	By 1/4 mile	ABC

LOC Approaches

Inoperative Component or Aid	Increase MDA	Increase Visibility	Approach Category
ALS, MM	None	By 1/4 mile	D

Fig. 8-15. Inoperative Components (AC 90-1A). Example only; check latest FAA (FAR) requirements.

LEGEND
INSTRUMENT APPROACH PROCEDURES (CHARTS)
GENERAL INFORMATION & ABBREVIATIONS

★ Indicates control tower operates non-continuously.
All distances in nautical miles (except Visibility Data which is in statute miles and Runway Visual Range which is in hundreds of feet).
Runway dimensions in feet.
Elevations in feet Mean Sea Level.
All radials/bearings are Magnetic.

ADF	Automatic Direction Finder	MALS/R	Medium Intensity Approach Light Systems /with RAIL
ALS	Approach Light System	MAP	Missed Approach Point
APP CON	Approach Control	MDA	Minimum Descent Altitude
ARR	Arrival	MIRL	Medium Intensity Runway Lights
ASR/PAR	Published Radar Minimums at this Aerodrome.	NA	Not Authorized
ATIS	Automatic Terminal Information Service	NDB	Non-directional Radio Beacon
		NoPT	No Procedure Turn Required (Procedure Turn shall not be executed without ATC clearance)
BC	Back Course	RA	Radio Altimeter Height
C	Circling	Radar Required	Radar vectoring requires for this approach
CAT	Category		
CHAN	Channel	Radar Vectoring	May be expected through any portion of the Nav Aid Approach, except final.
CLNC DEL	Clearance Delivery		
DH	Decision Height		
DME	Distance Measuring Equipment	RAIL	Runway Alignment Indicator Lights
DR	Dead Reckoning		
ELEV	Airport Elevation	RBn	Radio Beacon
FAF	Final Approach Fix	REIL	Runway End Identifier Lights
FM	Fan Marker	RCLS	Runway Centerline Light System
GPI	Ground Point of Interception	RNAV	Area Navigation
GS	Glide Slope	RRL	Runway Remaining Lights
HAA	Height Above Aerodrome	RTB	Return To Base
HAL	Height Above Landing	Runway Touchdown Zone	First 3000' of Runway.
HAT	Height Above Touchdown	RVR	Runway Visual Range
HIRL	High Intensity Runway Lights	S	Straight-in
IAF	Initial Approach Fix	SALS	Short Approach Light System
ICAO	International Civil Aviation Organization	(S) SALS/R	(Simplified) Short Approach Light System /with RAIL
Intcp	Intercept	SDF	Simplified Directional Facility
INT, INTXN	Intersection	TA	Transition Altitude
LDA	Localizer Type Directional Aid	TAC	TACAN
Ldg	Landing	TCH	Threshold Crossing Height (Height in feet Above Ground Level)
LIRL	Low Intensity Runway Lights		
LDIN	Lead in Light System	TDZ	Touchdown Zone
LOC	Localizer	TDZE	Touchdown Zone Elevation
LR	Lead Radial. Provides at least 2 NM (Copter 1 NM) of lead to assist in turning onto the intermediate/final course.	TDZL	Touchdown Zone Lights
		TLv	Transition Level
MALS	Medium Intensity Approach Light System	W/P	Waypoint (RNAV)

Fig. 8-16. Abbreviations.

The *Airman's Information Manual* (AIM) and the other publications here are available on annual subscription from the U.S. Government Printing Office; Washington, D.C. 20402.

AIM — BASIC FLIGHT INFORMATION AND ATC PROCEDURES

This publication contains instructional, educational, and training material — things that are basic and not often changed, such as:

Glossary of Aeronautical Terms — Definition of control zones and areas and other terms used in aviation.

Air Navigation Radio Aids — Theory and operations of such aids as LF/MF ranges, VOR's, radio beacons, Distance Measuring Equipment, Instrument Landing Systems (ILS), and marker beacons. The Frequency Utilization Plan (what frequencies are used where) and VHF/DF (VHF direction finding) are included.

Good Operating Practices — Some hints on operating the airplane in a safe manner and suggestions on use of the aircraft radio to avoid frequency congestion.

Airport and Air Navigation Lighting and Marking Aids — Information on airport beacons and runway lighting and marking. Information on enroute beacons and landmark lighting, including obstructions hazardous to flight, is in this part also.

Radar — General information on FAA and military radar plus specifics on precision and surveillance radar approaches.

Radiotelephone Phraseology and Techniques — Background on microphone technique and procedure words and phrases. Includes the phonetic alphabet and Morse code.

Altimetry — Background and use of the altimeter in the airways system.

Weather — The weather reporting aids available, in-flight weather safety advisories, weather radar, pilot reports (PIREPS). Information on thunderstorms, such as at which altitudes the most turbulent conditions exist, etc.

Wake Turbulence — Information on airplane wake turbulence (how it's developed, the places where it's most likely to be found, and suggested pilot action if it is encountered).

Medical Facts for Pilots — Discussions of the effects of hypoxia (lack of oxygen), use of alcohol with tips on how long to wait before flying after its use, plus sections on use of drugs, effects of vertigo and carbon monoxide on the pilot, and the danger of flying shortly after scuba diving.

Preflight — Weather briefing tips and how to file and cancel flight plans.

Departure IFR — Communications information (ground control and tower). Light signals. Clearances for taxi and take-off. Departure control procedures.

Enroute IFR — The airways systems and special control areas. Communications and operating procedures enroute (Instrument Flight Plan). Radar assistance to VFR (Visual Flight Rules) aircraft. Diagrams of cruising altitude requirements (VFR and IFR).

Arrival IFR — VFR advisory information procedures. Radar traffic information service and terminal radar service. Airport advisory service (Flight Service Stations). Unicom and Multicom information. Approach control. Instrument approach procedures.

Airport Operations — Traffic pattern standard procedures and traffic indicators (wind sock, tee, and segmented circle). Hand signals for taxi directors.

General — Airports of entry and departure and procedures. ADIZ (Air Defense Identification Zone) procedures.

Emergency Procedures — What to do under various emergency situations. Search and rescue procedures and visual emergency signals and codes. Radio communications failure procedures.

Pilot/Controller Glossary — ATC terms.

The Basic Flight Manual is issued every 112 days.

AIRPORT/FACILITY DIRECTORY

This publication is issued by the U.S. Department of Commerce (NOAA) every 56 days. There are seven regions or areas covered and you may subscribe to any or all of them. The major information contained is that of (as indicated by the title) airports and facilities for the various regions (Fig. 8-17).

The Airport/Facility Directory for each region

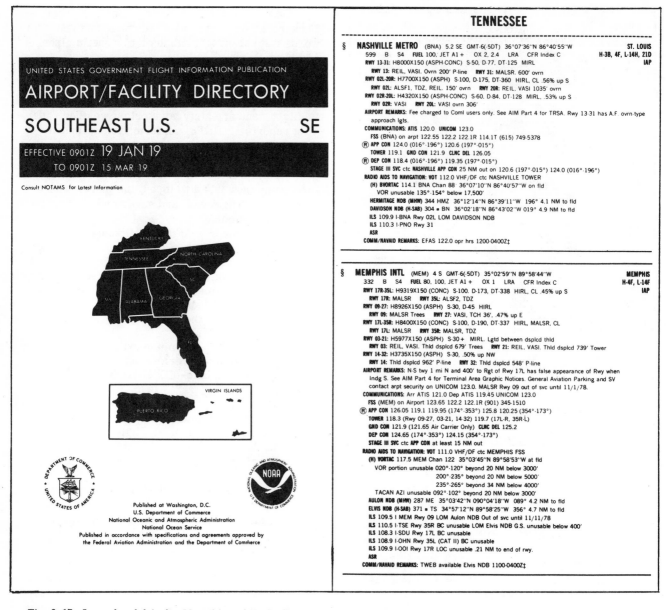

Fig. 8-17. Legend and data for Memphis and Nashville Metropolitan airports. A full explanation of the symbols is given in the Appendix of this book.

also contains information for that region such as:

Heliports and seaplane bases – A listing by state.

Special Notices – Information such as civil use of military fields, customs requirements and other data, including FSS and Weather Service Office telephone numbers, can be found in this section.

VOR Receiver Check Points – The facilities for that area having VOR check points with frequencies, locations and check point descriptions.

Air Route Traffic Control Centers – Locations of Centers and the remote transmitter sites and their frequencies for that area.

Preferred IFR Routes – Best routes through and within the area for High and Low IFR traffic.

Aeronautical Chart Bulletin – Major changes in aeronautical information sectional and area charts.

Enroute Flight Advisory Service (EFAS) – A diagram of the Flight Watch Control Stations and communications outlets for the states in that area. Get a current copy of the *Airport/Facility Directory* and check it carefully for the latest information included there.

NOTICES TO AIRMEN (CLASS II)

This is a publication containing current NOTAMS which are considered essential to flight as well as supplemental data affecting the other operational publications just discussed. It also includes current FDC NOTAMS, which are regulatory in nature, issued to establish restrictions to flight or amend charts or published Instrument Approach Procedures. This publication is issued every 14 days and is available through subscription from the Superintendent of Documents.

154

SUMMARY OF THE CHAPTER

Instrument charts and other government information sources are continually changing. This chapter has taken a general look at the types of information available, but you might find that some details or methods of presenting information may have changed before you read this. The point is to use this as an introduction to charts and other services. When you get the instrument rating, make sure that you are able to get the latest changes or corrections to the material you are *using*.

Chapter 9

PLANNING THE NAVIGATION

CHECKING THE ROUTE

MAYBE LATER you'll be able to grab an Enroute Chart and an Approach Procedures Chart and do a real good job of flying IFR (it's doubtful). But for now you'd better make sure that you've gone over the situation with a fine-tooth comb.

Look at the Enroute Chart. Check your proposed route, and it might not be a bad idea for your first few flights to mark along the route with a black or green pen. The VOR airways and radio data will be in blue on the map, so you'll want a contrasting color. While red would be great for daytime work, it would be hard to spot in red cockpit lighting at night. Take a few minutes to get a rough idea of distances and the VOR names and frequencies enroute. What about the minimum enroute altitudes? You can file for several different altitudes, but most pilots generally will file for the highest MEA (or the next 1000 feet higher) for the route unless there's a wide divergence in MEA's along the way. In planning for a trip from Memphis to Nashville, probably you would file for V-16, the most direct route. Look at the reduced section of the L-14 Low Altitude Enroute Chart in the back of this book.

The chances are good that V-16 would be what you'd get, but look at the Enroute Chart for other possible routes you *could* be assigned.

One in-flight problem is getting a clearance for an alternate (strange) route and then being unable to find it on the map right away. If you have your main route well marked you will at least know where it is as a starting point and can find more easily the alternate routes to either side. In congested areas there will be a maze of airways and possible alternates, so make it easy on yourself. Maybe you don't want to mark up your chart, but if you have subscribed to the service, you'll get a new one every 56 days anyway — and if you make the trip several times in that period, the route will already be marked. You'll find that after flying the same route a few times, you'll know every intersection along the way and won't be so astounded by the way the airline pilots seem to come back so quickly with clearances.

While you are planning a route into a new area of expected higher terrain, you should examine the sectional chart to get a general look at geographic and topographic points. It would be helpful to have some idea of possible obstruction problems if things don't go perfectly. You should carry a set of Sectionals or WAC charts of your flight route in case you have to do some flying by reference to the ground.

This doesn't exactly tie in with planning the navigation, but you should consider oxygen equipment. If you have your own airplane, or are flying a company airplane all the time, why not have an oxygen bottle and masks available? Not only might you have to fly at higher than usual altitudes, but what if one of your passengers needs oxygen when you are on solid IFR and can't get down immediately?

FLIGHT LOG

A flight log will be of particular value during your training. Making such a log will assure your having some advance knowledge of the route.

By the time you get to instrument training, it's likely that you'll have a good idea of cruising speeds (T.A.S.) at expected cross-country altitudes for your airplane. Or, if you are the thorough type, you may check the power setting chart for T.A.S. for the chosen altitudes for 65 (or 75) per cent power. Then you could feed in wind information for that altitude as it would apply.

In flying IFR and using radio aids, the wind side of the computer will be of less value than before, and you won't usually work out wind triangles. It's still an aid in figuring out your estimated groundspeed, however. If you have a transponder, the wind side of the computer will be even less important. But you should be conscientious in planning, as you'll want to be within the three-minute allowance at your reporting points if not in radar contact with Center. With radar contact, the Center will be keeping right up with you and will know your position better than you do.

Back to the wind problem: In dead reckoning, precomputing of the drift correction is necessary. Here it is not — you'll take care of that with the omni left-right needle (Course Deviation Indicator, to be more technical). You *are* interested in the component of wind acting along the route (for or against), and for quick estimate you can use the following:

The wind at your altitude will be given in knots and *true* direction. Your course is *magnetic;* and in areas where large magnetic variation exists, a correction is necessary. If you are flying in an area where the variation is 5° or less either way, forget it. Consider the true direction as being magnetic in getting a quick estimate.

For wind directly on the tail use full value (25 knots, etc.) to get groundspeed. Figure 9-1 shows the idea.

As an example, suppose your course is 065° magnetic, T.A.S. is 150 knots, the wind is from 280° (true) at 30 knots at the altitude chosen. Assume that variation is small and can be ignored. If the wind were right on the tail it would be from 245°. Here it is 35° off the (left) tail, the component of wind acting along the course (picking the nearest value) is 0.9 of 30, or 27 knots. Your groundspeed at that altitude will be 150 + 27 = 177 knots. To check the possible inaccuracies, reference to a trigonometric table would show that a wind at a 35° angle to the tail would have a component (along the course) of 0.81915 (the cosine of the angle) or about 82 per cent of the wind value. This would mean a component of 24.5 knots, or a groundspeed of 174.5 knots. On a 400-nautical-mile trip this would mean a difference of about *3 minutes total time.* You could split the difference between 30° (0.7) and 45° (0.9) for 35°, and get 0.8, which would make it even more accurate. Only minor errors would result on this — or any trip —

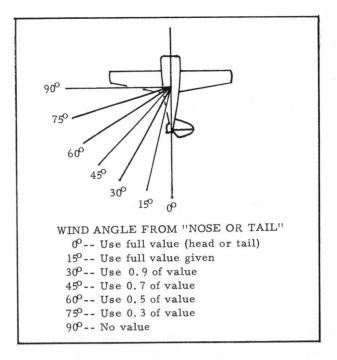

WIND ANGLE FROM "NOSE OR TAIL"

  0°-- Use full value (head or tail)
 15°-- Use full value given
 30°-- Use 0.9 of value
 45°-- Use 0.7 of value
 60°-- Use 0.5 of value
 75°-- Use 0.3 of value
 90°-- No value

Fig. 9-1. Multipliers for head- or tailwind components at various angles to the course line.

with the normally expected distance between check points by using this method.

How much time do you allow if you are climbing enroute to the next reporting point?

For fixed gear airplanes add 2/3 of a minute per thousand feet to be climbed. For single-engine, retractable gear airplanes and light twins add 1/2 minute per thousand feet. As an example, suppose you are flying a single-engine, retractable gear airplane and plan to fly at an altitude of 8000 feet (MSL). The airport elevation is 2000 feet, so you'll have to climb 6000 feet to get to that altitude. The first leg is 50 nautical miles, and you compute that at *cruise speed* it would take 20 minutes. For your climb you would add 3 minutes (6 × 1/2 minute) to get a total of 23 minutes ETE (estimated time enroute).

For the same leg in a fixed gear type you might get an ETE at cruise of 23 minutes. Adding the 2/3 minute per thousand you'd get (6 × 2/3 = 4), or 4 + 23 = 27 minutes for the leg, including climb. Expect these thumb rules to be accurate up to assigned altitudes of about 12,000 feet MSL.

On the flight log you can take into account the time required to fly from the fix serving the destination airport (VOR, etc.) to the destination airport *for your own purposes. However, when you file the flight plan you'll include only the flying time from the take-off to the enroute navigation aid serving the destination airport and then to the final approach fix.*

As a part of your preflight planning and flight log work look over the Approach Charts for the *destination airport, alternate airport,* and the *airport of departure.* Be sure to have these latter Approach Charts accessible during take-off — you might have problems and have to return to the airport of departure.

Alternate Airport

In earlier times an alternate airport was required for *all* IFR flights; and your airplane had to carry enough fuel to *complete the flight* to the first intended point of landing, to fly from that point to an alternate airport, and to fly thereafter for 45 minutes at normal cruising speeds. In Chapter 8 the weather minimums for alternate airports as listed on Approach Charts were discussed.

(FAR 91 states that the alternate airport requirement considers weather reports and forecasts and weather conditions.)

You are allowed to omit the designation of an alternate airport on the IFR flight plan provided the first airport of intended landing has a standard instrument approach procedure and, for at least one hour before and one hour after the estimated time of arrival, the weather reports or forecasts or any combination of them, indicate that the ceiling will be at least 2000 feet above the airport elevation and the visibility will be at least 3 miles.

If an alternate airport is required be sure to *plan the flight from the destination to the alternate*. It's disconcerting to get to the destination and discover that it's gone below minimums, and you are faced with flying a route to the alternate that you haven't really checked out. Later, with more experience you'll be able to pick the figures right off the Enroute Chart.

ACTUALLY PLANNING THE TRIP

For this sample flight you will be using a Zephyr Six, a four-place single-engine, retractable gear, high-performance airplane. Following are its specifications:

Engine — Lycoming 0-540, 250 hp
Gross Weight — 2900 lbs
Basic Empty Weight — 1744 lbs
Total Fuel — 60 gals
Usable Fuel — 56 gals (6 lbs per gal)
Baggage Capacity — 200 lbs

You weigh 160 pounds and have 2 passengers weighing 190 and 210 pounds respectively. The baggage to be carried weighs 150 pounds.

Since the unusable fuel (4 gallons) and oil (23 pounds) is already included in the basic empty weight of 1744 pounds, you'll only add the weight of 56 gallons of usable fuel. The arms given here are the same as those used back in Chapter 3 for the first example of weight and balance and would be given on the Weight and Balance Form of the airplane.

Item	Weight (lbs)	Arm (in)	Moment
Basic Empty Weight	1744		142,927
Fuel (56 gals)	336	90.0	30,240
Pilot	160	84.8	13,568
Passenger (front)	190	84.8	16,112
Passenger (rear)	210	118.5	24,885
Baggage	150	142.0	21,300
Total	2790 lbs		249,032 lb-in

Adding the 1266 pound-inches as required back in Figure 3-20, the total moment is 250,298 pound-inches.

You can see that you have 110 pounds to spare in weight, but the C.G. should be checked by dividing the total moment by the total weight and getting an answer of 89.7 inches (rounded off) aft of the datum. A check of Figure 3-20, which is the weight and balance envelope for this airplane, shows that the C.G. is within the limits. The heavy passenger was placed in the rear seat to give the worst loading combination for this example.

Looking at the Enroute Chart and planning to fly (for training purposes) to Nashville from Memphis via V-54S to Muscle Shoals, V-7 to Graham, V-7 to Nashville at 5000 feet, you can add up the mileage to get the total distance. In addition, assume that Chattanooga will be your alternate, and the distance from Nashville to Chattanooga should also be taken into account. For this example the wind at 5000 feet (your planned altitude) is from 230 (true) at 20 knots.

The total distance from the Memphis airport to the Nashville VOR over the route planned is 243 nautical miles. You'll have a basis for computing the time and fuel required. The figure of 243 nautical miles was obtained by assuming that you will fly from the Memphis airport direct to the Memphis VOR and then follow the Victor route. The chances of this being necessary are slim, but you could be asked to do it. Besides, it is a conservative approach to the problem.

You will have a tailwind component on all legs but should get a no-wind estimate to the destination and on to the alternate. From the Nashville airport to the Chattanooga VOR is 102 nautical miles, for a total of 345 nautical miles to be flown. Looking back to Figure 3-11, you'll note that at 5000 feet (density altitude) the T.A.S. at that altitude at 65 per cent power is 148 knots. You plan on using 65 per cent because it is much easier on the engine, with only a few knots cost in speed. Figure 3-12 indicates that this airplane uses 12.3 gallons of fuel per hour when properly leaned at 65 per cent.

Using your computer, the no-wind (and conservative) time to make the trip of 2 hours and 20 minutes is found (if it is necessary to go to Chattanooga). Allowing 1/2 minute per thousand feet to climb from MEM airport (elevation 331 feet) to 5000 feet MSL would be another 2.5 minutes (call it three). The no-wind time could be figured as 2 hours and 23 minutes. Again, referring to a computer with this knowledge, you'll find that the fuel required will be 29.7 gallons. Add another 5 gallons for taxiing, waiting, and extra fuel used in the climb and figure on about 35 gallons required to fly from Memphis to Nashville to Chattanooga. The requirement is to have a 45-minute reserve at normal cruising speed after reaching the alternate. The fuel burned in 45 minutes, at 12.3 gallons per hour, is 9.2 gallons (call it 9). The fuel now required is 35 + 9 = 44 gallons. Better also allow another 5 gallons to assume you make an approach to Nashville, miss it, and have to climb back up to proceed to the alternate. This makes a total of 49 gallons to cover all requirements, and you have 56 gallons of usable fuel. You can make it, even being on the conservative side. This calculating takes less time than would

appear and is merely done to get a check.

If there had been a headwind, the rough check should have been done using the expected headwind component for all legs, again a conservative approach. If the rough check shows that the trip can't be made in one hop as first supposed, or if it looks very close, the final, more accurate check and flight log will be completed before commitment to the flight. (There will be no commitment, naturally, until the weather and NOTAMS are taken into account.)

Notice at the top of the Enroute Chart that the magnetic variation is 5° East in the vicinity of Memphis and tapers off to about 3° East in the vicinity of Nashville. For simplicity it can be averaged as 4° East for wind computations.

Figure 9-2 is a flight log made out for the flight from MEM to BNA via the route just discussed. In

TRUE AIRSPEED 148 KNOTS			AIRPLANE NO. N 3456 J							T.O. TIME _____
WINDS ALOFT 230/20			ALTITUDE 5000							DATE _____

FROM	IDENT / FREQ	TO	IDENT / FREQ	M. C.	DIST. MILES	E.G.S.* / ETE*	TIME OVER	ATE*	G.S.	ETA	REMARKS
MEMPHIS AIRPORT	/	MEM VOR ▲	MEM 115.5	166°	8	100 / 06					AT CLIMB SPEED
MEM VOR	MEM 115.5	HOLLY SPRINGS VOR △	HLI 112.4	109°	25	158 / 10					
HOLLY SPRINGS VOR	HLI 112.4	STAMM *△	HLI/JKS 112.4/109.4	094°	17	161 / 06					STAMM * 213R - JKS
STAMM *	HLI/JKS 112.4/109.4	ALLSO *△	MSL/HAB 116.5/110.4	094°	54	161 / 20					ALLSO 348R - HAB
ALLSO *	/	MUSCLE SHOALS VOR ▲	MSL 116.5	072°	30	165 / 11					
MUSCLE SHOALS VOR	MSL 116.5	GILLE *△	MSL/HSV 116.5/112.2	359°	15	161 / 06					GILLE * 281R HSV
GILLE *	/	GORDONSB'G *△	GHM/JKS 111.6/109.4	359°	36	161 / 13					GORDONSBURG * 090R - JKS
GORDONSB'G *	/	GRAHAM VOR ▲	GHM 111.6	359°	17	161 / 06					
GRAHAM VOR	GHM 111.6	TIDWELL *△	GHM/BNA 111.6/114.1	062°	16	166 / 06					TIDWELL * 302R - SYI
TIDWELL *	/	NASHVILLE VOR ▲	BNA 114.1	064°	25	166 / 09					TOTAL DISTANCE - 243 TOTAL TIME (EST.) 1:33
NASHVILLE VOR	BNA 114.1	CHATT'OOGA VOR ▲	CHA 115.8	131°	102						BNA VOR TO CHA VOR — ALTERNATE —

POSITION REPORTS						CLEARANCES N 3456 J CLEARED		
POS.	TIME	ALT.	TYPE	EST.	NEXT	TO	BNA METRO AP	
						DEPT.OR SID	MEM COM 1	
						ROUTE	V545 V7 GHM, BNA	
						ALTITUDE	9000 -10 M 5000	
						HOLDING		
						SPECIAL		
						FREQ. CODE	DPT 124.15	4365

Fig. 9-2. A flight log for an IFR flight from Memphis to Nashville. (E.G.S. — Estimated Groundspeed, ETE — Estimated Time Enroute, ATE — Actual Time Enroute.) Times have been rounded off to the minute.

the early part of your training, you might list every intersection, but for later actual flights the best idea is to use *all* VOR's as check points (compulsory reporting points or not) and any intersections that are compulsory.

The symbols ▲ and △ aid in telling from the flight log whether the particular fix is a compulsory reporting point. You may use your own shorthand; for instance, an **X** or an asterisk (*) can be used as a symbol for "intersection."

Working out the flight log beforehand is a good way to avoid the problem of misreading map distances in the air. Remember that the Enroute Charts give the *total* distance between compulsory reporting points or VOR's in a box (check it), and the other distances (between noncompulsory intersections, etc.) are in the open. It seems that distracted pilots sometimes add the series of distances between compulsory points and leave out maybe a 10- or 15-mile section somehow. Needless to say, this also affects the *minutes* of a new estimate. As a check after completing the flight log, you should add the minutes required to fly individual legs. They should add up very close to the *total time* required to fly the *total distance*. Probably you'll be a couple of minutes off because of rounding-off times on various legs, but at least you have a double check of distances and times.

Included in Figure 9-2 are position report and clearance forms for easier organization in flight. If you were not in radar contact and had to give position reports, you could write that information out before transmitting, to save some stumbling. It would also serve as a permanent record if desired. The clearance box in Figure 9-2 contains the information in the order that it will be given to you. The example here might be the clearance you'd get before departing Memphis and states that "N3456J IS CLEARED TO THE NASHVILLE METRO AIRPORT WITH A COMMON ONE DEPARTURE" (Fig. 9-3). They might give you specific departure instructions here. The route is next: "VIA V-54S, V7" to Nashville, with altitude data as flown, such as "MAINTAIN 4000 UNTIL 10 MINUTES FROM MEMPHIS, THEN MAINTAIN 5000." Holding instructions, if any, would be then given ("EXPECT TO HOLD AT HOLLY SPRINGS, etc."), followed by any special instructions. The last item would be frequency and transponder code information: You are to contact Memphis departure control on 124.15 MHz and are assigned a discrete beacon code of 4365.

FLIGHT PLAN

Figure 9-4 is a sample instrument flight plan for the trip from Memphis to Nashville just planned.

Taking the items on the flight plan one by one:

Type of Flight Plan — You'd check IFR, naturally.

Aircraft Identification — The full identification (registration number of the airplane.

Aircraft Type — If you know the "official"

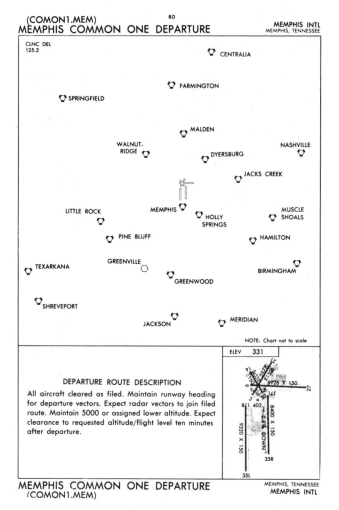

Fig. 9-3. The Memphis Common One Departure (SID).

designation of your airplane, use it. For instance, "PA-23-250" is the designator for the *Piper Aztec*, rather than "Aztec." The Flight Service Station will have a list of official designators for various airplanes. If you don't know the model number, they will help you in this regard. Here you would put the information concerning DME or transponder or other equipment aboard. Shown as an example is the designator for a 4096-Code transponder with altitude encoding ability and DME. For practice purposes, the flight in the following chapters will not be so equipped except at isolated situations as needed to make a point. (On paper, you can throw equipment in or out as desired, with little trouble or expense.)

True Airspeed (knots) — If heretofore you've been shying away from knots you'll soon find that for IFR work you'll have to *think* in knots. All distances on the Enroute Charts, Area Charts, and Approach Charts (except visibility minimums) are in nautical miles. If you aren't used to knots, don't make the mistake of talking in terms of "knots per hour." The term "knot" means *one nautical mile per hour* and is a short way of saying it. Another error made by the knot neophytes is to speak of knots as a distance. They'll look at the chart and say, for instance, "I measure it to be 163 knots between A and

DEPARTMENT OF TRANSPORTATION—FEDERAL AVIATION ADMINISTRATION **FLIGHT PLAN**					Form Approved OMB No. 04-R0072	

| 1. TYPE
VFR ☒ IFR
DVFR | 2. AIRCRAFT
IDENTIFICATION
N3456J | 3. AIRCRAFT TYPE/
SPECIAL EQUIPMENT
ZA-6/A | 4. TRUE
AIRSPEED
148 KTS | 5. DEPARTURE POINT
MEM INTL. | 6. DEPARTURE TIME
PROPOSED (Z) 2040 ACTUAL (Z) | 7. CRUISING
ALTITUDE
5000 |

8. ROUTE OF FLIGHT

V54S MSL, V7 GHM, BNA.

9. DESTINATION (Name of airport and city) NASHVILLE METRO	10. EST. TIME ENROUTE HOURS 1 MINUTES 35	11. REMARKS

| 12. FUEL ON BOARD
HOURS 4 MINUTES 40 | 13. ALTERNATE AIRPORT (S)
LOVELL FIELD
CHATTANOOGA | 14. PILOT'S NAME, ADDRESS & TELEPHONE NUMBER & AIRCRAFT HOME BASE
STEWART G. DICKSON
598-5318, NASHVILLE METRO | 15. NUMBER ABOARD
3 |

| 16. COLOR OF AIRCRAFT
BLUE AND WHITE | CLOSE VFR FLIGHT PLAN WITH_____FSS ON ARRIVAL |

Fig. 9-4. The flight plan as given to the Flight Service Station. The time enroute has been rounded off to the nearest 5 minutes.

B." What they should say is that it's 163 nautical miles between A and B.

One other confusion that arises is that the new pilot wants to put groundspeed in that item. *Don't.* ATC wants to know your true airspeed, which will give them comparative speeds of airplanes operating in the same airspace. They will have the winds at your altitude and can come right back to you if you give a bum estimate to your next reporting point (such estimate by *you* was based on *your* computation of groundspeed).

If your true airspeed varies more than 10 knots during the flight, let ATC know about it if you are in a non-radar environment. It won't be necessary if you are under radar surveillance. *Don't* get mixed up and discover that your *groundspeed* has changed that amount and let ATC know about it — except as a revised estimate to a fix. In other words, they'll have access to wind info but can't read your mind if you decide to do something radical about indicated (and calibrated and true) airspeed.

Point of Departure — If you are phoning the flight plan in or for some reason don't plan to fly from the particular airport where you are filing the flight plan in person, you'd better make this point clear.

Departure Time — You'll put the proposed time of departure in Zulu time. Needless to say, you should know by now to add 5 hours to Eastern *Standard* Time, 6 hours to Central *Standard*, etc., to get Zulu (Greenwich) time or the time it happens to be in Greenwich, England, even if you don't care what time it is over there.

The actual departure time will be recorded after you've departed. Remember, if at all possible, file the flight plan at least thirty minutes before the proposed time of departure.

Cruising Altitude — This will be the requested level enroute flight altitude. Normally file odd altitudes eastbound ($0°$-$179°$ magnetic) and even altitudes when westbound ($180°$-$359°$). However, you may want to take advantage of being able to get a handy low altitude just above the MEA (minimum enroute altitude) at an even altitude (say 4000 feet) when eastbound. You may ask for it but may not get it.

Route of Flight — In this case, a complicated route was picked to show a point. When you have a number of Victor airways, which may be the case on some long flights, it's best to name the fix at the end of each different airway to avoid possible errors. The example flight plan states that you are filing for V-54S (South) to Muscle Shoals VOR, V-7 through Graham VOR to Nashville.

Destination — You'll put the airport name and city here because some cities have more than one airport capable of handling IFR traffic.

Estimated Time Enroute — This is based on the time required from take-off to the final approach fix to be used for the approach to the destination airport. This is done in case of communications failure enroute. If you lose communications, you'll be expected to continue the flight at the last assigned altitude or the MEA (whichever is higher) and route, to the navaid serving the destination airport and thence to the facility or fix to be used for the approach. You will depart that facility to start an approach (ILS, VOR, etc.) at the time of arrival, based on the last estimate to that point, whether it be from the flight plan (you lost communications before reaching the first enroute fix to set up estimates with ATC) or from later estimates given to ATC, based on known groundspeeds.

161

This will be covered in more detail in later chapters.

Fuel on Board— This is *usable fuel available* at take-off. If you've been flying and haven't refueled, be sure to put the actual hours of fuel left at the cruising speed you used in Item 4. Make sure that you will have enough to get to the destination, to the alternate (if required) *plus* forty-five minutes.

Remarks — You can make snide remarks such as "I hope that this time I get an altitude and route somewhat near that filed for" and other such statements guaranteed to tickle the ATC dragon. (More likely you'll use this space to list your passengers' names or other such serious information.)

Alternate Airport — If an alternate is necessary, put it down here.

Name of Pilot — You should be able to handle this with no trouble.

Address of Pilot or Aircraft Home Base — Usually it's best to list the aircraft home base if there's a difference.

Number of Persons on Board — Another obvious one.

Color of Aircraft — Self-explanatory.

SUMMARY OF THE CHAPTER

Good preflight planning can make the difference between a no-sweat situation and everything turning to worms. You'll work out your own shorthand for use on the flight log and clearance copying. A small clipboard can be modified to have an elastic band (with a snap fastener) for strapping around your leg or on the control wheel, or you may want to buy one of the custom-made clipboards you see in ads in aviation magazines.

Figure 9-5 shows the probable order of your

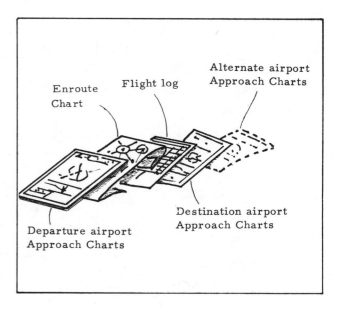

Fig. 9-5. Probable order of paperwork on the kneeboard.

paperwork on the clipboard.

The flight planning ideas offering in this chapter may make the instrument flight seem very complicated; this is not the case at all in the vast majority of IFR trips. As the ATC system improves, the headaches of "reporting points" will be completely phased out. In the future it's unlikely that you will make *any* routine reports after the departure to the approach. However, since reporting points are still required on many flights (in certain areas) this book will try to cover different possibilities.

Part of preflight planning is assuring that if you should lose all radios you know the route to better (VFR) weather. Later in your training you will probable dispense with the flight log and will use the Enroute Chart for that purpose.

Part Four

THE INSTRUMENT FLIGHT

Chapter 10

BEFORE THE TAKE-OFF

PREFLIGHT INSPECTION

IT'S IMPORTANT that you make a thorough pre-flight inspection anytime you fly, but for IFR work it's vital. For instance, you just get leveled off at your assigned altitude and are solidly on the gages, when one of the passengers says, "Hey, there's a solid stream of gas pouring off the back of both wings." You were in a kind of a hurry to get off. You saw the line boy fueling, so were sure that the tanks were full and didn't bother to check that the caps were on properly or secured. Now you have a choice: (1) Continue on your way and hope to complete the trip before you lose all the fuel (that's asking for it), or (2) let ATC in on the act and get a clearance to the nearest airport that has an instrument letdown procedure. Well, in that case, by not using fifteen seconds of your valuable time, you've caused the reshuffling of IFR traffic clear back to What Cheer, Iowa. Assuming you make it safely, the least you've done is to have caused a lot of people a lot of trouble.

Figure 10-1 is a preflight inspection for a light twin-engine airplane showing special IFR items to consider. (As far as checking the rest of the airplane is concerned, if you don't have a check system set up already, then it's too late to bring it up here.)

The following check is meant to bring up some ideas rather than give a specific system. There are too many airplane types and variations in antennas and other equipment for this to hit your situation exactly.

1. As always, the first thing to do is to make sure all the *switches are OFF*.

2. Shown at (2) is a combination *Navigation-Communications* (broad band) *antenna*. Check it for security of attachment and general condition.

3. This is a *broad band communications antenna*. Check it for security and general condition. (Of course, *your* airplane may be using whip antennas and you should check them carefully.)

4. The *marker beacon antenna* may not be in the exact spot on the belly as shown in Figure 10-1 and may be the older type, but you should check it for security and possible dirt and oil on it. (The dirt and oil problem is usually worse on a single-engine airplane.)

5. Check the *Automatic Direction Finder loop housing* and the *sensing antenna* for security and general condition.

6. If your airplane has the "separate pitot tube and static inlet" system make sure that the *static vents* are clear. Some airplanes have these static vents nearer the tail cone, while others may have them farther forward than shown.

7. This number covers *de-icer boots*. You would check the boots as you came to them in your clockwise (or counterclockwise) check. The statements made about the tail de-icers will stand for the wing boots, hence the same reference numbers.

Check for the condition of the rubber to assure that there are no large cracks that could cause leaks. Check the attachment of the boots to the airframe. Some have screws attaching the assembly

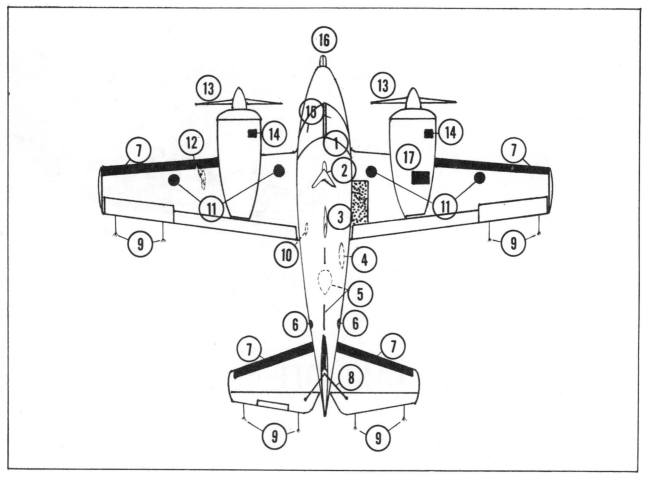

Fig. 10-1. A preflight inspection showing items of special interest for IFR operations.

and others are permanently (or semipermanently) glued or bonded to the airframe.

8. Shown here on the top of the fin is a *whip* (V type) *VOR/LOC antenna*. If your plane has such an antenna, check it for security and general condition.

9. If your plane has *static wicks,* which are designed to gradually discharge static electricity buildups, check them for security. Their job is to keep radio static down when the plane is flying through rain and other precipitation (or dust in extreme cases). You'll get the wing static wicks as you come to them.

10. The *DME antenna* is likely to be on the belly in the area of the spot indicated. Look it over. If you are not sure of the various types of antennas on your plane, you might review Chapter 5 or talk to some of the old hands around the airport. There may be some pretty exotic looking antennas. If you have a transponder, the antenna will be quite similar (in the majority of types) to the small DME blade antenna and will be on the belly in or near the center section.

11. You should have *full tanks and secured fuel tank caps*. Unless weight and balance considerations for a particular passenger and baggage loading require otherwise, *always* have full tanks when going IFR. In a partially filled tank situation you'd better *know* exactly how much fuel the plane has and how

much will be needed. Remember, you may have to hold or go to an alternate airport (or both).

12. The *pitot tube* is shown as being attached to the bottom of the left wing on this sample airplane. Check the pitot tube opening for obstructions. If this is a pitot-static combination, check the static vents for obstructions. Make sure the assembly is firmly attached, whether left or right wing or fuselage.

13. If the plane has *electric prop de-icers* check the slip ring and brush-block assemblies. (The slip ring has three concentric copper rings which turn with the propeller and spinner. The brush-block assembly is in contact with them, furnishing continuous electrical connection as needed.) The brush-blocks are very fragile on some models. Be careful not to break them. (Admittedly, this is more of a problem in changing them than during the preflight check.) Make sure that the three (or whatever number) brush-blocks are in contact with the slip rings. One problem found is that one (or more) of the brush-blocks may have slipped out of alignment with its respective ring and have been unevenly worn. If in doubt have a mechanic check them.

The propeller heating elements are wires encased in an oil- and abrasion-resistant pad of rubber bonded to each propeller blade. Check these for general conditions, including bonding integrity and worn or torn rubber.

166

14. Make sure that the oil is at the proper level and the *oil cap is secure.*

15. A *clean windshield* is a lot of help when you break out on final. Check the windshield wiper blades if the airplane is so equipped.

16. Shown in the figure is the older type of *glide slope antenna.* Make sure that it is on securely. (Well, it certainly looks like a handle to be used to pull the airplane and may have been used for that.)

17. If the airplane uses a propeller *fluid anti-icing system,* check the fluid level, make sure the outlets aren't clogged, and that the cap and flap are secure after you've done it.

During the check make sure that the fuel sumps and strainers are drained. Water in the sumps may freeze if the airplane is left out overnight in the winter, and the quick-drains may be "stuck" to where they cannot be moved. It *may* be just a drop or so of water frozen in the assembly, or there may be enough built-up ice in there to cut off part of the fuel flow. Everything is fine while you taxi, and maybe nothing shows up during the run-up; but when full power is applied the engine(s) is just not getting enough fuel to pull it off. This could happen just as you lifted off into the murk (and drained the carburetor float chamber). Even if you discover this problem during the run-up (or earlier), it means taxiing back to take care of it, with a resulting time loss. If the drains won't drain, find out why not and thaw them as necessary. Don't, however, be like Archibald Zorp, instrument pilot, who expedited the thawing process with an acetylene torch. Not only was his airplane completely thawed all over, but several other airplanes and one-half the hangar received the benefit of his efforts. Archibald has yet to make the instrument flight he planned so carefully. (He is now busily growing back his eyebrows.)

Check the alternator(s) belt(s) if you can see in that area.

When you get into the cabin, make sure that your checklist includes moving the fuel selectors to verify ease of switching tanks. It could cause trouble if you fly all fuel out of a tank (or tanks) and then discover, some miles from an airport, that the selector can't be moved to another tank.

STARTING

There is little to be added in this regard. You have the word on starting your particular airplane. Make sure all of the radios are off for starting, to save the battery. Also, the sudden surge of power through the system on starting doesn't do radios *any good at all,* so leave them off.

After the start check the oil pressure(s). As you make the run-up check, make sure that you have the proper vacuum pressure. For full-fledged IFR work it's best to have an engine-driven vacuum pump rather than venturi-driven instruments. With the pump you'll have gyro instruments operating *before* take-off and can tell if one of the instruments is not up to snuff. With the venturi you could be pretty well committed during take-off before discovering problems. In heavy icing conditions the venturi(s), being exposed, could ice over, with a resulting loss of gyro instruments.

TAXIING

Before taxiing you should check ATIS (Automatic Terminal Information Service) for the latest weather and runway information. ATIS information is normally updated each hour, even when no significant changes have occurred, but is redone anytime as required to report important changes.

Get your clearance on the clearance delivery frequency if available, or, if not, alert ground control that you are on an IFR flight plan to Nashville so that they'll have your clearance ready for you when you request it.

Here is a good time to check your VOR receivers with the VOT if available.

Now you can call for taxi instructions.

(Wait until the frequency is clear and then . . .)

You: MEMPHIS GROUND (CONTROL) THIS IS ZEPHYR THREE FOUR FIVE SIX JULIET, JONES FLYING SERVICE (RAMP) INFORMATION UNIFORM TAXI.

Memphis Ground: ZEPHYR FIVE SIX JULIET, MEMPHIS GROUND CONTROL, TAXI TO RUNWAY ONE SEVEN (followed by wind, altimeter setting and time, plus any necessary instructions for taxi). The clearance delivery frequency is listed just after the tower and ground control frequencies in the *Airport/Facility Directory,* as shown by Figure 10-2 for Dulles. It was mentioned in Chapter 6 that you could get your clearance on this frequency before taxiing, at some places, and you would have looked it up beforehand. (Memphis has regular clearance delivery.)

Fig. 10-2. Dulles International, like other busier airports, has pre-taxi clearance procedures.

PRETAKE-OFF CHECK

You will, of course, check the controls, trim settings, manifold heat, magnetos, and prop controls.

If you have electric prop de-icers, you should check them for operation (the prop de-icer ammeter will show surges as the equipment works). You should cycle the wing and tail de-icer boots if installed.

The turn and slip or turn coordinator should be checked as you taxi. The heading indicator and attitude gyro should be holding. (Set the H/I to the compass; set the attitude indicator as near to the actual attitude of the airplane as possible.) The vacuum pressure should be normal. In the twin both vacuum pumps should be working, as indicated by either the manual selector on the panel or the lack of red indicators on either of the vacuum gages (whichever type of equipment you have). Check pitot heat (ammeter). Tap the altimeter to make sure it has settled down.

The radios will be of "great interest" and you should check all such equipment. (You should, however, keep the transponder on standby until starting the take-off roll.) The latest transponder equipment will have a press-to-test operations check or test switch to be used while on the ground. The radio check should include all communications and navigation equipment. Sometimes you may have to wait for a clearance, so don't sit there with all radios on draining the battery. The use of alternators has alleviated this problem somewhat, but you still may have to run the engine(s) up to a high rpm.

While we're on the subject of radios, you'd better have a headset aboard in case the cabin speaker malfunctions. An extra microphone is a good idea, too.

An added point: If it's night, you wouldn't want to sit at the warm-up area with taxi lights (and/or landing lights) and white cabin lights on. (White cabin lights raise havoc with night vision.) Some pilots also turn off the rotating beacon and/or strobes if they are sitting at the approach end of the runway. Not only does the beacon use electric power, but it could also distract landing pilots as they cross the threshold. *Leave the navigation (position) lights on,* unless you want a 747 or something in the cabin with you.

Speaking generally about pretake-off checks, it would be very well for you to make up a new check list for your particular airplane and include the special checks for IFR flight in the places where they would apply. (For instance, the vacuum pressure check might fall logically right after the magneto check, etc.) Check the de-icer boot operation (if you have them), as outlined in the Supplement to the *Pilot's Operating Handbook.* You might want to underline the special IFR checks to separate them from the usual VFR check items. If you are getting formal instrument training with a flight school, such a check list may already be available. There are too many different airplane models to set up a pre-take-off check here, and you'll probably have plenty of VFR experience in the airplane before taking IFR training.

CLEARANCES

Your clearance will be issued in the following order (also see Fig. 9-2).

A. *Aircraft Identification* (N 3456 Juliet)

B. *Clearance limit* (CLEARED TO THE NASHVILLE AIRPORT)

C. *Departure procedure* or SID (MEMPHIS COMON ONE SID, MAINTAIN 4000, EXPECT 5000 TEN MINUTES AFTER DEPARTURE)

D. *Route of flight* (VICTOR 54 SOUTH, VICTOR 7) If a route given by a previous clearance is amended it may be given to you thusly:

(1) CHANGE VICTOR 7 TO VICTOR 21 (for example) or (2) VICTOR 54 NORTH TO MUSCLE SHOALS, REST OF ROUTE UNCHANGED. Or they may issue the full route.

E. *Altitude data in the order flown* (MAINTAIN 4000 UNTIL 10 MINUTES FROM MEMPHIS, MAINTAIN 5000) When a route or altitude in a previously issued clearance is amended, clearance delivery will restate all applicable altitude restrictions or state that altitude restrictions are cancelled.

F. *Holding instructions* (if applicable). If a delay is anticipated enroute and the holding pattern is not depicted you'll be issued either a general or detailed holding procedure (see Chapter 12 for details on holding).

G. *Any special information.*

H. *Frequency and beacon code information* (DEPARTURE CONTROL FREQUENCY 124.15, SQUAWK 4365)

You may get a "cleared as filed" clearance, which means that there were no complications or changes in the *route.* The *altitude* will always be stated by ATC, and you will confirm it.

If you are a refresher instrument pilot you'll note that ATC facilities do not say ATC CLEARS N3456 JULIET anymore. They'll say N 3456 JULIET (IS) CLEARED (TO, FROM, FOR etc.) However, Flight Service Stations or other sources *relaying* ATC information to you will prefix the clearance with ATC CLEARS (or ADVISES, REQUESTS, etc.).

If a clearance is one with which you can't comply, let them know and ask for a new one. (Expect a possible delay.)

A suggestion is to write down beforehand, widely spaced, the route and altitude you filed for, so that changes can be made without having to do a lot of writing of the symbols. Write the new route(s) and altitude above the old. Figure 10-3 shows this method for a different routing and altitude to Nashville. (You filed for V-54S, V-7 and for 5000 feet, but ATC wants you to go another way.) After copying, mark out the original figures as shown in Figure 10-3 and check the Enroute Chart.

But, going back to the example trip, we'll say that you're still going the V-54S, V-7 route.

The clearance you'll get before departure *normally* will be to the airport of intended landing. Under some conditions, at certain locations, a short range clearance procedure is used; and you would be advised of the frequency with which to contact the Center for the long range clearance.

When you've accepted a clearance, you are expected to follow it. Any clearance in which the time of execution by the pilot is optional will

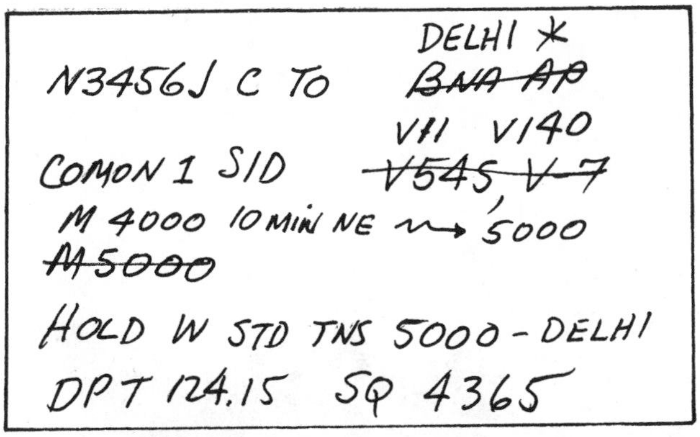

Fig. 10-3. Revising the expected clearance. Leave plenty of space between lines for inserting new routes and climb instructions.

state, "At pilot's discretion."

If you get a clearance enroute, for instance, "(YOUR NUMBER) CLIMB AND MAINTAIN SIX THOUSAND *IMMEDIATELY*," you'd better clear your present altitude for six thousand *immediately,* if not sooner. ATC doesn't throw "immediately" around without a reason.

Amendments to the initial clearance may be given enroute. You may have been cleared to the destination airport via a certain route before takeoff, but you may be given an amendment, or amendments, enroute to avoid conflict with other traffic. It can include holding, change of altitude, and rerouting. Unless it will exercise an extreme hardship, don't argue but accept the clearance and comply. *Don't* tie up valuable time and frequencies by chatter unless you are going to be placed in a dangerous situation. The controllers don't like to upset the status quo any more than you do unless it is absolutely necessary. If you have a beef, write or call the Center *after* the flight.

If you've been cleared to a fix short of the destination airport, it's the responsibility of ATC to give additional clearance at least *five minutes prior* to the time the flight arrives at the clearance limit. The new clearance may authorize flight beyond the limit or contain holding instructions. However, if

for some reason you don't get the clearance by the time you're three minutes from the fix, you'll be expected to reduce speed so as to cross the limit initially at or below maximum holding speed (unless further clearance comes through while you're slowing). If no clearance has come through, you will establish a *standard* holding pattern — which would mean that you would make right-hand turns and one-minute inbound legs — on the course on which you approach the fix. When you get clearance to another fix, you will acknowledge — giving the time of departure — and will depart.

The ARTCC Sector controller may be watching your antics on radar, will know you're holding, and will try to get you on your way. So it won't do any good to mention casually over the radio, "Well, here I am, still holding at Zilch intersection, well, well." They'll get you going as soon as possible.

You *always* should write down your clearances; it will help in a read-back and can serve as a record if you should need it later.

Normally, when given an altitude, you'll be told to *maintain* that altitude. This means you must maintain that altitude until cleared by ATC for another. On short flights in uncongested areas, you may be given "Cruise at such and such an altitude." *Cruise* means that you may ascend to or descend from cruising altitude and make an (approved) approach at the destination without further clearance from ATC. *This does not clear you to descend below minimum enroute altitude or MOCA, if applicable, or other minimum altitudes unless in VFR conditions.* If you've been cleared to cruise at a certain altitude and descend from it, you can't change your mind and go back up to it without a new clearance.

It's your responsibility to notify ATC immediately if your radio equipment cannot receive the type of signals required to comply with the clearance.

ATC will not issue a clearance specifying that a climb or descent on any portion of the flight be made under "VFR conditions" on any IFR flight, unless specifically requested by the pilot. You can sometimes save time at the departure if conditions are VFR by requesting this — rather than waiting for the full treatment.

Chapter 11

TAKE-OFF AND DEPARTURE

THE PRETAKE-OFF CHECK is complete, you've read the clearance back and gotten "CLEARANCE CORRECT" from ground control or clearance delivery, so you'll switch to tower frequency for take-off. Make sure all necessary radios (and lights, if applicable) are on. Have one VOR receiver on one of the VHF approach facilities serving the airport of departure, and have the local Approach Charts available just in case you have to make an unscheduled return. The other VOR receiver should be tuned to the first enroute VOR.

It was said earlier and will be said again: If conditions are such that a full-fledged instrument take-off is required, how do you plan to get back in if something goes wrong? If the bad conditions are local and other airports are available — and flyable — it might be feasible, but if the lousy weather is extensive, you'd better look closely before busting off.

You: MEMPHIS TOWER, (ZEPHYR THREE FOUR) FIVE SIX JULIET READY FOR TAKE-OFF ON (RUNWAY) ONE SEVEN RIGHT.

Memphis Tower: ZEPHYR FIVE SIX JULIET, MEMPHIS TOWER, CLEARED FOR TAKE-OFF. (The tower may give headings to turn to after take-off or other instructions; it depends on local and current conditions.)

You: ZEPHYR FIVE SIX JULIET.

You would turn the transponder from standby *on* to the assigned discrete code as late as possible before taking off.

After you've reached a point about one-half mile past the end of the runway, you will be switched to

departure control; heavy chatter on the radio or the traffic situation could delay this slightly.

Memphis Tower: ZEPHYR FIVE SIX JULIET, CONTACT DEPARTURE CONTROL, OVER. (No frequency given; you had that earlier and should have written it down.) As soon as the tower tells you to make the switch, acknowledge and do so.

You: ZEPHYR FIVE SIX JULIET.

DEPARTURE CONTROL

You: MEMPHIS DEPARTURE (CONTROL) THIS IS ZEPHYR THREE FOUR FIVE SIX JULIET, If your airplane has a working encoder altimeter you would give an altimeter check. For instance say, LEAVING 800 FEET, or whatever. Report to the nearest 100 feet. If your airplane has the 3/A mode (or no transponder) you could give departure control a general altitude check as noted shortly (PASSING TWO THOUSAND). OVER.

Memphis Departure Control: ZEPHYR FIVE SIX JULIET, MEMPHIS DEPARTURE RADAR CONTACT. TURN LEFT HEADING ONE FOUR ZERO TO JOIN VICTOR FIVE FOUR SOUTH, MAINTAIN FOUR THOUSAND (FEET).

You: ZEPHYR FIVE SIX JULIET, MAINTAIN FOUR THOUSAND, PASSING TWO (THOUSAND).

The Memphis tower computer has passed your time off to the Center.

After reaching a safe altitude, you'll set climbing power, turn off the boost pumps, and do the other required cockpit chores, as you finish the turn and continue to climb. Your tendency may be to get so engrossed in voice reports and the other jobs that the instrument scan is neglected, and you turn past the heading. (The bank sneaks over more steeply, and the climb may stop or turn into a descending turn for a few seconds.)

While you're getting things under control, it would be well to discuss the coordination between the tower (departure control) and Center. Unknown to you, the tower and Center have coordinated their actions and established a "Letter of Agreement" between the two facilities. Copies are on file at both places.

This letter and diagram outlined the jurisdiction of the two facilities, *but variations can be made from the so-called "rigid rules" at any time with coordination between the two.*

The tower has control of you at altitudes of up to 11,000 feet MSL and below and will notify the Center of your take-off. It's likely that you'll be picked up on radar by Center shortly after crossing the airport boundary.

Unless coordination has been effected with the Center, departure control will retain departures from the airport to tower altitudes (in this case 5000 feet MSL and below), until the aircraft is established in the departure area along the assigned route or (in the case of Memphis) until 20 miles from the Memphis airport on the assigned route.

If the altitude conflict is cleared up while you are still in the area of the tower's (departure control) jurisdiction, you may be told by departure control to "delete the 4000 restriction" or be given clearance to climb to your assigned altitude. If you don't have DME, you may ask departure control to let you know. Normally, you will be handed off to Center before reaching the 20-mile (or other) limit.

Memphis Departure Control: ZEPHYR FIVE SIX JULIET, CONTACT MEMPHIS CENTER (ON) ONE TWO FOUR POINT THREE FIVE (124.35 MHz).

You: ZEPHYR FIVE SIX JULIET, CENTER (ON) ONE TWO FOUR POINT THREE FIVE, OUT.

Write down the frequency.

You may also be given a time or point at which to contact the Center (CONTACT CENTER AT HOLLY SPRINGS or CONTACT CENTER AT FIVE FOUR — 54 minutes past the hour).

You: MEMPHIS CENTER, (THIS IS) ZEPHYR THREE FOUR FIVE SIX JULIET, FOUR THOUSAND, OVER.

Center: ZEPHYR FIVE SIX JULIET, MEMPHIS CENTER, MAINTAIN FIVE THOUSAND, OVER. Assume that once you are in radar contact you are remaining so, unless otherwise stated. More about this in the next chapter.

You: ZEPHYR FIVE SIX JULIET, CLEARED TO FIVE (THOUSAND). LEAVING FOUR (THOUSAND) AT (glance at the clock) FIVE TWO. (This is for time check purposes only.) You'll then leave 4000 immediately for 5000 or will start to climb as you acknowledge the clearance.

Looking back at Figure 6-4, the V-54S, V-7 heavy line, or the planned route to Nashville, you can see that you are in Sector 8 and the VHF frequency is 124.35 MHz. The UHF frequency is 239.3 MHz, but this is probably of no interest to you at this stage. Your route is marked as a heavy line in Figures 6-3 and 6-4, and you can see that you'll have several enroute frequency changes to make before getting to Nashville. You'll pass through Sectors 8 (124.35), 13 (128.5), 9 (120.8) and 23 (125.85) before being handed off to Nashville approach control. Your pertinent flight strips will have been passed to these Sectors, following the same idea as shown back in Figure 6-5.

Obviously, you're not going to carry a Sector chart of every Center and will change frequencies when, and as, requested.

You made a report upon leaving your last assigned altitude. This is required without a special request from ATC.

You are not required automatically to report reaching the new assigned altitude but may be asked by ATC to "report reaching five thousand." Or, you may have to report at interim altitudes on your way to the assigned altitude, such as CLIMB AND MAINTAIN EIGHT THOUSAND, REPORT PASSING SIX AND SEVEN THOUSAND. There is traffic that might possibly cause problems at the interim altitudes and ATC wants to know when you've cleared them.

When you are assigned a new altitude, climb (or descend) as rapidly as practicable to within 1000 feet of the assigned altitude and then climb (or descend) to the assigned altitude at a rate of no more than 500 feet per minute. It would be best to limit your *"as rapidly as practicable" to no more than 1000 feet per minute to lessen chances of loss of control.* In climbing, this probably will be no problem — in descending, however, you might overdo the rate.

And the most important thing of all to remember is this (it was said before). . . .

NEVER SACRIFICE CONTROL OF THE AIRPLANE TO MAKE VOICE TRANSMISSIONS. The man on the ground can't see that you might be having problems with turbulence, icing, and other such situations, so he may call at an "inopportune moment." Let him wait until you have things under control.

Chapter 12

ENROUTE

REFER TO the Enroute Chart for your route to Nashville. After you reach the assigned altitude (5000 feet MSL), leave the power at climb setting to expedite reaching cruise speed as you level off. Adjust power, lean the mixture(s), and switch tanks as required. Make sure the airplane is well trimmed. Keep your scan going. Later, in smooth air or *light* turbulence, you'll find an autopilot to be a great aid.

POSITION REPORTS

The enroute portion of the flight will bring up the subject of position reports, and they should be discussed here before you get too far from Memphis.

As far as enroute, constant altitude flight is concerned, you'll make position reports as follows:

1. At a compulsory reporting point as shown on the Enroute Chart. Holly Springs VOR is *not* a compulsory reporting point; Muscle Shoals and Graham VOR's are, as shown by the black triangle (▲).

2. When requested by ATC. Center may just want a report over Holly Springs VOR or Stamm intersection (△). These are "on request" reporting points.

Remember, if you are radar identified and will be remaining under radar surveillance, you will discontinue position reports, even over compulsory reporting points. If you are in radar contact, the controller, at the time radar service is lost or terminated, will say "RADAR CONTACT LOST" or "RADAR SERVICE TERMINATED." This sometimes catches

pilots napping. They've put the flight log away and only glance at the chart to get courses for following segments — now it's back to work. (One of the controllers' biggest gripes is that the pilot does not get back to work with his estimates.) Another term for resuming position reporting is, "Radar contact lost." That's why preflight planning is important.

As far as the exact *time* to give a position report (if required) you might note the following:

Over a VOR — The time reported should be the time at which the TO-FROM indicator makes its first complete reversal.

Over an ADF — The time reported should be the time when the needle makes a complete reversal.

Over a Z-Marker or Fan Marker — The time should be noted when the signal begins (aural or light) and when it ends. The mean of the two times should be taken as the actual time over the fix.

If you are giving a position with respect to a bearing and distance from a reporting point, be as accurate as possible.

OTHER REPORTS

The following reports will be made to ATC (including Centers, FSS, towers, or approach control, as applicable). These are in addition to the two mentioned in the last section.

1. *The time and altitude/flight level upon reaching a holding fix or point to which cleared.*

2. *When vacating any previously assigned*

altitude/flight level for a newly assigned altitude/ flight level.

3. *When leaving any assigned holding fix or point.*

4. *When leaving final approach fix inbound on final approach.*

5. *When an approach has been missed.* (You'll then request clearance for specific action, such as another approach, clearance to the alternate airport, etc.)

6. *A corrected estimate* at anytime it becomes apparent that an estimate as previously submitted is in error in excess of three minutes.

7. *That an altitude change will be made if oper*ating on a clearance specifying "VFR conditions-on-top."

Items 1, 3, 4 and 6 are not required when in "Radar Contact."

To repeat what was said about Pilot Reports in Chapter 7, if you encounter weather conditions that have not been forecast, or hazardous conditions that have been forecast, report it.

ENROUTE

(Refer to the Enroute Chart unless a specific Figure number is given.)

You took off at 1440 CST (2040Z) and at 1456 (2056Z) CST you are over Holly Springs as indicated by your VOR receiver. You set the omnibearing selector to 094° and turn to track outbound from that station. By waiting until the **TO-FROM** indicator reads a definite **FROM**, you may overshoot the turn slightly and have to turn further to the left (to a heading of, say, 084°) to get on the new course. That's expected.

When you make the turn and reset the OBS, it's probable that the needle will be well deflected to the left (maybe a full deflection). Remember that you are very close to the VOR at this point. Some pilots want to get back on *right now* and would turn back, taking a 45-degree cut. They'd find that the radial would be overshot in short order, and a turn back to the right would be necessary.

You hit Holly Springs within a minute of your ETA and feel that at this point the flight log estimated time enroute will hold. Figure 12-1 is the flight log for the trip, with the estimated and actual times of arrival for Holly Springs VOR and Stamm intersection.

Assume, for example purposes, that at this point radar contact is lost temporarily, and ATC needs certain information for traffic separation ahead.

Center: ZEPHYR FIVE SIX JULIET, GIVE ALLSO ESTIMATE AND REPORT PASSING THE TWO ZERO SIX (206) RADIAL OF JACK'S CREEK VOR.

You: REPORT TWO ZERO SIX RADIAL JACK'S CREEK VOR, ALLSO ESTIMATE, ZEPHYR FIVE SIX JULIET.

You've passed Stamm intersection you know but wonder if you've crossed the radial in question. You

were over HLI at 1456. Looking at the chart, you see that it's 71 nautical miles to Allso (17 to Stamm, plus 54 from there to Allso — or 101 miles, between HLI and MSL, minus 30 is also 71 miles). Your best groundspeed estimate is 161 knots which is set up on your computer (or should be), so you read off that for 71 miles you'll require 26 minutes.

You: ZEPHYR FIVE SIX JULIET, ALLSO ESTIMATE TWO TWO, OVER.

Center: ALLSO ESTIMATE TWO TWO, OUT.

Now, have you passed the 206 radial of Jack's Creek VOR? You tune in Jack's Creek (109.4 MHz) and *identify* it. You could check by one of the two ways discussed in Chapter 5: (1) setting the OBS to the cross-bearing station inbound bearing (026) and "turning" the airplane to that heading in your mind or (2) setting 206 on the OBS (FROM) and noting whether the relative bearing of the station and needle match. (In this example they do — left and left — so you haven't passed it yet.)

It's best always to use one VOR receiver as the enroute or course receiver and the other (if there are two) for any cross-bearing work. Usually the top (or No. 1 set) would be used for on-course indications and the lower set used for cross-bearings (or set up whatever is the most convenient for you). If you don't have a routine established and are slightly off-course, under pressure you could read the wrong VOR indication.

When the cross-bearing needle centers, you are on the 206 radial of JKS and would duly report this fact:

You: MEMPHIS CENTER, ZEPHYR FIVE SIX JULIET CROSSING TWO ZERO SIX RADIAL JACK'S CREEK AT ZERO FIVE (five minutes past the hour), FIVE THOUSAND, OVER.

Center: ZEPHYR FIVE SIX JULIET. REPORT PASSING ALLSO INTERSECTION, OUT. (Sometimes if you want a time check you would say as you crossed, "...CROSSING TWO ZERO SIX RADIAL *NOW*, AT ZERO FIVE.") If you forgot to set the clock or set it wrong before taxi, Center might come back with: FIVE SIX JULIET, TIME NOW ZERO SEVEN. If you had given them an estimate earlier based on your clock time, you'd better add two minutes to the earlier estimate, so you'll be there at "Center Time" (the correct time). Allso estimate would now be "two four" instead of "two two" as given earlier. Assume here, however, your clock is correct.

Speaking of time, one problem with pilots is the simple misreading of the clock when giving the actual time over a fix. The most troublesome times for this are between 20 and 29 minutes past the hour and 34 to 50 minutes past. Sometimes the pilot will read "28" instead of "23" or vice versa. If ATC comes back with a time correction, double check the clock before resetting.

At a point about midway between Holly Springs VOR and ALLSO intersection, you'll enter a new Sector (Fig. 6-4) and will hear:

Center: ZEPHYR FIVE SIX JULIET, CONTACT

TRUE AIRSPEED 148 KNOTS	AIRPLANE NO. N 3456 J	T.O. TIME 1440CST-2040Z
WINDS ALOFT 230/20	ALTITUDE 5000	DATE DEC 19-

FROM	IDENT / FREQ	TO	IDENT / FREQ	M.C.	DIST. MILES	E.G.S.* / ETE*	TIME OVER	ATE*	G.S.	ETA NEXT	REMARKS
MEMPHIS AIRPORT	/	MEM VOR ▲	MEM 115.5	166°	8	100 / 06	2046	06		—	AT CLIMB SPEED
MEM VOR	MEM 115.5	HOLLY SPRINGS VOR △	HLI 112.4	109°	25	158 / 10	2056	10	162	2102	
HOLLY SPRINGS VOR	HLI 112.4	STAMM ✳ △	HLI/JKS 112.4/109.4	094°	17	161 / 06	2102	06	161	2122	STAMM ✳ 213R - JKS
STAMM ✳	HLI/JKS 112.4/109.4	ALLSO ✳ △	MSL/HAB 116.5/110.4	094°	54	161 / 20	2122	20	161	2133	ALLSO 348R - HAB
ALLSO ✳	/	MUSCLE SHOALS VOR ▲	MSL 116.5	072°	30	165 / 11	2132				
MUSCLE SHOALS VOR	MSL 116.5	GILLE ✳ △	MSL/HSV 116.5/112.?	359°	15	161 / 06					GILLE ✳ 281R HSV
GILLE ✳	/	GORDON'SBG ✳ △	GHM/JKS 111.6/109.4	359°	36	161 / 13					GORDONSBURG ✳ 090R - JKS
GORDON'SBG ✳	/	GRAHAM VOR ▲	GHM 111.6	359°	17	161 / 06					
GRAHAM VOR	GHM 111.6	TIDWELL ✳ △	GHM/BNA 111.6/114.1	062°	16	166 / 06					TIDWELL ✳ 302R - SYI
TIDWELL ✳	/	NASHVILLE VOR ▲	BNA 114.1	064°	25	166 / 09					TOTAL DISTANCE - 243 TOTAL TIME (EST.) 1:33
NASHVILLE VOR	BNA 114.1	CHATT'OOGA VOR ▲	CHA 115.8	131°	102						BNA VOR TO CHA VOR - ALTERNATE -

POSITION REPORTS

POS.	TIME	ALT.	TYPE	EST.	NEXT

CLEARANCES

N 3456J CLEARED

TO	BNA METRO AP	GORD ✳
DEPT. OR SID	MEM COM 1	—
ROUTE	V545, V7	V545, V7
ALTITUDE	4000 -10/5000	5000
HOLDING		SOUTH R.H. TURNS
SPECIAL		EFC 01
FREQ / CODE	DPT 124.15	4365

Fig. 12-1. Flight log. As soon as you reach each fix, a new ETA would be worked out for the next one, based on the latest groundspeed information. The times over on the log don't exactly coincide with the estimate given in Figure 6-8 but you will find such discrepancies on actual flights, too.

CENTER ON ONE TWO EIGHT POINT FIVE (128.5 MHz) NOW.

You: ZEPHYR FIVE SIX JULIET, ONE TWO EIGHT POINT FIVE.

You: (on 128.5 MHz) MEMPHIS CENTER, ZEPHYR THREE FOUR FIVE SIX JULIET, FIVE THOUSAND, OVER.

Center: ROGER, ZEPHYR FIVE SIX JULIET, I HAVE A CLEARANCE FOR YOU, OVER.

You: ZEPHYR FIVE SIX JULIET, READY TO COPY, OVER.

Center: ZEPHYR THREE FOUR FIVE SIX JULIET (IS) CLEARED OVER THE MUSCLE SHOALS VOR TO GORDONSBURG INTERSECTION VIA VICTOR FIVE FOUR SOUTH, VICTOR SEVEN, MAINTAIN FIVE THOUSAND, EXPECT FURTHER CLEARANCE BEFORE REACHING GORDONSBURG, OVER.

You write down and read back the clearance as given, as you look for Gordonsburg intersection on the map. You find it and make sure that there are no gaps in the clearance. Gordonsburg intersection is now your clearance limit. If there are gaps, ask for a readback. And, of course, as always, if you are unable to comply for some reason, let ATC know about it.

Assuming that the clearance is acceptable, you continue and pass over Muscle Shoals VOR at 1532 (2132Z), about one minute ahead of time.

When talking directly to the Center and *not* in radar contact, you would alert them to a pending position report as follows:

You: MEMPHIS CENTER, ZEPHYR THREE FOUR FIVE SIX JULIET, MUSCLE SHOALS, OVER. (By naming a *fix* you alert the Center for the position report to follow.)

Center: GO AHEAD, ZEPHYR FIVE SIX JULIET, OVER.

(You would then proceed with your position report.)

If you had lost contact with the Center before reaching MSL but after getting the clearance to Gordonsburg and still had 122.1, 122.6 or 123.6 MHz (or 121.5 MHz) left, as well as being able to receive on VOR, 122.2, 122.4 or 123.6 MHz, you would call Muscle Shoals (or other facilities as necessary) to pass the word to the Center about your MSL passage, since it is a compulsory reporting point; and you aren't in voice contact with the Center and so don't know whether you're back in radar contact or not. You would note that Gordonsburg is your clearance limit, of course, using the PTA-TEN method of reporting.

You should add remarks such as "moderate turbulence" or "light icing" at the end of the position report — it will help other pilots in the area. Such remarks as "what's going on down there" or "get me down from here" are not considered cricket and can lose you points as well as causing you to move back three fixes (or result in elimination from the game).

In a fast airplane a turn such as required at MSL, up V-7, would mean that the airplane would move out of the airway route boundary or protected airspace if the turn was started at, or after, fix passage. If you have DME, it's perfectly legal to lead the turn enough to insure staying within the boundaries. (If you lead too soon, you'll crash the boundary on the *inside* of the turn and this isn't good either.) This is usually only a factor for planes with groundspeeds approaching 300 knots, so probably you'll not have any worries along this line yet.

TOTAL LOSS OF COMMUNICATIONS
(FAR 91.127)

Let's backtrack a little and look at the Enroute Chart again. Before take-off you were cleared via V-54S, V-7 to Nashville. You got another clearance at 2115Z, halfway between Holly Springs VOR and ALLSO intersection, as shown on the chart. If you had a total communications loss before this clearance was confirmed by you, even after ATC had broadcast it (your radios went out the instant ATC finished delivering the clearance — you didn't get to acknowledge), the pretake-off clearance still holds. You'd fly on to MSL, make a sharp left turn up V-7 to GHM, and continue to the Nashville VOR (which is the enroute navigation aid serving the destination airport).

You would depart the Nashville VOR, fly to the facility to be used for the approach (unless you plan to make a VOR approach). You would hold on the procedure turn side of the approach course at the estimated time of arrival for the route and altitude of the last clearance, which was the one received and acknowledged at the warm-up spot before take-off.

As far as altitude requirements are concerned, you would fly at the last assigned altitude or the minimum enroute altitude, *whichever is higher*. In this case, your last assigned altitude was 5000 feet and the minimum altitudes are well below this. (The highest MEA will be 3500 feet, so you *would maintain the assigned 5000 feet*. Figure 12-2 shows a hypothetical situation that could arise.

ATC will be expecting you to do as shown in Figure 12-2. If you decide "what the heck" and stay at 8000 for a couple more segments to save "all that stair-stepping down," you could find that you'd be in the airspace of other airplanes (and there can be few worse feelings than flying into wake turbulence when you're solidly on the gages).

In the situation of the flight to Nashville, you would maintain 5000 feet until over the facility to be used for the approach. If you arrive *before* the time based on an estimate along the route assigned, you would set up a holding pattern at 5000 feet on the procedure turn side of the final approach course. When the estimated time has elapsed, you'd shuttle down in the holding pattern on the procedure-turn side of the leg to the initial approach altitude and commence your approach. If you arrive later than the ETA as worked out earlier for the latest clearance, you would shuttle down immediately and start your approach. All approaches for the destination airport will be held clear for thirty minutes past your ETA without question and may be held longer if the pilots of other aircraft awaiting approach (who've been cleared out of the way — in a non-radar situation) agree to it. You checked wind direction and velocity at the destination before leaving but, unless an extra strong wind made it out of reason, you'd likely opt for an ILS front course approach. Listen to every possible available source of communication (VOR's, LF/MF, etc.); ATC may give you further word or clear you for an earlier approach.

If you lost communications after receiving the airborne clearance to Gordonsburg intersection, you would proceed as cleared and set up a holding pattern, if necessary, to insure arriving at the facility to be used for the approach at the destination airport either at the time (1) given as the estimated time of arrival on the flight plan or (2) as amended with ATC. The second situation is the more likely. Only if you were assigned the same route and altitude and hit every fix as predicted would (1) be valid.

IFR conditions. If the failure occurs in IFR conditions, or if paragraph (b) of this section cannot be complied with, each pilot shall continue the flight according to the following:

(1) Route.

(i) By the route assigned in the last ATC clearance received;

(ii) If being radar vectored, by the direct route from the point of radio failure to the fix, route, or airway specified in the vector clearance;

(iii) In the absence of an assigned route, by the route that ATC has advised may be expected in a further clearance; or

(iv) In the absence of an assigned route or a route that ATC has advised may be expected in a further clearance, by the route filed in the flight plan.

(2) Altitude. At the HIGHEST of the following altitudes or flight levels FOR THE ROUTE SEGMENT BEING FLOWN:

(i) The altitude or flight level assigned in the last ATC clearance received except that the altitude in (ii) shall apply for only the segment/s of the route where the minimum altitude is higher than the ATC assigned altitude.

(ii) The minimum altitude (converted, if appropriate, to minimum flight level as prescribed in § 91.81 (c)) for IFR operations; or

(iii) The altitude or flight level ATC has advised may be expected in a further clearance except that the altitude in (ii) shall apply for only the segment/s of the route where the minimum altitude is higher than the expected altitude.

(iv) The intent of the rule is that a pilot who has experienced two-way radio failure should, during any segment of his route, fly at the appropriate altitude specified in the rule for that *particular segment*. The appropriate altitude is whichever of the three is *highest* in each given phase of flight: (1) the altitude or flight level last assigned; (2) the MEA; or (3) the altitude or flight level the pilot has been advised to expect in a further clearance.

(3) Leave holding fix. If holding instructions have been received, leave the holding fix at the expect-further-clearance time received, or, if an expected approach clearance time has been received, leave the holding fix in order to arrive over the fix from which the approach begins as close as possible to the expected approach clearance time.

If holding instructions have not been received and the aircraft is ahead of its ETA, the pilot is expected to hold at the fix from which the approach begins. If more than one approach fix is available, it is pilot choice and ATC protects airspace at all of them. Descent for approach begins at the ETA shown in the flight plan, as amended with ATC.

(4) Descent for approach. Begin descent from the en route altitude or flight level upon reaching the fix from which the approach begins, but not before—

(i) The expect-approach-clearance time (if received); or

(ii) If no expect-approach-clearance time has been received, at the estimated time of arrival, shown on the flight plan, as amended with ATC.

2. In the event of two-way radio communications failure, ATC service will be provided on the basis that the pilot is operating in accordance with FAR 91.127.

3. In addition to monitoring the NAVAID voice feature, the pilot should attempt to reestablish communications by attempting contact:

a. on the previously assigned frequency, or

b. with an FSS or ARINC.

If communications are established with an FSS or ARINC, the pilot should advise of the aircraft's position, altitude, last assigned frequency and then request further clearance from the controlling facility. The preceding does not preclude the use of 121.5 MHz. There is no priority on which action should be attempted first. If the capability exists, do all at the same time.

NOTE.—ARINC is a commercial communications corporation which designs, constructs, operates, leases or otherwise engages in radio activities serving the aviation community. ARINC has the capability of relaying information to/from ATC facilities throughout the country.

4. Should the pilot of an aircraft equipped with a coded radar beacon transponder experience a loss of two-way radio capability he should:

(a) adjust his transponder to reply on Mode A/3, Code 7700 for a period of 1 minute,

(b) then change to Code 7600 and remain on 7600 for a period of 15 minutes or the remainder of the flight, whichever occurs first.

(c) repeat steps a and b, as practicable.

The pilot should understand that he may not be in an area of radar coverage. Many radar facilities are also not presently equipped to automatically display Code 7600 and will interrogate 7600 only when the aircraft is under direct radar control at the time of radio failure. However, replying on Code 7700 first increases the probability of early detection of a radio failure condition.

5. It is virtually impossible to provide regulations and procedures applicable to all possible situations associated with two-way radio communications failure. During two-way radio communications failure when confronted by a situation not covered in the regulation, pilots are expected to exercise good judgment in whatever action they elect to take. Should the situation so dictate, they should not be reluctant to use the emergency action contained in FAR 91.3(b).

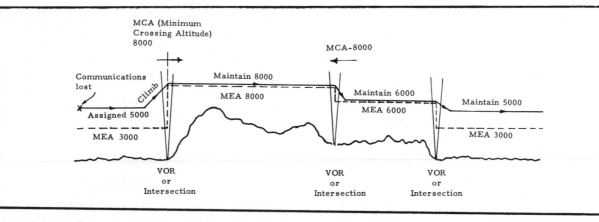

Fig. 12-2. Loss of communications procedures (*Airman's Information Manual*). If in VFR conditions, maintain VFR and land as soon as practicable.

If you are flying in the soup when the communications loss occurs, and have a transponder, squawk 7700 for one minute, then squawk 7600. This will alert ATC to your situation — see Chapter 5.

If you are in VFR conditions when the communications failure occurs, or fly into VFR after the failure, remain in VFR conditions and land as soon as practicable. This doesn't mean "land as soon as possible" (the pastures down there might be a little soft or short — and this might be the same situation at the closest airports). If the destination airport is within a few minutes, then go on to it (you'd have to get light signals to land). Get to a phone and let the nearest ATC facility know what happened.

Using common sense and the knowledge that nearly all (if not all) airplanes flying IFR have two communications transceivers, plus two navigation receivers, the chances of not being able to receive *any* instructions are slim indeed. Center will pass the word to all Flight Service Stations to call you on navaid frequencies. One possibility is that Center can call you on the normal frequencies (you can receive but not transmit) and tell you to acknowledge instructions and clearances by having you "Ident" or squawking other codes on the transponder. Or, if you aren't transponder equipped, they may ask you to make turns to acknowledge instructions.

If you have lost all communications the controllers will normally expect you to make a transition to the approach fix from the VOR (or applicable enroute navaid). As was said earlier, however, they will protect *all* approaches during the 30 minute grace period.

Practically speaking, just to lose both sets of communications and lose nothing else is indeed a remote possibility, but it could happen.

THE HOLDING PATTERN

If it became necessary for you to hold at Gordonsburg intersection, either because of waiting for an ATC clearance or the emergency situation of having to "kill time," you would set up a particular pattern as shown in Figure 12-3.

The standard holding pattern consists of right-hand turns with a 1-minute inbound leg (below 14,000 feet MSL). Each pattern will take exactly 4 minutes to complete in a no-wind condition. (Each of the two 180° — standard-rate — turns will require 1 minute and the two legs are 1 minute each.) ATC may issue holding instructions.

Holding is a pretty simple matter in the situation of holding at an enroute fix such as Gordonsburg intersection. You'd start slowing up to the holding airspeed within 3 minutes of the estimated initial time over the holding fix. It then becomes a matter of getting definite indication of a station (fix) passage and commencing a 180° right (or left, if instructed) standard-rate turn. As soon as you entered holding, which would be at the initial fix passage, you would normally report to ATC, giving the time and altitude/flight level upon reaching the holding fix.

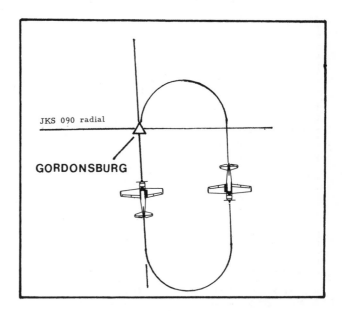

Fig. 12-3. Holding at Gordonsburg intersection.

Unless ATC says otherwise, when you arrive at a clearance limit such as Gordonsburg, hold in a standard right pattern on the course on which you approached the fix, until further clearance is received. You'll also be expected to hold at your last assigned altitude.

You: ZEPHYR FIVE SIX JULIET. COMMENCED HOLDING AT FIVE ZERO, FIVE THOUSAND, OVER.

Center: (Acknowledges.) EXPECT FURTHER CLEARANCE AT ZERO ONE, OUT.

The "expected further clearance" time is given, so that you'll have something to work with should communications be lost *while you are in the holding pattern.* When the expected further clearance time approaches and you have lost communications, be prepared to depart the fix at that time. The 4-minute no-wind pattern doesn't always work out evenly with this expected further clearance time, so you will modify your holding pattern to be at the fix at that time. For instance, you arrive over the fix in one of your patterns 3 minutes before the expected further clearance time. You don't have time to make a complete pattern and be back at the fix at the "go" time. Figure 12-4 shows the way to do it.

Note that it will require 2 minutes to make the two 180° turns, leaving 1 minute to be divided between the two legs as shown in Figure 12-4.

The previous discussion was based on no-wind conditions, an unlikely situation. ATC requires that the inbound leg be 1 minute at altitudes below 14,000 MSL (1 1/2 minutes above that altitude), and your *initial* outbound leg should be 1 minute to check what's up. Suppose you enter holding, fly outbound for 1 minute, and find you reach the fix 45 seconds after rolling out inbound. (The inbound leg is measured from roll-out from the outer 180° turn in to the fix and should be timed accordingly.) Okay, you arrived 15 seconds early so next time make that outbound leg 15 seconds longer, or for 1 minute and 15

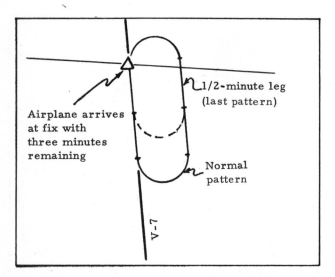

Fig. 12-4. The legs of the last pattern are shortened in order to hit the fix at the expected further clearance time.

seconds (you have a headwind outbound and a tailwind inbound). The purists will argue that this is not the case; you'll have to fly outbound slightly longer than 15 seconds in order to add the 15 seconds to the inbound leg. This is true and can be worked out on a computer (since you have nothing else to do but fly the airplane on instruments, keep up with the clock, talk to ATC, etc.). You will get settled down at about the second pattern and can add what is necessary to get the required 1-minute inbound leg. The same theory applies to a reversed wind situation.

Practically speaking, although it is very easy to work everything out nicely while sitting at a desk, it becomes a different matter in the airplane. As veteran instrument pilots often put it, "The holding pattern is a situation where you are holding somewhere in the general vicinity of a fix — you think."

The crosswind correction for the holding pattern as recommended by the FAA is to hold twice the wind correction angle outbound as was used to stay on the inbound course. This allows both turns to be

standard rate but the inbound and outbound legs will not be parallel (which is okay).

The straight-in or direct entry is the most usual case, but you may have variations on the theme. Holding-pattern entries can be quite confusing, and it's best when given holding instructions actually to sketch the pattern on the chart as shown by Figure 12-5.

Figure 12-6 indicates one way to enter the pattern. When the fix is passed, turn to a heading 30° (or less) to the holding side of the pattern as shown in Figure 12-6 (105° bearing). Set your OBS to the outbound bearing 105. Setting the OBS for the outbound bearing this first time is a good idea for orientation. Fly the outbound leg for one minute. Start a standard-rate turn and reset the OBS for the *inbound* bearing of 315°. It is suggested that you leave the OBS on this setting for the rest of the holding pattern, now that you are established.

One other simple method of entry in this case would be to turn parallel to the outbound leg after passing the fix and hold this for one minute, then turn right to intercept the course. Your instructor may have some suggestions on this.

The numbered items in Figure 12-6 give left-right needle indications and headings at the points mentioned.

Your job will be to visualize the airplane's relationship to the holding pattern. If you don't "see" where you are, the problem of entry is very difficult.

After the end of one minute, start a standard-rate turn to the right. When you get to Point 2, which is 45° from the inbound heading, the moment of truth will arrive. Do you speed up the turn? (The needle is centering now.) Or do you roll out at that heading and fly straight until the needle gets nearly centered? It's a good move, if the needle isn't moving toward the center position as you think it should, to stop at that heading and hold it until the needle is nudging toward the center at what you have found (through practice) to be the proper rate. At this 45° entry

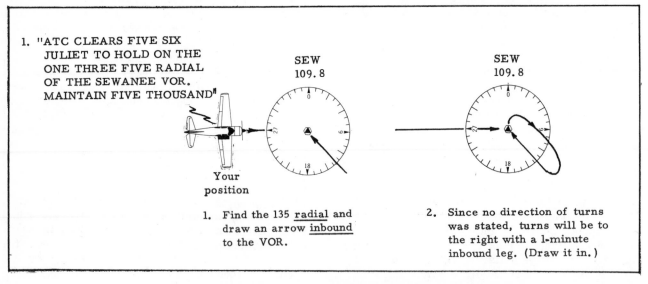

1. "ATC CLEARS FIVE SIX JULIET TO HOLD ON THE ONE THREE FIVE RADIAL OF THE SEWANEE VOR. MAINTAIN FIVE THOUSAND"

Your position

SEW 109.8

SEW 109.8

1. Find the 135 <u>radial</u> and draw an arrow <u>inbound</u> to the VOR.

2. Since no direction of turns was stated, turns will be to the right with a 1-minute inbound leg. (Draw it in.)

Fig. 12-5. Sketch the holding pattern; it makes the entry easier to accomplish.

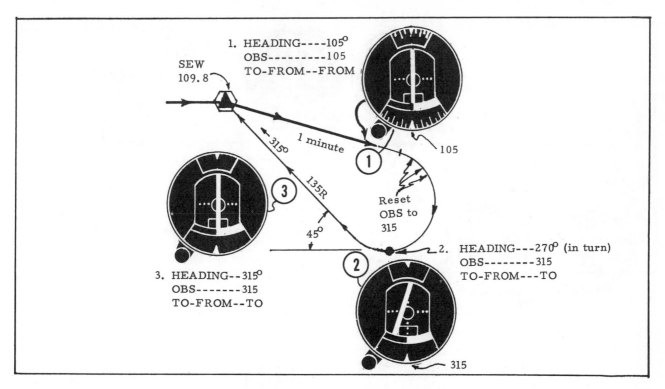

Fig. 12-6. Entering the holding pattern of Figure 12-5.

angle on a *localizer*, if the needle is moving toward the center at all, you should keep turning, because the localizer is *four* times as sensitive as the VOR, as far as your receiver is concerned.

The actions you have to take in turning on that first inbound leg can tell you a lot about corrections to be required on the legs.

The FAA recommends a standard entry for holding patterns as shown in Figure 12-7.

This method has not had wide acceptance because of the problem of computing 70° from an odd heading (313°, for instance). It's really a matter of common sense on your holding pattern entries. There is a "buffer zone" around the racetrack pattern, but you would want to insure that your entry did not cause you to fly outside the allowable area.

Holding at a VOR intersection is the toughest problem. Always use the same VOR indicator for on-course indications (usually the No. 1 set) and the other for cross-bearing information.

If you have only one VOR receiver, you'll be pretty busy switching back and forth. Holding at an LF/MF intersection in precipitation static with only one coffee grinder is an experience old-time instrument pilots turn pale thinking about — so maybe *one* crystal-controlled VOR receiver isn't so bad after all.

Distance Measuring Equipment (DME) Holding Pattern

This equipment has simplified intersection holding to the point where it's as easy (or easier) than holding at a VOR. For instance, you are instructed to hold at an intersection which is 10 nautical miles

out from a VORTAC. You've been told by ATC to use an outbound leg length of 5 miles. You are holding *away* from the VORTAC, so your pattern and instrument indications will be as shown in Figure 12-8.

General and Detailed Holding Instructions

1. General holding instructions.
   a. The direction to hold *from* holding point.
   b. The holding fix.
   c. On (specified) radial, course, magnetic bearing airway number or jet route.

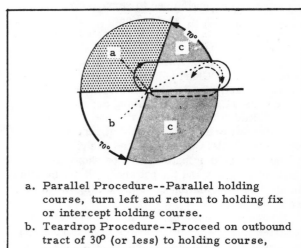

a. Parallel Procedure--Parallel holding course, turn left and return to holding fix or intercept holding course.
b. Teardrop Procedure--Proceed on outbound tract of 30° (or less) to holding course, turn right to intercept holding course.
c. Direct Entry Procedure--Turn right and fly the pattern.

Fig. 12-7.

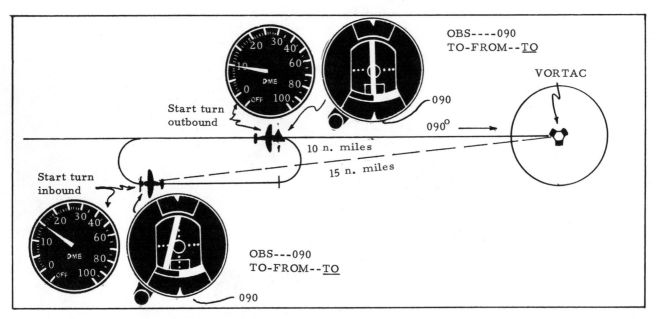

Fig. 12-8. DME holding.

d. Outbound leg length in nautical miles if DME is to be used.

e. Left turns, if nonstandard pattern is to be used.

f. Time to expect further clearance, or time to expect approach clearance.

2. Detailed holding instructions – same as 1(a), (b), and (c) above with the following additions:

a. To (d) – or minute/s if DME is not to be used.

b. To (e) – or right turns if standard pattern is to be used.

Depiction of Holding Patterns on Charts

Holding patterns (standard or nonstandard) at fixes most consistently used to serve a terminal area/airport by either an Air Route Traffic Control Center or a terminal facility will be charted.

The holding patterns will be charted on either or both U.S. Government Enroute High/Low Altitude and appropriate Area Charts. A particular pattern may be shown on both the Enroute and Area charts if the fix is consistently used for holding enroute and terminal traffic.

If aircraft are generally held at particular enroute fixes by ARTCC the patterns are charted.

Only one holding pattern will be shown at a fix on an individual chart and the patterns will not be labeled with altitude information or letter coding for any special purposes.

If you're required to hold at a fix where a pattern is charted, ATC will *not* issue holding instructions. You'll be expected to hold in the pattern shown, unless otherwise advised by ATC.

If you're required to hold at a fix where the pattern is not charted, you will be given holding instructions by ATC at least five minutes before you are estimated to reach the clearance limit.

If you don't receive a clearance beyond the fix before arrival over the fix and the holding pattern is charted, maintain the assigned altitude and use the depicted holding pattern. If this happens at a fix *without* a charted holding pattern, use a standard right hand pattern on the course on which you approached the fix.

If you are in doubt about holding, get instructions from ATC. Remember, below 14,000 feet MSL the inbound legs are 1 minute.

Points on Holding

1. When holding at a VOR, you should begin the turn to the outbound leg at the time of the first complete reversal of the TO-FROM indicator.

2. The direction to hold with relation to the holding fix will be specified as one of the eight general points of the compass, i.e., North, Northeast, East, etc. Your instructions to hold at Gordonsburg intersection could be: HOLD SOUTH OF GORDONSBURG INTERSECTION ON VICTOR SEVEN, (or nautical miles will be given if DME is to be used). In some situations the argument could arise whether the holding pattern is closer for instance, to "South" or "Southeast" (here it's obviously closer to South), so the airway is inserted. You'll hold on the airway, but on the southern side of the fix rather than on the northern side. Holding on an enroute VOR could possibly entail *not* holding on the inbound airway, as covered in Figure 12-5. For holding at an intersection enroute, there's little choice except whether you're to hold on "this side" or "the other side" and normally it will be on "this side."

Holding Speeds

The maximum holding speed for propeller-driven aircraft is 175 knots I.A.S.; for civil Turbojet planes it is 200 knots to 6000 MSL, from 6000 to 14,000 the maximum is 210 knots, and above 14,000 MSL, 230 knots (all speeds I.A.S.).

At what speed should you hold? In theory, you should think in terms of the speed at which the minimum fuel is required as shown in Figure 12-9.

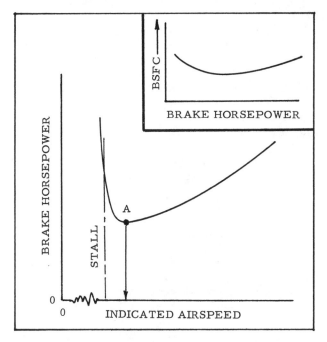

Fig. 12-9. Minimum power required to fly the airplane at a certain weight and altitude.

Point A shows the indicated airspeed at which the minimum power is required. If the engine specific fuel consumption is constant (which it isn't, as shown by the inset), then Point A would indicate the speed for holding. Brake specific fuel consumption (BSFC) is the *pounds of fuel per hour being used by each brake horsepower*. You'll find that for most general aviation type engines it works out to be about 0.45 pounds per brake horsepower per hour, leaned in the area of cruising power. The Lycoming O-540 is rated at 250 brake horsepower. The manual states that at 75 per cent power (188 hp), leaned, the fuel consumption is approximately 14.0 gallons per hour. This can be examined by the following:

Brake horsepower = 188; 0.45 × 188 = 84.6 pounds of fuel used per hour. Basing the fuel weight on the average of 6 pounds per gallon, 84.6 divided by 6 = 14.1 gallons per hour. This is slightly higher than noted by the manufacturer but is within reason. You can also use the rule of thumb of multiplying the horsepower being used (188) by 0.075 to get the same answer in gallons per hour (0.075 × 188 = 14.1).

All right, this is very interesting, you say — but what do you use for a holding speed? Because you don't have power curves or brake specific fuel consumption charts (and would be busy enough flying the airplane without setting up a research project), thumb rules may be substituted. Look at the bottom of the green arc on your airspeed indicator and note the value. That's the flaps up, power off, stall speed

at maximum certificated weight. The following thumb rules are based on that value.

For single-engine airplanes with fixed gear, multiply that figure (using 60 knots as an example) by the figure 1.2 (1.2 × 60 = 72 knots).

For retractable gear singles and twins, use a factor of 1.3 (1.3 × 60 = 78 knots). These factors apply whether the speeds are given in knots, mph, kilometers per hour, or feet per second.

The holding speed has been discussed in terms of theory. But what about a practical situation? Turbulence may make it advisable to use a higher airspeed than given by the thumb rule. A speed of 20 per cent (or 30 per cent) above the stall doesn't give you much to play with in turbulent air, and control could be questionable, particularly in a situation where the airplane was loaded in a rearward C.G. condition. Add a few knots for better control, but don't exceed a speed of 1.6 times the stall speed as marked on the airspeed indicator — you might overstress the airplane if a strong vertical gust is encountered. (These rules are based on *calibrated* airspeeds.)

The airplane type you'll be flying will probably have a recommended holding speed of somewhere between 70 to 115 knots, so you'll have no problem of exceeding the maximum allowable speed of 175 knots for prop planes.

To set up holding, you'll slow the airplane to the recommended speed by throttling back and maintaining altitude. You have no idea how long holding will be required, even though you've been given an "expect further clearance" time. Normally, this will be the time you *will* be cleared, but ATC has extended it on occasion and you'll want to economize. Pull the rpm back (if a controllable pitch prop is being used) to the lowest value you can get without the prop "hunting." Use whatever manifold pressure is necessary to maintain altitude at the chosen airspeed. Check on further leaning, but don't damage the engine. Obviously the airplane should be as clean as possible — flaps up and gear up (if possible). You'll maintain a constant altitude in the pattern, unless instructed by ATC to change altitudes — this technique was covered in Chapter 4.

DEPARTING GORDONSBURG

At 2201Z you receive the following clearance:
Center: ZEPHYR FIVE SIX JULIET IS CLEARED TO THE TIDWELL INTERSECTION VIA VICTOR SEVEN, DESCEND AND MAINTAIN FOUR THOUSAND. CONTACT NASHVILLE APPROACH CONTROL ONE TWO ZERO POINT SIX APPROACHING TIDWELL, OVER.
You: (Read back clearance and add): LEAVING FIVE FOR FOUR THOUSAND AT ONE ZERO. (You leave 5000 at once.)

You've been tuned into Graham VOR and pass it at 2207Z. (You resume cruise speed immediately upon leaving Gordonsburg.)
You: GRAHAM AT ZERO SEVEN, FOUR THOUSAND, NASHVILLE TWO TWO. (Center acknowledges.)

It's 16 miles to Tidwell, and if you haven't already done so, you'd take a last-minute glance over the Approach Charts for the type of approach planned. Your first choice (and that of Nashville approach control) would be the ILS.

Tune in the ATIS in the Graham area and get the latest information and use the code when initially contacting approach control or they will have to give you all the information.

At this point you'd probably tune the No. 2 VOR receiver to BNA (114.1 MHz) and have the DME tuned in also. The No. 1 receiver has the glide slope here, so this would be set on the frequency of 109.9 MHz for the ILS. The glide slope power switch should be turned on if necessary for your airplane.

The ADF should be tuned to the LOM (BN — 304 kHz). (You probably won't pick it up clearly until well past Tidwell.)

The marker beacon receiver should be *on*. (Press to test the lights.)

The Center hand-off to approach control will be by computer. When Center equipment indicates that you've been accepted by approach control, you'll be given the frequency change. In most cases (ARTS) you'll stay on the discrete transponder code you used throughout the flight. *Then* the Center will give you the frequency change to approach control.

As you approach Tidwell, you'll call Nashville approach control on 120.6 MHz.

You: NASHVILLE APPROACH (CONTROL), ZEPHYR FIVE SIX JULIET, TIDWELL AT ONE THREE, FOUR THOUSAND, INFORMATION DELTA, OVER.

Approach Control: ZEPHYR FIVE SIX JULIET. Approach control may give turn or descent clearance like the following:

Approach Control: ZEPHYR FIVE SIX JULIET, NASHVILLE APPROACH CONTROL, TURN RIGHT, HEADING ZERO EIGHT ZERO FOR VECTOR TO THE FINAL COURSE.

You: (Acknowledge.)

Approach Control: NASHVILLE WEATHER THREE HUNDRED OVERCAST, VISIBILITY ONE MILE, DRIZZLE AND FOG. NASHVILLE ALTIMETER TWO NINE EIGHT NINE.

You: (Acknowledge.)

Controller instructions are (ATC Manual 7110.65):

"Issue approach information by including the following, except omit information currently contained in the ATIS broadcast if the pilot states the appropriate ATIS code or says he has received it from the center or another source:

1. Approach clearance or type of approach to be expected if two or more approaches are published and the clearance limit does not indicate which will be used.

2. Runway in use if different than that to which the instrument approach is made

3. Surface wind

4. Ceiling and visibility if the ceiling at the airport of intended landing is reported below 1,000 feet or below the highest circling minimum, whichever is greater, or the visibility is less than 3 miles

Altimeter setting at the airport of intended landing.

Issue any known changes classified as special weather observations as soon as the volume of traffic, controller workload, and communications frequency congestion permit. Special weather observations need not be issued after they are included in the ATIS broadcast and the pilot states the appropriate ATIS code."

Complete your landing check list except for gear, flaps, and prop. Note if pitot heat or de-icing (airframe and propeller) is necessary.

You are now very glad that you took the time to study the Approach Charts thoroughly before the flight . . . .

Chapter 13

THE INSTRUMENT APPROACH AND LANDING

THE APPROACH (and landing), particularly when the weather is at minimums, requires complete attention to the work at hand and is the part of the flight where precision is most required. Unfortunately, after a tough enroute session, it is also the point where the pilot is most likely to be suffering from fatigue and get-on-the-ground fever. Sneaking down below minimums to "see if you can break out" can result in your carrying added weight in the form of television towers, smokestacks, or transmission or telephone lines. Sure, it's a pain in the neck to have to make a missed approach and fly to an alternate (you have no clean clothes, or maybe there's a big neighborhood shindig tonight). The controllers will want your intentions as to an alternate, etc.

This chapter will cover the expected procedures for the most common approaches made in instrument flight. The approach and missed approach will be covered for the ILS (both back and front course), VOR, ADF, and Airport Surveillance Radar plus a general look at other approaches at BNA.

ILS FRONT COURSE APPROACH

In Chapter 12, approach control was vectoring you to intercept the front course of the Nashville ILS. You'll be further worked by radar, so that you will intercept at an angle of no more than 30° and at a far enough distance out from the outer marker, so that you won't be rushed. Figure 13-1 shows the Approach Chart for the Nashville ILS front course.

Your airplane will be in Approach Category "A".

Be sure that you are ready to take over and complete the approach at any time. Being vectored by radar is pretty easy living, and sometimes the pilot is caught short when approach control says, "Take over and complete the approach." There may be a great deal of confusion and delay while navigation receivers are turned on.

Set the OBS to 016 on the No. 1 omni as a *reminder* of your base course on the ILS approach. (Remember the OBS setting is not a factor when using the localizer frequencies.)

Radar can descend you to the interception altitude at a safe point and thereby dispense with the problem of flying to the outer marker and shuttling down or turning back outbound and making a procedure turn.

Notice in Figure 13-1 that you are expected to stay above 2800 feet until out of procedure turn, and the Minimum Safe Altitude for the route from Tidwell is 3000 feet. When you are being radar vectored in the approach control area Minimum Vector Altitudes will apply. Each radar approach control has an MVA chart or charts showing the various Sectors and minimum altitudes for each. Each Sector boundary is at least three miles from the Obstruction determining the MVA. Figure 13-2 shows a typical MVA chart.

You are approaching Point 1 on the approach in Figure 13-3.

Radar will descend you as necessary after obstructions are passed.

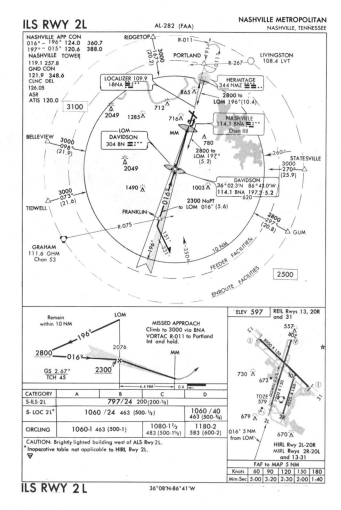

ILS RWY 2L

Fig. 13-1. Front course ILS Approach Procedures Chart for Nashville Metropolitan Airport. Note that there is a "brightly" lighted building west of ALS Rwy 2L.

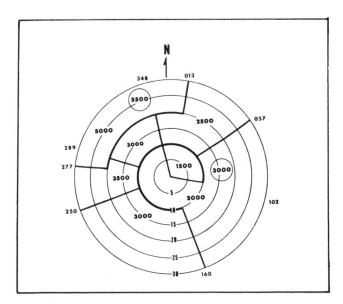

Fig. 13-2. A typical MVA Chart.

Approach Control: ZEPHYR FIVE SIX JULIET, NOW FIVE AND ONE-HALF MILES SOUTHWEST OF THE OUTER MARKER. DESCEND AND MAINTAIN TWO THOUSAND FIVE HUNDRED, UNTIL ESTABLISHED ON THE FINAL APPROACH COURSE, TURN LEFT HEADING ZERO FOUR FIVE, CLEARED FOR AN ILS, RUNWAY TWO LEFT, APPROACH. CONTACT THE TOWER, ONE ONE NINE POINT ONE BEFORE CROSSING THE OUTER MARKER, OVER.

You: ZEPHYR FIVE SIX JULIET.

You would start your descent to 2300 feet as you intercept the localizer south of the OM, so as to cross it at that altitude.

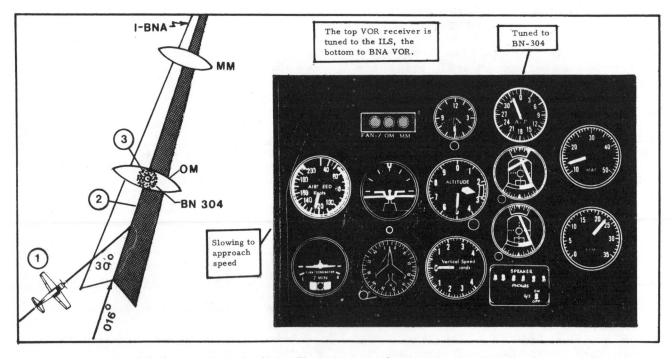

Fig. 13-3. Approaching the localizer. The instrument indications when the plane is at Point 1.

Remember that the localizer is four times as sensitive as a VOR, so as soon as the needle starts moving from the peg, you'd better think of turning the airplane. The rate of turn naturally will depend on the interception angle. If you are intercepting it at a 10° angle there would be no urgency in turning; a 90° interception could mean a fairly steep turn would be required to keep from overshooting the center line. (And rather than taking a chance on losing control of the airplane, you'd better work it so that such an angle is not necessary.) If you are being vectored by radar to the final approach course the maximum angle of intercept is 30°.

You are to call the tower now, according to your instructions from approach control earlier, as you intercept the localizer. Notice on the Low Altitude Enroute Chart that the tower's control jurisdiction — the control zone (outlined by dashed lines) — starts at the outer marker. Approach control could have coordinated with the tower and had you contact them when you first intercept the localizer — it's up to them. The tower at Nashville will be watching your airplane on their "bright display" a reproduction of the radar screen. Of course, at a non-radar situation you'd call passing the final approach fix (the outer marker).

You: NASHVILLE TOWER, ZEPHYR FIVE SIX JULIET, YOUR FREQUENCY, OVER.

Nashville Tower: ZEPHYR FIVE SIX JULIET, CLEARED TO LAND RUNWAY TWO LEFT, WIND THREE TWO ZERO DEGREES, ONE FIVE (320/15 knots), OVER.

You: (Acknowledge.)

If there is a possibility of traffic conflict, you may not get a landing clearance at this time. In conditions above VFR minimums (1000 feet ceiling and 3 miles visibility), there may be VFR traffic under the overcast that will have to be coordinated. (You are up at 2300 feet MSL and are on instruments.) With the ceiling of 300 feet and 1 mile visibility given here as an example, there would only be IFR traffic to coordinate — which will have been done — and landing clearance can often be given some distance out.

Figure 13-4 shows the instrument indications with the airplane has intercepted the localizer and is about 1/4 mile south of the outer marker at Point 2 in Figure 13-3. Note that the glide slope needle is beginning to show signs of life.

The outer marker is reached and the glide slope indicator centers. This is not always the case. The layout of a particular approach could require intercepting the glide slope before reaching the outer marker, but normally the systems are designed for G/S interception at the outer marker. You drop the gear. Or if you preferred to have the gear already down and change power for the descent, that's fine — and you may want to do both. But as covered in Chapter 4, know what power is required for an approximate descent of 500 fpm. The required rate of

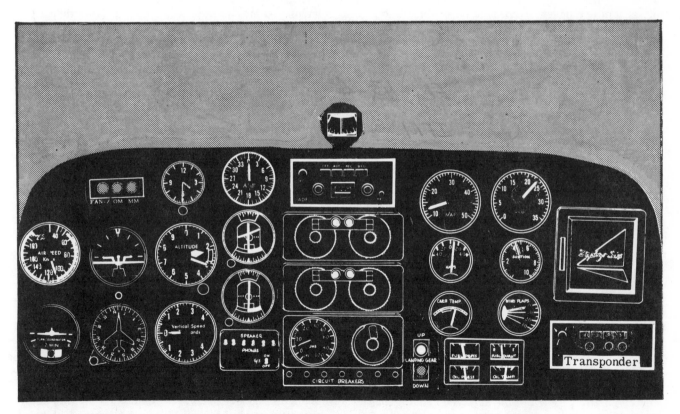

Fig. 13-4. The instrument indications at Point 2 in the approach. (On the localizer and approaching the LOM.) On actual instruments the airplane will intercept the glide slope about one mile from the outer marker and start its descent. If the glide slope (ground or air equipment) is inoperative, you would maintain an altitude of 2300 feet to the outer marker.

185

Fig. 13-5. At the outer marker and compass locator and on the glide slope. The gear is put down at this point (if not already down).

descent to stay on the glide slope varies with air-speed (or more properly, groundspeed). The rates of descent are based on no-wind conditions, and a head- or tailwind component would mean corrections in sink rate. The wind, as given, would result in a headwind component of about 10 knots, making your groundspeed about 100 knots.

Figure 13-5 shows all kinds of things happening as the plane reaches the outer marker (and locator). (Point 3 in Figure 13-3.)

The surface wind at Nashville indicated that you might expect both a crosswind and headwind component on the approach. Figure 13-6 shows that the airplane is low and to the right of course. The ILS head shows that the airplane is to fly up and to the left. The heading indicator indicates that the heading is reasonable (disregarding such things as precession — which will not be a factor in this illustration). The altimeter shows 1750 feet MSL at this point.

The common problem of new instrument pilots is that of "flying" the localizer needle rather than picturing the position of the airplane relative to the system and using the H/I for corrections as should be done. "Making a turn" using the needle and guessing can result in overshooting back and forth across the center line *and* the likelihood of a danger-ous condition or, at least, a missed approach.

Okay, you're off to the right and will turn left and hold the new heading until the needle is at the center position. *Turn to the selected heading and fly it.* Precise directional control is required. For most cases, a 10° cut should be considered about the maximum. In this case, a heading of 005° would be

used because it can be easily read, even if it is 1° past the "maximum" cut. In extreme cases of cross-winds you may have to make cuts greater than 10°.

In close (approaching the middle marker), 5° corrections could be considered pretty much the outside limit, once the drift correction heading is established. In other words, you know the proper heading for drift correction but have been careless and let the heading slip off.

What about altitude corrections? You are low at the position shown in Figure 13-6. By ramming open the throttle, you could gain altitude in short order (and probably fly on through the glide slope). Back in Chapter 4 in the section on descents, it was men= tioned that throttle jockeying can ruin an approach. If you are below the glide slope, you don't want to *gain* altitude. Level flight probably would be the most radical correction. Normally, you would just decrease the rate of descent, the amount of correc-tion depending on the error, of course. (If you get so low that the G/S needle is pegged at the top, you'd better add full power, clean up, and *climb for another shot at it.*

You might have the right combination of power and airspeed for a nonturbulent condition, but an up- or downdraft could result in a G/S needle deviation. On a hot bumpy day, correcting with throttle would be a chore. For very *minor* corrections use the eleva-tors to ease back onto the glide slope, then a power change is unnecessary. Depending on the airplane and airspeed used, you might set a limit of ±5 knots to be used for minor deviations from the glide slope. In other words, if you have to vary the airspeed more than, for instance, 5 knots to hold the glide slope, then you'd better do something with the power. *Try not to make radical power changes;* one inch of manifold pressure for the variable pitch prop or 100-rpm change for a fixed pitch prop should be sufficient for minor variations, and you may take off part of this after the glide slope is "wired" again.

Concerning the idea of using elevators to correct back to the glide slope: This will work if the airplane is operating on the "front side" of the power curve. In discussing the power in Chapter 4, it was noted that the rate of descent was proportional to the deficit power existing at a particular airspeed. In Chapter 12 it was noted that the point of minimum power was found at 1.2 times the flaps-up, power-off stall speed for fixed gear (but otherwise clean) airplanes and 1.3 for retractables, both singles and twins.

Figure 13-7 shows the instruments just as the airplane crosses the middle marker. You'll reach the minimum (797 MSL) right after passing the middle marker.

Don't be surprised if, on the final approach, varying winds are encountered. The surface wind may be reported as right down the runway, but you have quite a correction plugged in to hold the localizer needle centered during most of the approach. There may be a drop in wind velocity during the approach as you lose altitude, and you must adjust for it.

One problem (and it can be a hairy one for single-piloted planes) is that of flying half instruments, half VFR on the latter stages of the approach. Often there will be a low broken layer in the last part of the approach. The occasional glimpses of the ground are enticing enough to draw your eyes away from the instruments — just as you fly into the clouds again. It will take a second or two to get reoriented with the instruments, and at low altitudes this can be fatal. Of course, you have to look out, otherwise you'd make a missed approach when it was not necessary. On an actual instrument approach you'll be able to see out the windshield from the corners of your eyes, and any spectacular change of visibility will be readily apparent. With hood work or dual-piloted actual instruments, your safety pilot (or copilot) will be watching for the ground.

You might try some practice approaches at 1.3 times the landing flaps (bottom of the white arc) stall speed with the landing flaps extended and see how it grabs you. Remember that you won't want to make the approaches at too slow an airspeed because of the delaying of traffic. There's also the problem of control in turbulence at too low a speed.

Once you've broken out, double check the gear and other check list items and complete your landing. From here on, it's just like a VFR approach and landing. You'll be told to contact ground control after turning off and will taxi into the point of destination.

You don't have to cancel your instrument flight plan if you complete an approach and land — it's done automatically. However, if you start the approach and decide it can be done VFR, you would cancel it with approach control if this were your wish. A lot of times, even though you've broken out VFR, you'll want to continue IFR for practice purposes. Don't be too hasty cancelling your IFR flight plan.

Fig. 13-6. The airplane is to the right of the center line and low.

Fig. 13-7. Crossing the middle marker and compass locator at Nashville. The airplane broke out at 900 feet MSL. The airplane has just been turned to the runway heading for a wing-down approach. Flaps may be lowered now as required.

Sometimes things could deteriorate before you complete the approach.

You can always ask for a visual approach, if conditions allow it, rather than cancelling IFR.

Missed Approach

Check Figure 13-1 for the missed-approach procedure for your approach. If you got to the minimum altitude and did not have the runway in sight, you'd apply climb power, clean up the airplane, and climb to 3000 feet MSL outbound on the north course to Portland intersection and hold. The localizer needle will still point to the center line as you go out the north course, since you will still be flying "inbound" heading.

This is one of the required reports as mentioned in Chapter 12. *You'll* have to make up *your*

mind what the next move is to be. If the field has obviously gone below minimums, there's no use (and it is a waste of time and fuel) in trying another approach. It will depend on the situation; perhaps scud or rain showers moved across the field and momentarily caused the problem. You might check on the weather before going to the alternate.

Obviously, if the weather is below minimums for an ILS (with glide slope operative), you wouldn't be able to complete a VOR or ADF approach with their *higher minimums*.

If you're moving on to the alternate, you'll get a clearance and follow it. You'd notify the tower of your missed approach and would be switched to departure control at Nashville. If you listed the alternate in the REMARKS section of your flight plan, the information will be forwarded to the destination well before you get there.

VOR APPROACH

Figure 13-8 is the Approach Chart for a VOR approach at Nashville. Notice that the minimums are higher than for the ILS.

Radar (again) can be a big help in vectoring you so that a procedure turn is not required. As an example, however, assume that you will have to go through the entire process.

You'll be on V-7 on an inbound bearing of 064° (magnetic) and will have a minimum altitude of 3000 feet crossing the VOR. Hold it, you say. According to Figure 13-8, the minimum safe altitude is 3100 feet from an inbound magnetic course of 360° to 180°; now there's this talk about *3000 feet*. What gives? Okay, it's fairly simple; the 3100 feet (MSL) has to cover *180 degrees of approach*, not a planned airways approach. The 3100 feet must cover all approaches within a 180 degree azimuth. The 3000 foot figure covers a specific airway from a specific direction.

After the station passage, set your OBS to the outbound bearing of 107° and turn back to get on it. (A 30° cut should be sufficient.)

Figure 13-8 notes that the procedure turn or other type of reversal is to be made *within* 10 miles of the VOR. Usually three minutes outbound is

plenty of time (and later you'll probably knock this down to two minutes). You want to go far enough so that there will be enough straightaway inbound to allow you to get squared away.

Going back to the initial part of the approach, you'll have the airplane slowed up and will be descending from 3000 feet (you'll tell approach control of your actions — and they'll normally tell you to report to the tower coming out of the procedure turn, inbound).

The 2400 feet shown on the profile of the Approach Chart means that you must not be *lower than that* coming out of the procedure turn. You *can* be higher at this point.

After one minute, make the 180° turn to the right. Reset the OBS to the inbound bearing (287 here) and fly until the needle centers and turn inbound (use the proper amount of lead on the turn).

You may use a "90-270" type of turn (turn to the left 90° and immediately roll right back into a right turn of 270°), which should put you on the inbound heading. (It may not work out precisely as a "90-270," but you should roll out to get on the inbound heading anyway, since that's the prime object.) You can make any corrections necessary to get on or stay on the inbound bearing. Actually, the procedure turn as shown in Figure 13-8 can waste more time than is necessary. It's finally been made official that the pilot may use any procedure he chooses to reverse course. (The pilots were using their own reversal procedures, anyway.)

After you've completed the reversal, start the descent to the altitude for station crossing. Double check your tracking inbound.

As to when to get the gear down, it might be best for this approach to do this over the initial station passage. (You'll have to be slowed up enough at that point to do it.)

The VOR approach requires less concentration than the ILS, but it would probably be best to have the landing check list completed before the procedure turn is *started*. Or you may prefer to wait until you are out of the procedure turn; work out the best routine for your airplane and approach. Don't exceed a 500-fpm descent when on final.

Note the importance of a DME receiver in getting lower minimums on this approach. The DME will let you let down an extra 120 feet to the MDA (100 feet in the circling minima).

Circling Minima

The visibility is the deciding factor here in the difference in minima between the straight-in (within a 30° angle) and circling for landing at another runway.

Don't descend below the circling minima until you are in a position for final descent to land. You should maneuver in the shortest path to the base or downwind leg, as required. There is no restriction against flying over the airport or other runways. Watch out for VFR traffic, and use standard left hand turns or controller instructions for maneuvering.

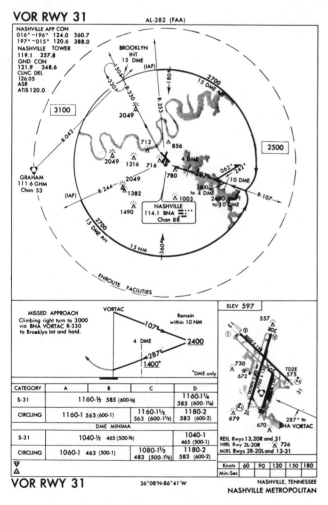

Fig. 13-8. VOR Approach Chart.

If there's no control tower you may want to fly over the airport to check out wind direction and traffic. If you are making an instrument approach to a military field you'd better have the full story on their procedures and requirements. Circling approaches should be avoided at night. In uncongested areas, particularly, loss of orientation could occur.

Missed Approach

Figure 13-8 shows the missed approach as a climbing turn to 3000 feet via the BNA VORTAC R-330 to Brooklyn intersection and hold.

Speaking generally, if you decide to make an early missed approach, fly the instrument approach as specified on the approach chart to the missed approach point at or above the MDA (or DH) before turning; the obstacle clearance is predicated on this. If you turn early it's possible that you wouldn't have proper clearance from obstructions.

Memorize the steps of the missed approach. It's no time to be fumbling around. One check pilot used to say "execute a missed approach" then he would cover up the missed approach procedure with his hand. Woe betide the trainee who hadn't already memorized it. Too often, a pilot is psychologically set to land and finish the flight and a missed approach requirement catches him short.

Under some conditions you might call approach control before starting the approach and say, "In case of a missed approach I'll do (thus and so)." Sometimes things may be too busy for this.

When you are going to an alternate remember WARP (weather, altitude, refile and procedures) or DRAFT (destination, route, altitude, fuel and time). But remember most of all (again) that *you are to fly the airplane first, voice procedures can wait.* If Radar Service is provided you should conform to radar vectors. After this missed approach you'll have to decide whether to take another try at it or go on to the alternate.

LOCALIZER BACK COURSE APPROACH

Figure 13-9 is the approach chart for a back course ILS approach. No glide slope is available.

Without radar contact you probably would ease on to the localizer at the outer marker, fly on past the airport, maintaining at least 2500 feet MSL. Approach control radar might save some time by getting you on at another spot. During the outbound part of the approach, the needle will read normally — fly the localizer needle (assuming that you have the ILS tuned in properly).

You would tune in the Nashville VORTAC on the other set. You'll then make the procedure turn within 10 miles of Hermitage NDB. As the approach chart indicates, the procedure turn is nonstandard (the turn off the course is to the *right*) because of obstructions.

Flying inbound on the back course, you'll have to keep close watch to fly "against the needle" unless

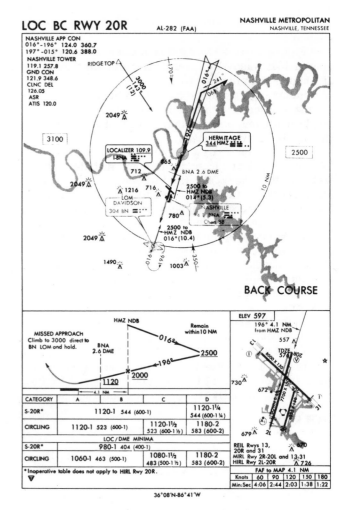

Fig. 13-9. The ILS back course approach at Nashville.

you have special equipment, such as a Course Director, and would descend to 2000 feet after leaving the procedure turn.

Note that you would maintain 2500 feet until out of procedure turn and descend to cross the Hermitage NDB at or above 2000. You would not descend below 1120 feet unless you had a DME operating. You could then descend to 980 feet after passing the BNA 2.6 DME spot. Note that DME allows lower straight-in and circling minima.

The Back Course approach requires timing from the final approach fix (Hermitage) to the MAP (missed approach point) if DME is not used. With a ground speed of 100 knots you would use 2 minutes, 30 seconds, in this distance of 4.1 nautical miles.

Assuming that it was necessary to land on runway 31 from this approach, a turn for a right downwind leg might be in order. Or if that weren't possible, the airplane could be flown over the airport and turned on a *left* downwind leg. There are as many possibilities as there are runways. You'll just have to work with the tower for the best deal for your situation.

On any approach if you have time, let the tower know when you are "contact" (or have broken out).

The missed-approach procedure is given in Figure 13-9 and is a straightforward explanation. Just remember that flying outbound on the south course with a standard localizer receiver (no Course Director), you'll have to fly against the needle because you're still going "the wrong way."

The voice reports ("inbound from final approach fix," "missed approach," and other requested reports) follow those of the earlier sections.

ADF APPROACH

Figure 13-10 is the NDB (ADF) Approach Chart for Nashville.

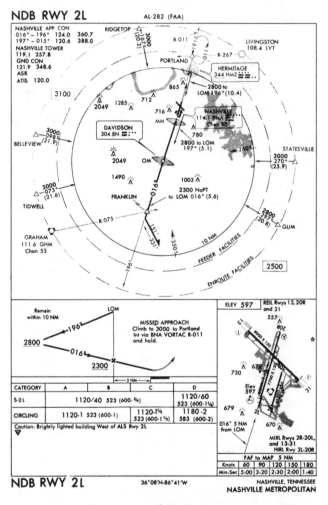

Fig. 13-10. NDB (ADF) approach.

Again there will be the requirement for timing the last portion of the approach from the LOM to the airport. For instance, as shown in Figure 13-10, if your groundspeed is 100 knots, in 3 minutes and 3 seconds you should be at the airport.

The missed-approach procedure is to climb to 3000 feet flying outbound on a course of 011° (VOR) to the Portland intersection and hold. You'd make the usual missed-approach voice report and decisions.

The NDB approach is being eliminated at many

places because it requires too much training compared to the VOR. The ADF is also subject to the precipitation static problems of LF/MF equipment.

The importance of your reporting leaving the Final Approach Fix on *any* approach when not in radar contact cannot be overemphasized. ATC may depend on this report for separation between arrivals and departures and between successive arrivals.

LOCALIZER APPROACH

This is included as another type of approach for you to consider. You would cross the Ayers intersection and outer marker and descend to no lower than 2700 feet out of procedure turn and crossing the outer marker. The missed approach procedure and minima are self-explanatory. (Ayers intersection was named after the late James Ayers, ATC specialist at Nashville, who was of great help in the first edition of this book.)

Other approaches at Nashville are shown in Figures 13-12 and 13-13. They are variations of those discussed earlier. You should study them for items not specifically mentioned earlier. You can save yourself some grief on the written and flight

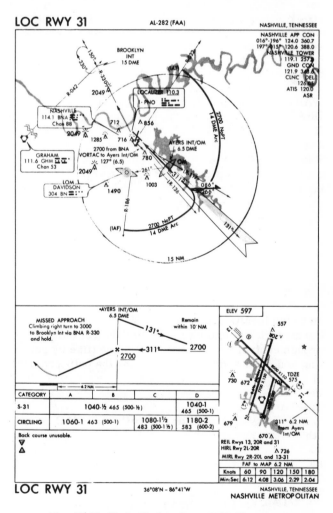

Fig. 13-11. Nashville Localizer RWY 31 approach.

tests and later by being familiar with the information shown on the approach charts.

RADAR CONTROLLED APPROACH

Figure 13-14 is a sample chart showing precision approach (PAR) and surveillance approach (ASR) information for the Nashville airport (which also stands for most airports). Nashville does not have precision approach radar (PAR). Charleston, S.C. has been included to show PAR minimums.

Read the information at the top of Figure 13-14.

Surveillance Approach

The surveillance approach at Nashville is good for the big runways day (d) or night (n) for both straight-in (s) and circling (c). Since it is a surveillance report with no glide slope information, the minimums are "Minimum Descent Altitudes" rather than "Decision Heights." Runway 2L has a lower straight-in visibility minimum because it is the ILS runway and has the better approach lighting.

The surveillance approach (called ASR approach or airport surveillance radar approach) must have

higher minimums than the precision approach radar because it does not have any means of keeping up with the altitude of the airplane. Altitude advisories are given to the pilot on final for each mile *upon request* and you will correct as necessary to get back on the "glide slope." You'll be asked before starting final what your approach speed will be (give it in knots, preferably, to save time). You will be told the approximate rate of descent needed to approach at the correct angle. The wind can affect this, of course, and you may have to make power adjustments to get the right rate of descent for headwind or tailwind conditions. The same thing applies on this approach as for the front course ILS -- the throttle will have to be eased back to keep the manifold pressure from getting too high and slowing down or stopping the rate of descent. This applies to throttling back the fixed pitch propeller airplane.

You'll be given heading information as necessary to keep you lined up with the runway center line. The controller will soon establish the proper crab if there is a crosswind. On a nonhooded practice approach, you'll find yourself using body English to help him get you lined up. One problem sometimes encountered is that 2° heading changes may be called for by the radar controller — on days of very

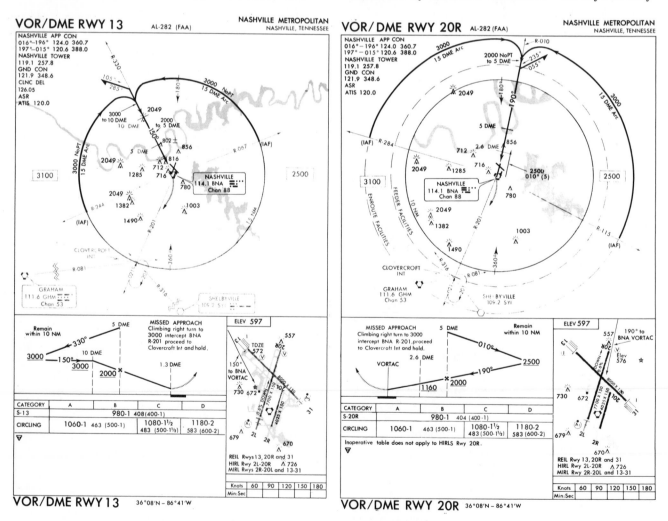

Fig. 13-12. Other approaches at Nashville.

192

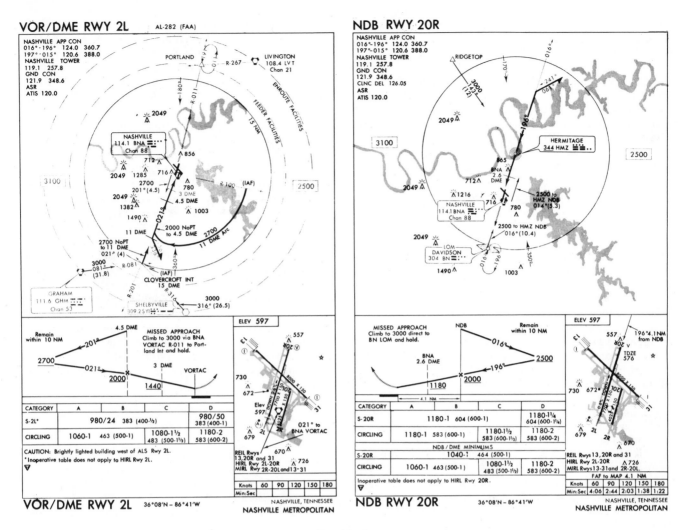

Fig. 13-13.

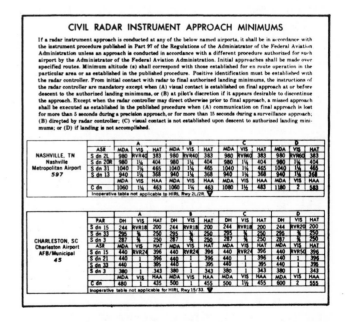

Fig. 13-14. Surveillance approach minimums for Nashville
and PAR minimums for Charleston, S.C.

choppy air conditions you may be lucky to hold it
within 5° of the required heading on approach.

When you reach Missed Approach Point, the controller will say: ONE (or the applicable distance)
MILE FROM RUNWAY, TAKE OVER AND COM-
PLETE YOUR LANDING. If you were not in contact
with the ground and airport, a missed-approach procedure would be executed.

Acknowledge altitude and heading instructions
until on final. If you don't hear from the controller
for any 30-second period, execute a missed approach. You may want to hold the gear extension
until ready to start descending (he'll warn you, saying, "Prepare to commence descent in one mile"),
or you may want to use the technique of putting the
gear down on turning final and reducing power when
the descent is begun. Whatever you do, have the
method established during your hooded practice sessions, rather than getting caught short under actual
IFR conditions.

A method of establishing the required rate of
descent on this type of approach (and it works for a
3° slope ILS, too) is to add a zero to your approach
speed and divide it by two. (Approach speed 110
knots — add a zero to make it 1100 and divide by 2

193

for a required rate of descent of 550 fpm.) If there's a headwind component, subtract it from your approach speed (say a 15-knot wind as an example). Subtract the 15 from 110 knots to get 95 knots. Add the zero (950) and divide by 2 for a required 475-fpm descent. You may be guided on final to an intermediate step-down fix minimum crossing altitude, and subsequently told to descend to the MDA without giving the desired altitude per mile. Under usual conditions only your *distance* from the missed approach point will be given at each mile.

Precision Approach

Figure 13-14 shows that Charleston has a PAR approach with its lower minimums (using a "Decision Height") and an ASR approach as well. It's noted that both Nashville and Charleston have special take-off minimums.

The initial part of the approach pattern for the PAR will be close to that of the surveillance approach. As you approach the glide slope, as shown in Figure 13-15, the final controller (usually a different person) will tell you when to start the descent and will give your relative position to the slope ("On glide slope, twenty feet above," etc.). You'll adjust your rate of descent as necessary and will also be given heading instructions. Usually, the final controller will say, "This is your final controller, acknowledge no further transmissions" (in other words, *shut up*). "If no transmission is heard for a period of five seconds or more, execute a missed approach." He then starts talking at such a rate that you couldn't get a word in edgewise, if you wanted to. (Remember, he has to keep talking or you'll think you've lost him and will pull up.)

You'll be given weather or other information vital to operations ("You are cleared to land, etc.").

The military term for this is GCA (Ground Controlled Approach). Some pilots prefer it to the ILS because it leaves them free to give all their attention to the flight (and engine) instruments. The

sound of another human voice has a good psychological effect for some. At some airports where PAR is available, you may request a monitor on an ILS front course approach and receive advisories on the ILS (localizer) frequency. The localizer voice facilities are being eliminated at many airports.

No-Gyro Approach

As an emergency standby you may practice what is termed a "no-gyro approach." This is not a completely accurate title because it is assumed that the attitude gyro and H/I are inoperative, with the turn and slip or turn coordinator remaining. (The needle or small airplane is gyro operated, so it's not really a "no-gyro approach.")

Because it is assumed that you have no *accurate* means of turning to headings -- the magnetic compass will not be precise enough -- the controller will say, "Turn left . . . stop turn." You will be expected to use standard-rate turns by reference to the turn and slip for the pattern, except after turning final where half standard-rate turns will be expected.

It will keep you busy, and it would be good training to practice one or more of these approaches, if possible.

ADDED NOTES

You should have done as many types of approaches as possible during the training process. The radar controllers are always glad to allow you practice ASR and PAR approaches if traffic permits, since they want the practice also. Sometimes you'll be asked to comment on the approach and you should give your honest opinion -- it helps them and will help you also in analyzing possible future problems of your own.

Here are a few points you should keep in mind concerning approaches:

1. The missed-approach might be required anywhere during the approach; even while making the procedure turn it might be necessary to call it off. Don't be ashamed to start all over again if things start going to pot. Too many pilots try to salvage an approach that should be stopped in favor of a new try, but no, they fight it all the way down and make a deep impression — on some fixed object.

It's a good idea in that case to let the tower (or approach control, as applicable) know that the reason you want to make the approach again is because you "didn't like the way that one was going," or some other such subtle clue, so that the pilots following won't think the weather was the cause.

2. Anticipate that approach control will grab you for vectoring before you reach the enroute navaid serving the destination airport. This was the case in the ILS front course approach example and probably will be the case for *all* approaches at a place where approach control radar is available.

3. Forget the part of the trip that's behind you and concentrate on the approach. For some reason

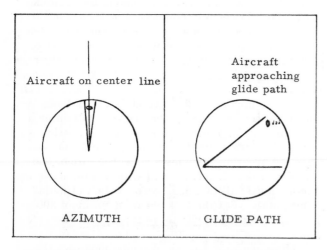

Fig. 13-15. Precision approach radar can check the airplane's position with reference to the glide path. The latest PAR presentations have azimuth and glide path information on the same scope.

Fig. 13-16. A precise instrument approach is a fitting end to a well-planned and well-executed instrument flight.

this is the time when passengers suddenly remember a lot of questions. Don't allow any outside distractions to interfere.

4. There have been a number of instances of aircraft striking the ground on approach when visual cues are lost during low visibility landings. Pilots have been known to continue the descent below DH or MDA after flying into a thin layer of fog (or snow or rain). So . . . if you lose the runway you'd better add full power and execute a missed approach, or you could break your airplane.

5. The approach is not complete until the airplane is locked in the hangar. *Don't* feel that you have it made as soon as you break out and have the field in sight. There's still some work for you to do.

MINIMUM SAFE ALTITUDE WARNING (MSAW)

To assist air traffic controllers to detect aircraft that are within or are approaching unsafe proximity to terrain/obstacles, the FAA has furnished ARTS III facilities with a computer function called Minimum Safe Altitude Warning (MSAW). The function generates an alert when a participating aircraft is, or is predicted to be, below a predetermined minimum safe altitude. Aircraft on an IFR flight plan, that are equipped with an operating altitude encoding transponder, automatically participate in the MSAW program. That is, no specific request is necessary. Pilots on VFR (or no flight plans) may, provided they are equipped with an operating altitude encoding transponder, participate by asking the air

traffic controller. The controller will evaluate any observed alerts and, when appropriate, issue a radar safety advisory.

Federal Aviation Regulations place responsibility for safe altitude management on the pilot. Minimum Safe Altitude Warning provides the controller information which, when judged to be significant, can be relayed to assist the pilot with that responsibility. Participation in the MSAW program does not relieve you, as the pilot, of responsibility for safe altitude management.

For general terrain altitude monitoring, MSAW maintains a computerized grid map of the terminal area. The grid map is comprised of 2-mile squares. The highest known obstacle in each grid or bin determines the minimum safe altitude for that location. The minimum safe altitude is 500 feet above the highest terrain/obstacle in each bin. The ARTS computer compares the current Mode C altitude of an aircraft against the minimum safe altitude. It then looks ahead 30 seconds to see if the aircraft will enter a bin below the minimum safe altitude if it continues its present heading, altitude, or rate of climb/descent. Then the program assumes a 5-degree climb, and computes to see if the aircraft will remain above the minimum safe altitude if it were to start climbing immediately. For the look ahead, a buffer of 300 feet (instead of 500 feet) above the highest obstacle is used.

Minimum Safe Altitude Warning monitors the final approach course from the final approach fix, and a point approximately 2 miles from the landing threshold. It first checks 100 feet below the minimum descent altitude (MDA)/stepdown fix altitude. It then looks ahead down the final approach course, using the computer-established descent rate to determine if the aircraft will be 200 feet below the MDA/stepdown fix altitude in 15 seconds.

If an aircraft is, or is predicted to be, below a minimum safe altitude, the computer alerts the controller by displaying "LA" in the aircraft's data tag. An aural alarm is also sounded to attract the controller's attention. The controller will evaluate the situation and, if appropriate, issue a radar safety advisory; i.e., "LOW ALTITUDE ALERT, CHECK YOUR ALTITUDE IMMEDIATELY."

It is the pilot's responsibility to evaluate the situation and determine what action may be necessary when an advisory is received. The pilot is expected to inform ATC immediately should any action be taken after receiving a radar safety advisory.

There are situations under which the controller will not receive an MSAW-generated low altitude alert. In some instances, the controller may not be aware of the condition. The situations include:

1. ATC radar beacon interrogator is not operating.

2. The ARTS III computer with the MSAW program is not operating.

3. The aircraft is not being tracked by the ARTS III.

4. The aircraft's Mode A or C transponder is sending garbled, weak, or erroneous signals. (Both Mode A and Mode C signals are required for MSAW processing.)

5. The aircraft is not within radar coverage because it is below line of sight or too far away from the radar site.

6. A departing aircraft is within 3 miles of the airport, or an arrival is on final approach to an instrument runway and within 2 miles of the airport or between the stepdown fix and the airport. (Because of the various types of activity in an airport traffic area it is not currently practical to continue MSAW processing within these areas.)

7. The aircraft has been inhibited from computer processing for low altitude alerts. Aircraft are sometimes purposely operated at low altitudes. Minimum Safe Altitude Warning processing of these flights will be inhibited because the controller would receive continuous alerts (false alarms) causing the intentionally low flying VFR pilot to be unnecessarily advised to check his altitude.

8. Due to radar antenna rotation time, the computer needs about 10 seconds to establish a definite course and/or altitude change. Consequently, there are two conditions which may result in low altitude alerts being issued too late to permit the pilot to take corrective action. These are:

a. An aircraft's projected track is clear of any known obstacle and an abrupt turn is made towards one.

b. An aircraft operating at an altitude just above the programed MSAW altitude makes an abrupt descent.

Chapter 14

THE INSTRUMENT WRITTEN AND FLIGHT TESTS AND BEYOND

WRITTEN TEST

YOU'VE HAD FAA written tests before, so you already have a general background on the procedures. FAR Part 61 ground instruction requirements were covered in Chapter 1, but you might look at the following suggestions about taking the written test:

NATURE OF THE TEST

The Instrument Rating Test Book (you can buy your own copy) contains questions that cover areas of necessary knowledge for an aspiring instrument pilot. You will be provided a question book by the testing center or examiner and "your" questions are taken from that book. (You cannot use your personal test book for obvious reasons.) The result is mailed to the applicant on the appropriate form. Planning materials, including charts, aircraft data, weather information, and *Airport/Facility Directory* excerpts are provided for taking the test. Similar materials, for use with the sample test in this guide, are provided in the Appendix.

The passing grade is 70 per cent. All answer sheets are graded by a computer which is programmed to indicate the areas missed. Later you will get an Airman Written Test Report that will indicate the test score and list the number of each of the questions missed. You can then refer to your

own question book to review the questions and material you were weak on.

TAKING THE TEST

1. Answer test items in accordance with the latest regulations and procedures.
2. Read every question thoroughly. Failure on the written test is frequently caused by not reading carefully, rather than lack of knowledge. Do not try to solve the problem before you understand the question.
3. Do not consider a complicated problem a "trick" question. Each test item has a specific objective. There is only one correct and complete answer.
4. Do not waste too much time on problems that stump you. Go on to the test items that you can answer readily, then return to those causing difficulty.
5. For a computer problem, select the answer closest to your own solution. The problem has been checked with various types of computers. If you've solved correctly, your answer will be closer to the correct answer than to any of the other choices.
6. Enter personal data in appropriate spaces on the test answer sheet, in a complete and legible manner, to aid in scoring. You'll be instructed to pick a certain 80 questions from a total of over 1000 questions. You will be given a Question Selection Sheet which one of these 1000+ questions are to be answered on *your* test.

RETESTING

An applicant who receives a failing grade may apply for retesting by presenting his Airman Written Test Report

1. after 30 days from the date he failed the test; or

2. sooner, by presenting a statement from an appropriately rated flight or ground instructor certifying that he has given the applicant additional instruction and considers him ready and prepared for retesting.

It is strongly suggested that you get a copy of the Instrument Rating Question Book (FAA-T-8080-7) to get an idea of the knowledge requirements and also to get used to reading the questions carefully if you haven't taken any tests lately.

FLIGHT TEST

Figure 14-1 is a suggested checklist for you when going up for the flight test.

Following are the items you will need to be aware of for the flight test:

The flight test guide contains an *Objective* for each required pilot operation. Under each pilot operation, pertinent procedures or maneuvers are listed with *Descriptions* and *Acceptable Performance Guidelines*.

1. The *Objective* states briefly the purpose of each pilot operation required on the flight test.

2. The *Description* provides information on what may be asked of you regarding the selected procedure or maneuver. The procedures or maneuvers listed have been found most effective in demonstrating the objective of that particular pilot operation.

3. *The Acceptable Performance Guidelines* include the factors which will be taken into account by the examiner in deciding whether the applicant (you) has met the objective of the pilot operation. The airspeed, altitude, and heading tolerances given represent the minimum performance expected in good flying conditions. However, consistently exceeding these tolerances before corrective action is initiated, or prematurely descending below DH or MDA, is indicative of an unsatisfactory performance. Any procedure or action, or the lack thereof, which requires the intervention of the examiner to maintain safe flight will be cause for disqualification.

In the event you take the instrument pilot flight test and the commercial pilot flight test simultaneously, the maneuvers selected by the examiner for each may be combined and evaluated together, where practicable.

GENERAL PROCEDURES FOR FLIGHT TESTS

For an instrument pilot airplane rating your ability to perform the required pilot operations is

---

APPLICANT'S FLIGHT TEST GUIDE
CHECKLIST

(Suggested)

APPOINTMENT WITH INSPECTOR

OR EXAMINER: Name _____

Time/Date _____

PROPERLY CERTIFICATED AIRPLANE WITH DUAL CONTROLS

☐ View-Limiting Device
☐ Aircraft Documents:
    Airworthiness Certificate
    Registration Certificate
    Operating Limitations
☐ Aircraft Maintenance Records:
    Airworthiness Inspections
    Required Systems Check
☐ FCC Station License
PERSONAL EQUIPMENT
☐ Enroute Charts, SIDS and STARS and
    Instrument Approach Charts
☐ Instrument Checklist
☐ Current Airman's Information Manual
☐ Computer and Plotter
☐ Flight Plan Form
☐ Flight Logs
PERSONAL RECORDS
☐ Pilot Certificate
☐ Medical Certificate
☐ Signed Recommendation
☐ Written Recommendation
☐ Written Test Results
☐ Logbook
☐ Notice of Disapproval
    (If applicable)
☐ Approved School Graduation Certificate
    (If applicable)
☐ FCC Radiotelephone Operator Permit
☐ Examiner's Fee
    If applicable)

Fig. 14-1. Flight test checklist.

based on the following:

1. completing a checklist for instrument flight operations appropriate to the airplane and equipment used

2. performing procedures and maneuvers within the airplane's performance capabilities and limitations, including use of the airplane's systems

3. performing emergency procedures and maneuvers appropriate to the airplane used

4. piloting the airplane with smoothness and accuracy

5. exercising judgment

6. applying your aeronautical knowledge

7. showing that you are master of the aircraft, with the successful outcome of a procedure or maneuver never seriously in doubt

If you fail any of the required pilot operations, you fail the flight test. The examiner or you may discontinue the test at any time when the failure of a required pilot operation makes you ineligible for the certificate or rating sought. If the test is discontinued, you are entitled to credit for only those entire pilot operations that you have successfully performed.

AIRPLANE AND EQUIPMENT REQUIREMENTS FOR FLIGHT TEST

You are required by revised FAR 61.45 to provide an airworthy airplane for the flight test. This airplane must be capable of, and its operating limitations must not prohibit, the pilot operations required on the flight test. Flight instruments required are those appropriate for controlling the airplane in instrument conditions. Appropriate flight instruments are considered to be those outlined in FAR Part 91 for flight under instrument flight rules. The required radio equipment is that necessary for communications with ATC and for the performance of VOR, ADF, and ILS (glide slope and localizer) approaches.

I. MANEUVERING BY REFERENCE TO INSTRUMENTS

Objective: To determine that you can safely and accurately maneuver the airplane in instrument conditions.

A. Straight-and-Level Flight

1. *Description.* You may be asked to demonstrate straight-and-level flight with changes in airspeed and airplane configuration. You will be expected to maintain altitude and heading to accurately control airspeed.

2. *Acceptable Performance Guidelines.* Your performance will be evaluated on the basis of your ability to maintain altitude within ± 100 ft., heading within $\pm 10°$, and airspeed within ± 10 kts. of that assigned.

B. Turns

1. *Description.* You may be asked to demonstrate heading changes using various means to determine rate and amount of turn. You should perform these turns in level, climbing, and descending flight. This may also include changes in airspeed and airplane configuration. Turns for this demonstration may be selected from the following:
   a. standard rate turns
   b. timed turns
   c. turns to predetermined headings
   d. magnetic compass turns
   e. steep turns
2. *Acceptable Performance Guidelines.* Your performance will be evaluated on the basis of your ability to complete turns within $\pm 10°$ of desired headings. You are to maintain altitude within ± 100 ft. and airspeed within ± 10 kts. of that assigned.

C. Climbs and Descents

1. *Description.* You may be asked to demonstrate changes of altitude including:
   a. constant airspeed climbs and descents
   b. rate climbs and descents
   c. climbs and descents to predetermined altitudes and headings

The examiner may request that the above demonstrations be performed in various airplane configurations.

2. *Acceptable Performance Guidelines.* Your performance will be evaluated on your ability to maintain airspeed within ± 10 kts. and vertical rate within ± 200 ft. per minute of that desired. Level-offs and rollouts shall be completed within ± 100 ft. and $\pm 10°$ of the altitude and heading assigned.

II. IFR NAVIGATION

Objective: To determine that you can safely and efficiently navigate in instrument conditions in the National Airspace System in compliance with Instrument Flight Rules and ATC clearances and instructions.

A. Time, Speed and Distance

1. *Description.* You may be asked to demonstrate preflight and inflight computations to determine ETE, ETA, wind correction angle, and groundspeed.

2. *Acceptable Performance Guidelines.* Your performance will be evaluated on the basis of your ability to make accurate and timely computations.

B. VOR Navigation

1. *Description.* You may be asked to demonstrate:
   a. intercepting a VOR radial at a predetermined angle
   b. tracking on a selected VOR radial
   c. determinating position using intersecting VOR radials
2. *Acceptable Performance Guidelines.* Your performance will be evaluated on the basis of your accuracy in determining your position by means of cross bearings, your interception procedures, and your ability to maintain orientation and the assigned flight path.

C. ADF Navigation

1. *Description.* You may be asked to use ADF for homing, intercepting, and tracking predetermined radio bearings to and from non-directional beacons, and for determining position by use of cross bearings.

2. *Acceptable Performance Guidelines.* Your performance will be evaluated on the basis of your accuracy in determining the position by means of cross bearings, your interception procedures, and your ability to maintain orientation and the assigned track.

D. Navigation by ATC Instructions

1. *Description.* You may be asked to show that you can comply with ATC instructions and

procedures. This includes navigation by adherence to radar vectors and specific instructions for headings and altitude changes.

2. *Acceptable Performance Guidelines*. Evaluation of your performance will be based on your promptness and accuracy in responding to and complying with ATC navigation instructions.

III. INSTRUMENT APPROACHES

Objective: To determine that you can execute safe and accurate instrument approaches to published minimums under instrument conditions.

A. VOR Approach

1. *Description*. You will be required to demonstrate a published VOR approach procedure.

2. *Acceptable Performance Guidelines*. You are to descend on a course so as to arrive at the MDA at or before the missed approach point, in a position from which a normal landing approach can be made, straight-in or circling, as appropriate. The missed approach point will be determined by accurate timing from the final approach fix. Deviations of more than ± 10 knots from the desired approach speed shall be disqualifying. Descent below minimum altitudes during any part of the approach or descent below the MDA prior to the examiner reporting the runway in sight, will be disqualifying. If a circling approach is made, exceeding the radius of turn dictated by published visibility minimums or descending below the MDA prior to reaching a position from which a normal approach to the landing runway can be made, will also be disqualifying.

B. ILS Approach

1. *Description*. You will be required to demonstrate a published ILS approach procedure.

2. *Acceptable Performance Guidelines*. As directed by the examiner, you will descend on a straight-in approach to the DH, or on a circling approach to the MDA, arriving at a position from which a normal landing approach can be made straight-in or circling, as appropriate. Deviations of more than ± 10 kts. from the desired approach speed will be disqualifying. Descent below minimum altitudes during any part of the approach, full scale deflection of the CDI or the glide slope indicator after glide slope interception, or descending below the DH or MDA prior to the examiner reporting the runway environment in sight, will be disqualifying. If a circling approach is made, exceeding the radius of turn dictated by published visibility minimums or descending below the MDA prior to reaching a position from which a normal approach to the landing runway can be made, will also be disqualifying.

C. ADF Approach

1. *Description*. You will be required to demonstrate an ADF approach using a published NDB (non-directional beacon) approach procedure.

2. *Acceptable Performance Guidelines*. You are to descend on a course so as to arrive at the MDA at or before the missed approach point, in a position from which a normal landing approach can be made, straight-in or circling, as appropriate. The missed approach point will be determined by accurate timing from the final approach fix. Deviations of more than ± 10 kts. from the desired approach speed will be disqualifying. Descending below minimum altitudes during any part of the approach, or descending below the MDA prior to the examiner reporting the runway environment in sight, will be disqualifying. If a circling approach is made, exceeding the radius of turn dictated by published visibility minimums or descending below the MDA prior to reaching a position from which a normal approach to the landing runway can be made, will also be disqualifying.

IV. CROSS-COUNTRY FLYING

Objective: To determine that you can competently conduct enroute and terminal operations within the National Airspace System in instrument conditions, using radio aids and complying with ATC instructions.

A. Selection of Route

1. *Description*. You may be asked to select a route for a 250 nautical mile IFR flight, based on information contained in the Airman's Information Manual, Enroute Charts, Instrument Approach Procedure Charts, and other appropriate sources of information. This includes facilities for all departures and arrivals.

2. *Acceptable Performance Guidelines*. Your performance will be evaluated on your ability to obtain and apply pertinent information for the selection of a suitable route. Failure to determine current status and usability of facilities will be disqualifying.

B. Procurement and Analysis of Weather Information

1. *Description*. You may be asked to procure and analyze weather reports and forecasts pertinent to the proposed flight. This information should provide (1) forecast weather conditions at destination, (2) the basis for selecting an alternate airport, and (3) the basis for selecting a route to avoid severe weather.

2. *Acceptable Performance Guidelines*. You will correctly analyze the weather reports and forecasts and understand their significance to the proposed flight. Failure to recognize conditions which would be hazardous to your flight will be disqualifying.

C. Development of Flight Log

1. *Description*. You may be asked to develop a flight log for the proposed flight. This log should include at least the enroute courses, estimated ground speeds, distances between checkpoints,

estimated time between checkpoints (ETEs), and amount of fuel required. On the basis of your log, you are expected to prepare an IFR flight plan for the examiner's review.

2. *Acceptable Performance Guidelines.* Your performance will be evaluated on the completeness and accuracy of your flight log and flight plan.

D. Aircraft Performance and Limitations

1. *Description.* You may be asked to apply the information contained in the airplane flight manual or manufacturer's published recommendations to determine the aircraft performance capabilities and weight and balance limitations.

2. *Acceptable Performance Guidelines.* Your performance will be evaluated on your proper application of aircraft performance and loading data in the conduct of the proposed flight.

E. Aircraft Systems and Equipment

1. *Description.* You may be asked to explain the use of the instruments, avionic equipment, and any special system installed in the airplane used, including indications of malfunctions and limitations of these units.

2. *Acceptable Performance Guidelines.* Your performance will be evaluated on your knowledge of the instruments and equipment which are installed in the airplane used for the flight test.

F. Preflight Check of Instruments and Equipment

1. *Description.* Prior to takeoff, you may be asked to perform a systematic operational check of engine instruments, flight instruments, and avionic equipment. All equipment should be appropriately set for your departure clearance.

2. *Acceptable Performance Guidelines.* Your performance will be evaluated on the thoroughness and accuracy of your checks and procedures. Failure to properly check and set instruments and equipment will be disqualifying.

G. Maintaining Airways or ATC Routes
(See IFR Navigation)

H. Use of Radio Communications

1. *Description.* You may be asked to demonstrate the use of two-way radio voice communication procedures for reports, ATC clearances, and other instructions. Radio communications may be simulated at the discretion of the examiner.

2. *Acceptable Performance Guidelines.* Your performance will be evaluated on the basis of your use of proper frequencies, correct phraseology, and the conciseness, clarity, and timeliness of your transmissions. Acceptance of clearances based on facilities or frequencies not appropriate to the equipment being used or to the aircraft performance capabilities, will be disqualifying.

I. Holding Procedures

1. *Description.* You may be directed, by ATC or the examiner, to hold in either a standard or a nonstandard pattern at a specified fix. You should make a proper entry as described in the Airman's Information Manual, remain within protected airspace, apply adequate wind correction, and accurately time the pattern so as to leave the fix at the time specified.

2. *Acceptable Performance Guidelines.* Your performance will be evaluated on your compliance with instructions, and your entry procedure, orientation, accuracy and timing. Deviations of more than ± 100 ft. from the prescribed altitude or more than ± 10 kts. from holding airspeed will be disqualifying.

J. Instrument Approach Procedures
(See Instrument Approaches)

V. EMERGENCIES

Objective: To determine that you can promptly recognize and take appropriate action for abnormal or emergency conditions and equipment malfunctions while in instrument conditions.

A. Recovery from Unusual Attitudes

1. *Description.* The examiner may place the airplane in unusual flight attitudes which may result from vertigo, wake turbulence, lapse of attention, or abnormal trim conditions. You should recover and return to the original altitude and heading. For this demonstration, the examiner may limit the use of flight instruments by simulating malfunctions of the attitude indicator and heading indicator.

2. *Acceptable Performance Guidelines.* Evaluation will be based on the promptness, smoothness, and accuracy demonstrated. All maneuvering will be conducted within the operating limitations for the airplanes used. Any loss of control which makes it necessary for the examiner to take over to avoid exceeding any operating limitation of the airplane will be disqualifying.

B. Equipment or Instrument Malfunctions

1. *Description.* You may be asked to demonstrate the emergency operation of the retractable gear, flaps, and the electrical, fuel, deicing, and hydraulic systems if operationally practical. Emergency operations such as the use of CO_2 pressure for gear extension, or the discharge of a pressure fire extinguisher system will be simulated only. Occasionally, during the performance of flight maneuvers described elsewhere in this guide, the examiner may simulate a partial or complete loss of flight instruments, navigation instruments, or equipment.

2. *Acceptable Performance Guidelines.* You are to respond to emergency situations in accordance with procedures outlined in the manufacturer's

published recommendations. Your performance will be evaluated on the basis of your competency in maintaining aircraft control, your knowledge of the emergency procedures, the judgment you display, and the accuracy of your operations.

C. Loss of Radio Communications

1. *Description.* The examiner may simulate loss of radio communications. You should know the actions required pertaining to altitudes, routes, holding procedures, and approaches.

2. *Acceptable Performance Guidelines.* Evaluation will be based on your knowledge of, and compliance with, the pertinent procedures required by Part 91 of the Federal Aviation Regulations and the emergency procedures outlined in the *Airman's Information Manual.* An explanation or simulation of the proper procedures for loss of radio communications is acceptable.

D. Engine-Out Procedures (Multiengine Airplane)

1. *Description.* You may be asked to demonstrate your ability to positively and accurately maneuver the airplane after one engine has been throttled to simulate the drag of a feathered propeller, or with one propeller feathered, as agreed upon by the applicant and examiner. Feathering of a propeller for flight test purposes will be performed only under such conditions and at such altitudes and positions that a safe landing can readily be accomplished if an emergency develops or difficulty is encountered in unfeathering.

2. *Acceptable Performance Guidelines.* Evaluation will be based on your ability to promptly identify the inoperative engine, and to follow the procedures outlined in the manufacturer's published recommendations. In cruising flight, you are to maintain your heading and altitude within $\pm 20^\circ$ and ± 100 ft. If the airplane is incapable of maintaining altitude with an engine inoperative under existing circumstances, you will maintain an airspeed within ± 5 kts. of the engine-out best rate-of-climb speed.

During approaches, you are to promptly correct any deviation from the desired flight path.

Any loss of control that makes it necessary for the examiner to take over, or any attempt at prolonged flight contrary to the single-engine operating limitations of the airplane, will be disqualifying.

E. Missed Approach Procedures

1. *Description.* At any time during an instrument approach, you may be asked to execute the missed approach procedure depicted on the approach chart being used. If the examiner does not specifically ask for the missed approach but he fails to report the runway in sight at the DH or MDA, you are to immediately initiate the missed approach procedure as described on the chart, or as directed by ATC.

2. *Acceptable Performance Guidelines.* The evaluation will be made on the basis of your judging when to execute the missed approach, the appropriateness of your communication and navigation procedures, your ability to maintain positive airplane control and to operate all airplane systems in accordance with applicable operating instructions for the airplane being used. Descent below the MDA or DH (as appropriate) prior to initiation of the missed approach procedure, will be disqualifying except in those instances where the runway environment was in sight at MDA or DH.

. . . AND BEYOND

A couple of notes:

1. As far as keeping your instrument experience up to date is concerned, you'll have to have had at least 6 hours of actual or simulated instrument flight in the last 6 months. *Not more* than 3 hours of the 6 may be obtained in a synthetic trainer and this trainer type has to be approved by the FAA administrator. You'll need six approaches in that time, too.

2. As far as what flight time may be logged as instrument time is concerned, only the pilot manipulating the controls of an aircraft during the time it is flown solely by reference to instruments under actual or simulated instrument flight conditions can get credit for it. (Except for instructors in actual conditions.)

Keep up your proficiency.

After you get the instrument rating try not to be too lordly with the noninstrument-rated pilots at the airport. Of course, you may be expected to do a little snowing, but hold it down to a dull roar. Remember that those VFR types have been sitting on the ground for a long time now, watching pilots like you take off into weather that has kept them haunting the Weather Service Office at Podunk Greater International Airport and other such well-known places. They squeaked in by the skin of their teeth (the airport went well below VFR minimums shortly after they got in and has been that way for days). The bitter part about it is that the tops are running only three or four thousand feet. It's CAVU above, and the weather at their destination is very fine VFR -- and there they sit. You think about that as you complete the filing of your IFR flight plan and move toward your airplane. The VFR pilots in the FSS watch you, and you have a pretty good idea of what they're thinking (". . . he's filing IFR and is going . . . and he doesn't look like he's got any more on the ball than I do.")

Since this sounds suspiciously like paragraph one of Chapter 1 of this book, it looks as if this is where we came in . . .

APPENDIX A
PILOT/CONTROLLER GLOSSARY

(Credit: *Airman's Information Manual*)

(Selected Terms)

AIRPORT LIGHTING — Various lighting aids that may be installed on an airport. Types of airport lighting include:

1. Approach Light System / ALS – An airport lighting facility which provides visual guidance to landing aircraft by radiating light beams in a directional pattern by which the pilot aligns the aircraft with the extended centerline of the runway on his final approach for landing.
   Condenser-Discharge Sequential Flashing Lights / Sequenced Flashing Lights may be installed in conjunction with the ALS at some airports.
   Types of Approach Light Systems are:

   a. ALSF-I – Approach Light System with Sequenced Flashing Lights in ILS CAT-I configuration,

   b. ALSF-II – Approach Light System with Sequenced Flashing Lights in ILS CAT-II configuration,

   c. SSALF – Simplified Short Approach Light System with Sequenced Flashing Lights,

   d. SSALR – Simplified Short Approach Light System with Runway Alignment Indicator Lights,

   e. MALSF – Medium Intensity Approach Light System with Sequenced Flashing Lights,

   f. MALSR – Medium Intensity Approach Light System with Runway Alignment Indicator Lights,

   g. LDIN – Sequenced Flashing Lead-in Lights,

   h. RAIL – Runway Alignment Indicator Lights (Sequenced Flashing Lights which are installed only in combination with other light systems).

2. Runway Lights / Runway Edge Lights – Lights having a prescribed angle of emission used to define the lateral limits of a runway. Runway lights are uniformly spaced at intervals of approximately 200 feet, and the intensity may be controlled or preset.

3. Touchdown Zone Lighting – Two rows of transverse light bars located symmetrically about the runway centerline normally at 100 foot intervals. The basic system extends 3,000 feet along the runway.

4. Runway Centerline Lighting – Flush centerline lights spaced at 50-foot intervals beginning 75 feet from the landing threshold and extending to within 75 feet of the opposite end of the runway.

5. Threshold Lights – Fixed green lights arranged symmetrically left and right of the runway centerline, identifying the runway threshold.

6. Runway End Identifier Lights / REIL – Two synchronized flashing lights, one on each side of the runway threshold, which provide rapid and positive identification of the approach end of a particular runway.

7. Visual Approach Slope Indicator / VASI – An airport lighting facility providing vertical visual approach slope guidance to aircraft during approach to landing by radiating a directional pattern of high intensity red and white focused light beams which indicate to the pilot that he is "on path" if he sees red/white, "above path" if white/white, and "below path" if red/red. Some airports serving large aircraft have three-bar VASIs which provide two visual glide paths to the same runway.

8. Boundary Lights – Lights defining the perimeter of an airport or landing area. (Refer to AIM Part 1)

AIRPORT ROTATING BEACON / ROTATING BEACON — A visual NAVAID operated at many airports. At civil airports alternating white and green flashes indicate the location of the airport. The total number of flashes are 12 to 15 per minute. At military airports, the beacons flash alternately white and green, but are differentiated from civil beacons by dualpeaked (two quick) white flashes between the green flashes. Normally, operation of an airport rotating beacon during the hours of daylight means that the reported ground visibility at the airport is less than three miles and/or the reported ceiling is less than 1,000 feet and, therefore, an ATC clearance is required for landing or takeoff. (See Control Zone, Special VFR Operations, Instrument Flight Rules) (Refer to AIM Part 1)
ICAO – AERODROME BEACON – Aeronautical beacon used to indicate the location of an aerodrome.

AIRPORT SURFACE DETECTION EQUIPMENT / ASDE — Radar equipment specifically designed to detect all principal features on the surface of an airport, including aircraft and vehicular traffic and to present the entire image on a radar indicator console in the control tower. Used to augment visual observation by tower personnel of aircraft and/or vehicular movements on runways and taxiways.

AIRPORT SURVEILLANCE RADAR / ASR — Approach control radar used to detect and display an aircraft's position in the terminal area. ASR provides range and azimuth information but does not provide elevation data. Coverage of the ASR can extend up to 60 miles.

AIRPORT TRAFFIC AREA — Unless otherwise specifically designated in FAR Part 93, that airspace within a horizontal radius of 5 statute miles from the geographical center of any airport at which a control tower is operating, extending from the surface up to, but not including, an altitude of 3,000 feet above the elevation of the airport. Unless otherwise authorized or required by ATC, no person may operate an aircraft within an airport traffic area except for the purpose of landing at, or taking off from, an airport within that area. ATC authorizations may be given as individual approval of specific operations or may be contained in written agreements between airport users and the tower concerned. (Refer to FAR Parts 1 and 91)

ALTITUDE READOUT / AUTOMATIC ALTITUDE REPORT — An aircraft's altitude, transmitted via the Mode C transponder feature, that is visually displayed in 100-foot increments on a radar scope having readout capability. (See Automatic Radar Terminal Systems, NAS Stage A, Alpha Numeric Display) (Refer to AIM Part 1)

ALTITUDE RESERVATION / ALTRV — Airspace utilization under prescribed conditions normally employed for the mass movement of aircraft or other special user requirements which cannot otherwise be accomplished. ALTRVs are approved by the appropriate FAA facility. (See Air Traffic Control Systems Command Center)

ALTITUDE RESTRICTION — An altitude or altitudes stated in the order flown, which are to be maintained until reaching a specific point or time. Altitude restrictions may be issued by ATC due to traffic, terrain or other airspace considerations.

ALTITUDE RESTRICTIONS ARE CANCELLED — Adherence to previously imposed altitude restrictions is no longer required during a climb or descent.

APPROACH CLEARANCE — Authorization by ATC for a pilot to conduct an instrument approach. The type of instrument approach for which cleared and other pertinent information is provided in the approach clearance when required. (See Instrument Approach Procedure, Cleared for Approach) (Refer to AIM Part 1, FAR Part 91)

APPROACH GATE — The point on the final approach course which is 1 mile from the final approach fix on the side away from the airport or 5 miles from landing threshold, whichever is farther from the landing threshold. This is an imaginary point used within ATC as a basis for final approach course interception for aircraft being vectored to the final approach course.

APPROACH LIGHT SYSTEM — (See Airport Lighting)

APPROACH SEQUENCE — The order in which aircraft are positioned while on approach or awaiting approach clearance. (See Landing Sequence)
ICAO – APPROACH SEQUENCE – The order in which two or more aircraft are cleared to approach to land at the aerodrome.

APPROACH SPEED — The recommended speed contained in aircraft operating manuals used by pilots when making an approach to landing. This speed will vary for different segments of an approach as well as for aircraft weight and configuration.

ARC — The track over the ground of an aircraft flying at a constant distance from a navigational aid by reference to distance measuring equipment (DME).

AREA NAVIGATION / RNAV — A method of navigation that permits aircraft operations on any desired course within the coverage of station-referenced navigation signals or within the limits of self-contained system capability. (Refer to AIM Parts 1 and 3, FAR Part 71)

1. Area Navigation Low Route – An area navigation route within the airspace extending upward from 1,200 feet above the surface of the earth to, but not including 18,000 feet MSL.
2. Area Navigation High Route – An area navigation route within the airspace extending upward from and including 18,000 feet MSL to flight level 450.
3. Random Area Navigation Routes / Random RNAV Routes – Direct routes, based on area navigation capability, between waypoints defined in terms of degree/distance fixes or offset from published or established routes/airways at specified distance and direction.
4. RNAV Waypoint / W/P – A predetermined geographical position used for route or instrument approach definition or progress reporting purposes that is defined relative to a VORTAC station position.

ICAO – AREA NAVIGATION/RNAV – A method of navigation which permits aircraft operation on any desired flight path within the coverage of station-referenced navigation aids or within the limits of the capability of self-contained aids or a combination of these.

AUTOMATED RADAR TERMINAL SYSTEMS / ARTS — The generic term for the ultimate in functional capability afforded by several automation systems. Each differs in functional capabilities and equipment. ARTS plus a suffix Roman Numeral denotes a specific system. A following letter indicates a major modification to that system. In general, an ARTS displays for the terminal controller aircraft identification, flight plan data, other flight associated information, e.g., altitude and speed, and aircraft position symbols in conjunction with his radar presentation. Normal radar co-exists with the alphanumeric display. In addition to enhancing visualization of the air traffic situation, ARTS facilitate intra/inter-facility transfer and coordination of flight information. These capabilities are enabled by specially designed computers and subsystems tailored to the radar and communications equipments and operational requirements of each automated facility. Modular design permits adoption of improvements in computer software and electronic technologies as they become available while retaining the characteristics unique to each system:

1. ARTS IA – The functional capabilities and equipment of the New York Common IFR Room Terminal Automation System. It tracks primary as well as secondary targets derived from two radar sources. The aircraft targets are displayed on a radar type console by means of an alphanumeric generator. Aircraft identity is depicted in association with the appropriate aircraft target. When the aircraft is equipped with an encoded altimeter (Mode C), its altitude is also displayed. The system can exchange flight plan information with the ARTCC.
2. ARTS II – A programmable non-tracking, computer aided display subsystem capable of modular expansion. ARTS II systems provide a level of automated air traffic control capability at terminals having low to medium activity. Flight identification and altitude

may be associated with the display of secondary radar targets. Also, flight plan information may be exchanged between the terminal and ARTCC.

3. ARTS III – The Beacon Tracking Level (BTL) of the modular programmable automated radar terminal system in use at medium to high activity terminals. ARTS III detects, tracks and predicts secondary radar derived aircraft targets. These are displayed by means of computer generated symbols and alphanumeric characters depicting flight identification, aircraft altitude, ground speed and flight plan data. Although it does not track primary targets, they are displayed coincident with the secondary radar as well as the symbols and alphanumerics. The system has the capability of communicating with ARTCCs and other ARTS III facilities.

4. ARTS IIIA – The Radar Tracking and Beacon Tracking Level (RT&BTL) of the modular, programmable automated radar terminal system. ARTS IIIA detects, tracks and predicts primary as well as secondary radar derived aircraft targets. An enhancement of the ARTS III, this more sophisticated computer driven system will eventually replace the ARTS IA system and upgrade about half of the existing ARTS III systems. The enhanced system will provide improved tracking, continuous data recording and fail-soft capabilities.

AUTOMATIC ALTITUDE REPORTING — That function of a transponder which responds to Mode C interrogations by transmitting the aircraft's altitude in 100-foot increments.

BRAKING ACTION (GOOD, MEDIUM OR FAIR, POOR, NIL) — A report of conditions on the airport movement area providing a pilot with a degree/quality of braking that he might expect. Braking action is reported in terms of good, medium (or fair), poor or nil. (See Runway Condition Reading)

CEILING — The height above the earth's surface of the lowest layer of clouds or obscuring phenomena that is reported as "broken," "overcast," or "obscuration," and not classified as "thin" or "partial".
ICAO – CEILING – The height above the ground or water of the base of the lowest layer of cloud below 6000 metres (20,000 feet) covering more than half the sky.

CENTER'S AREA — The specified airspace within which an air route traffic control center (ARTCC) provides air traffic control and advisory service. (See Air Route Traffic Control Center) (Refer to AIM Part 1)

CIRCLE TO LAND MANEUVER / CIRCLING MANEUVER — A maneuver initiated by the pilot to align the aircraft with a runway for landing when a straight-in landing from an instrument approach is not possible or is not desirable. This maneuver is made only after ATC authorization has been obtained and the pilot has established required visual reference to the airport (See Circle to Runway, Landing Minimums) (Refer to AIM Part 1)

CIRCLE TO RUNWAY (RUNWAY NUMBER) — Used by ATC to inform the pilot that he must circle to land because the runway in use is other than the runway aligned with the instrument approach procedure. When the direction of the circling maneuver in relation to the airport/runway is required, the controller will state the direction (eight cardinal compass points) and specify a left or right downwind or base leg as appropriate; e.g., "Cleared VOR Runway 36 approach circle to Runway 22" or "Circle northwest of the airport for a right downwind to Runway 22." (See Circle to Land Maneuver, Landing Minimums) (Refer to AIM Part 1)

CLEARANCE LIMIT — The fix, point, or location to which an aircraft is cleared when issued an air traffic clearance.
ICAO – CLEARANCE LIMIT – The point of which an aircraft is granted an air traffic control clearance.

CLEARANCE VOID IF NOT OFF BY (TIME) — Used by ATC to advise an aircraft that the departure clearance is automatically cancelled if takeoff is not made prior to a specified time. The pilot must obtain a new clearance or cancel his IFR flight plan if not off by the specified time.
ICAO – CLEARANCE VOID TIME – A time specified by an air traffic control unit at which a clearance ceases to be valid unless the aircraft concerned has already taken action to comply therewith.

CLEARED FOR THE OPTION — ATC authorization for an aircraft to make a touch-and-go, low approach, missed approach, stop and go, or full stop landing at the discretion of the pilot. It is normally used in training so that an instructor can evaluate a student's performance under changing situations. (See Option Approach) (Refer to AIM Part 1)

CONTACT APPROACH — An approach wherein an aircraft on an IFR flight plan, operating clear of clouds with at least 1 mile flight visibility and having received an air traffic control authorization, may deviate from the prescribed instrument approach procedure and proceed to the airport of destination by visual reference to the surface. This approach will only be authorized when requested by the pilot and the reported ground visibility at the destination airport is at least 1 statute mile. (See Visual Approach) (Refer to AIM Part 1)

CONTROL SECTOR — An airspace area of defined horizontal and vertical dimensions for which a controller, or group of controllers, has air traffic control responsibility, normally within an air route traffic control center or an approach control facility. Sectors are established based on predominant traffic flows, altitude strata, and controller workload. Pilot-controller communications during operations within a sector are normally maintained on discrete frequencies assigned to the sector. (See Discrete Frequency)

CONTROL SLASH — A radar beacon slash representing the actual position of the associated aircraft. Normally, the control slash is the one closest to the interrogating radar beacon site. When ARTCC radar is operating in narrowband (digitized) mode, the control slash is converted to a target symbol.

CRUISE — Used in an ATC clearance to authorize a pilot to conduct flight at any altitude from the minimum IFR altitude up to and including the altitude specified in the clearance. The pilot may level off at any intermediary altitude within this block of airspace. Climb/descent within the block is to be made at the discretion of the pilot. However, once the pilot starts descent and reports leaving an altitude in the block he may not return to that altitude without additional ATC clearance. Further, it is approval for the pilot to proceed to and make an approach at destination airport and can be used in conjunction with:

1. An airport clearance limit at locations with a standard/special instrument approach procedure. The FARs require that if an instrument letdown to an airport is necessary the pilot shall make the letdown in accordance with a standard/special instrument approach procedure for that airport, or

2. An airport clearance limit at locations that are within/below/outside controlled airspace and without a standard/special instrument approach procedure. Such a clearance is NOT AUTHORIZATION for the pilot to descend under IFR conditions below the applicable minimum IFR altitude nor does it imply that ATC is exercising control over aircraft in uncontrolled airspace; however, it provides a means for the aircraft to proceed to destination airport, descend and land in accordance with applicable FARs governing VFR flight operations. Also, this provides search and rescue protection until such time as the IFR flight plan is closed. (See Instrument Approach Procedure)

CRUISING ALTITUDE / LEVEL — An altitude or flight level maintained during en route level flight. This is a constant altitude and should not be confused with a cruise clearance. (See Altitude)

DELAY INDEFINITE (REASON IF KNOWN) EXPECT APPROACH/FURTHER CLEARANCE (TIME) — Used by ATC to inform a pilot when an accurate estimate of the delay time and the reason for the delay cannot immediately be determined; e.g., a disabled aircraft on the runway, terminal or center area saturation, weather below landing minimums. (See Expect Approach Clearance, Expect Further Clearance)

DISCRETE CODE / DISCRETE BEACON CODE — As used in the Air Traffic Control Radar Beacon System (ATCRBS), any one of the 4096 selectable Mode 3/A aircraft transponder codes except those ending in zero zero; e.g., discrete codes: 0010, 1201, 2317, 7777; non-discrete codes: 0100, 1200, 7700. Non-discrete codes are normally reserved for radar facilities that are not equipped with discrete decoding capability and for other purposes such as emergencies (7700), VFR aircraft (1200), etc. (See Radar) (Refer to AIM Part 1)

DISCRETE FREQUENCY — A separate radio frequency for use in direct pilot-controller communications in air traffic control which reduces frequency congestion by controlling the number of aircraft operating on a particular frequency at one time. Discrete frequencies are normally designated for each control sector in en route/terminal ATC facilities. Discrete frequencies are listed in the Airport/Facilities Directory, AIM Part 3, and DOD FLIP IFR En Route Supplement. (See Control Sector)

FINAL APPROACH – IFR — The flight path of an aircraft which is inbound to an airport on a final instrument approach course, beginning at the final approach fix or point and extending to the airport or the point where a circle to land maneuver or a missed approach is executed. (See Segments of an Instrument Approach Procedure, Final Approach Fix, Final Approach Course, Final Approach Point)

ICAO – FINAL APPROACH - That part of an instrument approach procedure from the time the aircraft has:

1. completed the last procedure turn or base turn, where one is specified, or
2. crossed a specified fix, or
3. intercepted the last track specified for the procedures; until it has crossed a point in the vicinity of an aerodrome from which:
   a. a landing can be made; or
   b. a missed approach procedure is initiated.

FINAL APPROACH COURSE — A straight line extension of a localizer, a final approach radial/bearing, or a runway centerline, all without regard to distance. (See Final Approach – IFR, Traffic Pattern)

FINAL APPROACH FIX / FAF — The designated fix from or over which the final approach (IFR) to an airport is executed. The FAF identifies the beginning of the final approach segment of the instrument approach. (See Final Approach Point, Segments of an Instrument Approach Procedure, Glide Slope Intercept Altitude)

FINAL APPROACH POINT — The point, within prescribed limits of an instrument approach procedure, where the aircraft is established on the final approach course and final approach descent may be commenced. A final approach point is applicable only in non-precision approaches where a final approach fix has not been established. In such instances, the point identifies the beginning of the final approach segment of the instrument approach. (See Final Approach Fix, Segments of an Instrument Approach Procedure, Glide Slope Intercept Altitude)

IFR TAKEOFF MINIMUMS AND DEPARTURE PROCEDURES — FAR, Part 91, prescribes standard takeoff rules for certain civil users. At some airports, obstructions or other factors require the establishment of nonstandard takeoff minimums, departure procedures, or both, to assist pilots in avoiding obstacles during climb to the minimum en route altitude. Those airports are listed in NOS/DOD Instrument Approach Charts (IAPs) under a section entitled "IFR Takeoff Minimums and Departure Procedures." The NOS/DOD IAP chart legend illustrates the symbol used to alert the pilot to nonstandard takeoff minimums and departure procedures. When departing IFR from such airports, or from any airports where there are no departure procedures, SIDs, or ATC facilities available, pilots should advise ATC of any departure limitations. Controllers may query a pilot to determine acceptable departure directions, turns, or headings after takeoff. Pilots should be familiar with the departure procedures and must assure that their aircraft can meet or exceed any specified climb gradients.

ILS CATEGORIES —

1. ILS Category I – An ILS approach procedure which provides for approach to a height above touchdown of not less than 200 feet and with runway visual range of not less than 1800 feet.

2. ILS Category II – An ILS approach procedure which provides for approach to a height above touchdown of not less than 100 feet and with runway visual range of not less than 1200 feet.

3. ILS Category III.
   a. IIIA – An ILS approach procedure which provides for approach without a decision height minimum and with runway visual range of not less than 700 feet.
   b. IIIB – An ILS approach procedure which provides for approach without a decision height minimum and with runway visual range of not less than 150 feet.
   c. IIIC – An ILS approach procedure which provides for approach without a decision height minimum and without runway visual range minimum.

INSTRUMENT RUNWAY — A runway equipped with electronic and visual navigation aids for which a precision or nonprecision approach procedure having straight-in landing minimums has been approved.

ICAO – INSTRUMENT RUNWAY – A runway intended for the operation of aircraft using nonvisual aids and comprising:

1. Instrument Approach Runway – An instrument runway served by a nonvisual aid providing at least directional guidance adequate for a straight-in approach.

2. Precision Approach Runway, Category I – An instrument runway served by ILS or GCA approach aids and visual aids intended for operations down to 60 metres (200 feet) decision height and down to an RVR of the order of 800 metres (2600 feet.)

3. Precision Approach Runway, Category II – An instrument runway served by ILS and visual aids intended for operations down to 30 metres (100 feet) decision height and down to an RVR of the order of 400 metres (1200 feet.)

4. Precision Approach Runway, Category III – An instrument runway served by ILS (no decision height being applicable) and:
   a. By visual aids intended for operations down to an RVR of the order of 200 metres (700 feet);
   b. By visual aids intended for operations down to an RVR of the order of 50 metres (150 feet);
   c. Intended for operations without reliance on external visual reference.

MAXIMUM AUTHORIZED ALTITUDE / MAA — A published altitude representing the maximum usable altitude or flight level for an airspace structure or route segment. It is the highest altitude on a Federal airway, Jet route, area navigation low or high route, or other direct route for which an MEA is designated in FAR Part 95, at which adequate reception of navigation and signals is assured.

MINIMUM CROSSING ALTITUDE / MCA — The lowest altitude at certain fixes at which an aircraft must cross when preceding in the direction of a higher minimum en route IFR altitude (MEA). (See Minimum En Route IFR Altitude)

MINIMUM DESCENT ALTITUDE / MDA — The lowest altitude, expressed in feet above mean sea level, to which descent is authorized on final approach or during circle-to-land maneuvering in execution of a standard instrument approach procedure where no electronic glide slope is provided. (See Nonprecision Approach Procedure)

MINIMUM EN ROUTE IFR ALTITUDE / MEA — The lowest published altitude between radio fixes which assures acceptable navigational signal coverage and meets obstacle clearance requirements between those fixes. The MEA prescribed for a Federal airway or segment thereof, area navigation low or high route, or other direct route, applies to the entire width of the airway, segment or route between the radio fixes defining the airway, segment or route. (Refer to FAR Parts 91 and 95, AIM Part 1)

MINIMUM HOLDING ALTITUDE / MHA — The lowest altitude prescribed for a holding pattern which assures navigational signal coverage, communications, and meets obstacle clearance requirements.

MINIMUM IFR ALTITUDES — Minimum altitudes for IFR operations as prescribed in FAR Part 91. These altitudes are published on aeronautical charts and prescribed in FAR Part 95, for airways and routes, and FAR Part 97, for standard instrument approach procedures. If no applicable minimum altitude is prescribed in FAR Parts 95 or 97, the following minimum IFR altitude applies:

1. In designated mountainous areas, 2000 feet above the highest obstacle within a horizontal distance of 5 statute miles from the course to be flown; or

2. Other than mountainous areas, 1000 feet above the highest obstacle within a horizontal distance of 5 statute miles from the course to be flown; or

3. As otherwise authorized by the Administrator or assigned by ATC. (See Minimum En Route IFR Altitude, Minimum Obstruction Clearance Altitude, Minimum Crossing Altitude, Minimum Safe Altitude, Minimum Vectoring Altitude) (Refer to FAR Part 91)

MINIMUM OBSTRUCTION CLEARANCE ALTITUDE / MOCA — The lowest published altitude in effect between radio fixes on VOR airways, off-airway routes, or route segments which meets obstacle clearance requirements for the entire route segment and which assures acceptable navigational signal coverage only within 22 nautical miles of a VOR. (Refer to FAR Part 91 and 95)

MINIMUM RECEPTION ALTITUDE / MRA — The lowest altitude at which an intersection can be determined. (Refer to FAR Part 95)

MINIMUM SAFE ALTITUDE / MSA —

1. The minimum altitudes specified in FAR Part 91, for various aircraft operations.

2. Altitudes depicted on instrument approach charts and identified as minimum sector altitudes or emergency safe altitudes which provide a minimim of 1000 feet obstacle clearance within a specified distance from the navigation facility upon which an instrument approach procedure is predicated. These altitudes are for EMERGENCY USE ONLY and do not necessarily guarantee NAVAID reception. Minimum sector altitudes are established for all procedures (except localizers without an NDB) within a 25 nautical mile radius of the navigation facility. Emergency safe altitudes are established for some military procedures within a 100 nautical mile radius of the navigation facility.

ICAO – MINIMUM SECTOR ALTITUDE - The lowest altitude which may be used under emergency conditions which will provide a minimum clearance of 300 meters (1000 feet) above all obstacles located in an area contained within a sector of a circle of 25 nautical miles radius centred on a radio aid to navigation.

MINIMUM VECTORING ALTITUDE / MVA —
The lowest MSL altitude at which an IFR aircraft will be vectored by a radar controller, except as otherwise authorized for radar approaches, departures and missed approaches. The altitude meets IFR obstacle clearance criteria. It may be lower than the published MEA along an airway or J-route segment. It may be utilized for radar vectoring only upon the controllers' determination that an adequate radar return is being received from the aircraft being controlled. Charts depicting minimum vectoring altitudes are normally available only to the controllers and not to pilots. (Refer to AIM Part 1)

MINIMUMS / MINIMA —
Weather condition requirements established for a particular operation or type of operation; e.g., IFR takeoff or landing, alternate airport for IFR flight plans, VFR flight. (See Landing Minimums, IFR Takeoff Minimums, VFR Conditions, IFR Conditions) (Refer to FAR Part 91, AIM Part 1)

MISSED APPROACH —

1. A maneuver conducted by a pilot when an instrument approach cannot be completed to a landing. The route of flight and altitude are shown on instrument approach procedure charts. A pilot executing a missed approach prior to the Missed Approach Point (MAP) must continue along the final approach to the MAP. The pilot may climb immediately to the altitude specified in the missed approach procedure.

2. A term used by the pilot to inform ATC that he is executing the missed approach.

3. At locations where ATC radar service is provided the pilot should conform to radar vectors, when provided by ATC, in lieu of the published missed approach procedure. (See Missed Approach Point) (Refer to AIM Part 1)

ICAO – MISSED APPROACH PROCEDURE - The procedure to be followed if, after an instrument approach, a landing is not effected, and occuring normally:

1. When the aircraft has descended to the decision height and has not established visual contact, or

2. When directed by air traffic control to pull up or to go around again.

OUTER FIX — A general term used within ATC to describe fixes in the terminal area, other than the final approach fix. Aircraft are normally cleared to these fixes by an Air Route Traffic Control Center or an Approach Control Facility. Aircraft are normally cleared from these fixes to the final approach fix or final approach course.

PILOT'S DISCRETION — When used in conjunction with altitude assignments, means that ATC has offered the pilot the option of starting climb or descent whenever he wishes and conducting the climb or descent at any rate he wishes. He may temporarily level off at any intermediary altitude. However, once he has vacated an altitude he may not return to that altitude.

PROCEDURE TURN INBOUND — That point of a procedure turn maneuver where course reversal has been completed and an aircraft is established inbound on the intermediate approach segment or final approach course. A report of "procedure turn inbound" is normally used by ATC as a position report for separation purposes. (See Final Approach Course, Procedure Turn, Segments of an Instrument Approach Procedure)

RADAR SAFETY ADVISORY — A radar advisory issued by ATC to radar identified aircraft under their control if an aircraft is observed to be at an altitude which, in the controller's judgment, places the aircraft in unsafe proximity to terrain, obstructions or other aircraft. The controller may discontinue the issuance of further advisories if the pilot advises he is taking action to correct the situation or has the other aircraft in sight.

1. Terrain/Obstruction Advisory – A Radar advisory issued by ATC to aircraft under their control if an aircraft is observed at an altitude which, in the controller's judgment, places the aircraft in unsafe proximity to terrain/obstructions; e.g., "Low Altitude Alert, advise you climb immediately to three thousand."

2. Aircraft Conflict Advisory – A radar advisory issued by ATC to aircraft under their control if an aircraft not under their control is observed at an altitude which, in the controller's judgment, places both aircraft in unsafe proximity to each other. With the alert, ATC will offer the pilot an alternate course of action when feasible, e.g., – "Conflict Alert, advise you turn right heading zero niner zero" or "climb to eight thousand immediately."

The issuance of a radar safety advisory is contingent upon the capability of the controller to observe an unsafe condition. The course of action provided will be predicated on other traffic under ATC control. Once the advisory is issued, it is solely the pilot's prerogative to determine what course of action, if any, he will take. (See Radar Advisory) (Refer to AIM Part 1)

RADAR SERVICE TERMINATED — Used by ATC to inform a pilot that he will no longer be provided any of the services that could be received while under radar contact. Radar service is automatically terminated and the pilot is not advised in the following cases:

1. When the aircraft cancels its IFR flight plan.
2. At the completion of a radar approach.
3. When an arriving aircraft receiving Stage I, II, or III service is advised to contact the tower.
4. When an aircraft conducting a visual approach is advised to contact the tower.
5. When an aircraft vectored to a final approach course for an instrument approach has landed or the tower has the aircraft in sight, whichever occurs first.

REMOTE COMMUNICATIONS AIR/GROUND FACILITY / RCAG — An unmanned VHF/UHF transmitter/receiver facility which is used to expand ARTCC air/ground communications coverage and to facilitate direct contact between pilots and controllers. RCAG facilities are sometimes not equipped with emergency frequencies 121.5 MHz and 243.0 MHz. (Refer to AIM Part 1)

SEGMENTS OF AN INSTRUMENT APPROACH PROCEDURE — An instrument approach procedure may have as many as four separate segments depending on how the approach procedure is structured.

1. Initial Approach – The segment between the initial approach fix and the intermediate fix or the point where the aircraft is established on the intermediate course or final approach course.
2. Intermediate Approach – The segment between the intermediate fix or point and the final approach fix.
3. Final Approach – The segment between the final approach fix or point and the runway, airport or missed approach point.
4. Missed Approach – The segment between the missed approach point, or point of arrival at decision height, and the missed approach fix at the prescribed altitude. (Refer to FAR Part 97)

ICAO

1. *Initial Approach* – That part of an instrument approach procedure consisting of the first approach to the first navigational facility associated with the procedure, or to a predetermined fix.
2. *Intermediate Approach* – That part of an instrument approach procedure from the first arrival at the first navigational facility or predetermined fix, to the beginning of the final approach.
3. *Final Approach* – That part of an instrument approach procedure from the time the aircraft has:
   a. Completed the last procedure turn or base turn where one is specified, or
   b. crossed a specified fix, or
   c. intercepted the last track specified for the procedures; until it has crossed a point in the vicinity of an aerodrome from which:
      (1) a landing can be made; or
      (2) a missed approach procedure is initiated.

4. *Missed Approach Procedure* – The procedure to be followed if, after an instrument approach, a landing is not effected and occurring normally:
   a. When the aircraft has descended to the decision height and has not established visual contact, or
   b. when directed by air traffic control to pull up or to go around again.

TERMINAL RADAR PROGRAM — A national program instituted to extend the terminal radar services provided IFR aircraft to VFR aircraft. Pilot participation in the program is urged but is not mandatory. The progressive stages of the program are referred to as Stage I, Stage II and Stage III. The stage service provided at a particular location is contained in AIM, Part 3.

1. Stage I / Radar Advisory Service for VFR Aircraft – Provides traffic information and limited vectoring to VFR aircraft on a workload permitting basis.
2. Stage II / Radar Advisory and Sequencing for VFR Aircraft – Provides, in addition to Stage I service, vectoring and sequencing on a full-time basis to arriving VFR aircraft. The purpose is to adjust the flow of arriving IFR and VFR aircraft into the traffic pattern in a safe and orderly manner and to provide traffic advisory to departing VFR aircraft.
3. Stage III / Radar Sequencing and Separation Service for VFR Aircraft – Provides, in addition to Stage II services, separation between all participating aircraft. The purpose is to provide separation between all participating VFR aircraft and all IFR aircraft operating within the airspace defined as a Terminal Radar Service Area (TRSA), or Terminal Control Area (TCA). (See Terminal Radar Service Area, Controlled Airspace) (Refer to AIM Parts 1, 3 and 4)

VISUAL APPROACH — An approach wherein an aircraft on an IFR flight plan, operating in VFR conditions under the control of an air traffic control facility and having an air traffic control authorization, may deviate from the prescribed instrument approach procedure and proceed to the airport of destination under VFR conditions.

ICAO – VISUAL APPROACH – An approach by an IFR flight when either part or all of an instrument approach procedure is not completed and the approach is executed in visual reference to terrain.

WORDS TWICE —

1. As a request: "Communication is difficult. Please say every phrase twice."
2. As information: "Since communications are difficult, every phrase in this message will be spoken twice."

APPENDIX B

AIRPORT/FACILITY DIRECTORY

DIRECTORY LEGEND
SAMPLE

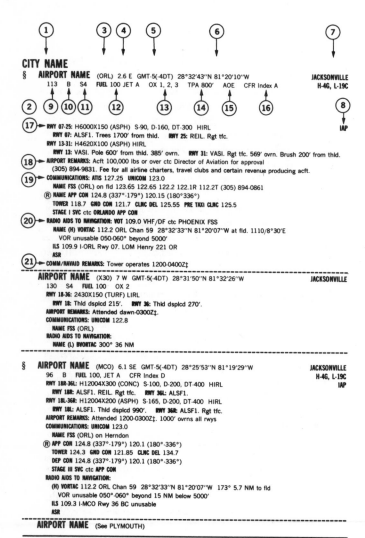

DIRECTORY LEGEND

LEGEND

This Directory is an alphabetical listing of data on record with the FAA on all airports that are open to the public, associated terminal control facilities, air route traffic control centers and radio aids to navigation within the conterminous United States, Puerto Rico and the Virgin Islands. Airports are listed alphabetically by associated city name and cross referenced by airport name. Facilities associated with an airport, but with a different name, are listed individually under their own name, as well as under the airport with which they are associated.

The listing of an airport in this directory merely indicates the airport operator's willingness to accommodate transient aircraft, and does not represent that the facility conforms with any Federal or local standards, or that it has been approved for use on the part of the general public.

The information on obstructions is taken from reports submitted to the FAA. It has not been verified in all cases. Pilots are cautioned that objects not indicated in this tabulation (or on charts) may exist which can create a hazard to flight operation.

Detailed specifics concerning services and facilities tabulated within this directory are contained in Airman's Information Manual, Basic Flight Information and ATC Procedures.

The legend items that follow explain in detail the contents of this Directory and are keyed to the circled numbers on the sample on the preceding page.

① CITY/AIRPORT NAME

Airports and facilities in this directory are listed alphabetically by associated city and state. Where the city name is different than the airport name the city name will appear on the line above the airport name. Airports with the same associated city name will be listed alphabetically by airport name and will be separated by a dashed rule line. All others will be separated by a solid rule line.

② NOTAM SERVICE

§—NOTAM "D" (Distant teletype dissemination) and NOTAM "L" (Local dissemination) service is provided for airport. Absence of annotation § indicates NOTAM "L" (Local dissemination) only is provided for airport. See AIM, Basic Flight Information and ATC Procedures for detailed description of NOTAM.

③ LOCATION IDENTIFIER

A three or four character code assigned to airports. These identifiers are used by ATC in lieu of the airport name in flight plans, flight strips and other written records and computer operations.

④ AIRPORT LOCATION

Airport location is expressed as distance and direction from the center of the associated city in nautical miles and cardinal points, i.e., 3.5 NE.

⑤ TIME CONVERSION

Hours of operation of all facilities are expressed in Universal Coordinated Time (UTC) and shown as "Z" time. The directory indicates the number of hours to be subtracted from UTC to obtain local standard time and local daylight saving time UTC—5(—4DT). The symbol ‡ indicates that during periods of daylight saving time effective hours will be one hour earlier than shown. In those areas where daylight saving time is not observed that (—4DT) and ‡ will not be shown.

⑥ GEOGRAPHIC POSITION OF AIRPORT

⑦ CHARTS

The Sectional Chart and Low and High Altitude Enroute Chart and panel on which the airport or facility is located.

⑧ INSTRUMENT APPROACH PROCEDURES

IAP indicates an airport for which a prescribed (Public Use) FAA Instrument Approach Procedure has been published.

⑨ ELEVATION

Elevation is given in feet above mean sea level and is the highest point on the landing surface. When elevation is sea level it will be indicated as (00). When elevation is below sea level a minus (–) sign will precede the figure.

⑩ ROTATING LIGHT BEACON

B indicates rotating beacon is available. Rotating beacons operate dusk to dawn unless otherwise indicated in AIRPORT REMARKS.

⑪ SERVICING

S1: Minor airframe repairs.
S2: Minor airframe and minor powerplant repairs.
S3: Major airframe and minor powerplant repairs.
S4: Major airframe and major powerplant repairs.

DIRECTORY LEGEND

⑫ FUEL

FUEL

CODE		PRODUCT
80	Grade 80 gasoline (Red)	
100	Grade 100 gasoline (Green)	
100LL	Grade 100LL gasoline (low lead) (Blue)	
115	Grade 115 gasoline	
A	Jet A—Kerosene freeze point—40° C.	
A1	Jet A-1—Kerosene, freeze point—50° C.	
A1+	Jet A-1—Kerosene with icing inhibitor, freeze point—50° C.	
B	Jet B—Wide-cut turbine fuel, freeze point—50° C.	
B+	Jet B—Wide-cut turbine fuel with icing inhibitor, freeze point—50° C.	

⑬ OXYGEN

OX 1   High Pressure
OX 2   Low Pressure
OX 3   High Pressure—Replacement Bottles
OX 4   Low Pressure—Replacement Bottles

⑭ TRAFFIC PATTERN ALTITUDE

TPA—Traffic Pattern Altitude is provided only for those airports without a 24 hour operating control tower. "Altitudes shown are Above Ground Level (AGL)"

⑮ AIRPORT OF ENTRY AND LANDING RIGHTS AIRPORTS

AOE—Airport of Entry-A customs Airport of Entry where permission from U.S. Customs is not required, however, at least one hour advance notice of arrival must be furnished.
LRA—Landing Rights Airport-Application for permission to land must be submitted in advance to U.S. Customs. At least one hour advance notice of arrival must be furnished.
NOTE: Advance notice of arrival at both an AOE and LRA airport may be included in the flight plan when filed in Canada or Mexico, if destination is an airport where flight notification service is available. This notice will also be treated as an application for permission to land in the case of an LRA. (See Customs, Immigration and Naturalization, Public Health and Agriculture Department requirements in the International Flight Information Manual for further details.)

⑯ CERTIFICATED AIRPORT (FAR 139) and FAA INSPECTION

Airports serving Civil Aeronautics Board certified carriers and certified under FAR, Part 139, are indicated by the CFR index; i.e., CFR Index A, which relates to the availability of crash, fire, rescue equipment.
All airports not inspected by FAA will be identified by the note: Not insp. This indicates that the airport information has been provided by the owner or operator of the field.

FAR—PART 139 CERTIFICATED AIRPORTS
INDICES AND FIRE FIGHTING AND RESCUE EQUIPMENT REQUIREMENTS

Airport Index	Required No. Vehicles	Aircraft Length	Scheduled Departures	Agent + Water for Protein Foam
A	1	≧90′	≦1	500#DC or 450#DC + 50 gal H²0
		〉90′, ≦126′	〈5	300#DC + 500 gal H²0
B	2	〉90′, ≦126′	≦5	Index A + 1500 gal H²0
		〉126′, ≦160′	〈5	
C	3	〉126′, ≦160′	≦5	Index A + 3000 gal H²0
		〉160′, ≦200′	〈5	
D	3	〉160′, ≦200′	≦5	Index A + 4000 gal H²0
		〉200′	〈5	
E	3	〉200′	≦5	Index A + 6000 gal H²0
Ltd.	Vehicle and capacity requirements for airports limited operating certificates are determined on a case by case basis.			

〉 Greater Than: 〈 Less Than: ≦ Equal or Greater Than: ≧ Equal or Less Than;   H²0—Water; DC—Dry Chemical.

NOTE: If AFFF (Aqueous Film Forming Foam) is used in lieu of Protein Foam, the water quantities listed for Indices A thru E can be reduced 33-1/3%.

⑰ RUNWAY DATA

Runway information is shown on two lines. That information common to the entire runway is shown on the first line while information concerning the runway ends are shown on the second or following line. Lengthy information will be footnoted and placed in the Airport Remarks.
Runway direction, surface, length, width, weight bearing capacity, lighting, gradient (when gradient exceeds 0.3 percent) and appropriate remarks are shown for each runway. Direction, length, width, lighting and remarks are shown for sealanes.

DIRECTORY LEGEND

RUNWAY SURFACE AND LENGTH

Runway lengths prefixed by the letter "H" indicate that the runways are hard surfaced (concrete, asphalt). If the runway length is not prefixed, the surface is sod, clay, etc. The runway surface composition is indicated in parentheses after runway length as follows:

(ASPH)—Asphalt   (GRVL)—Gravel, or cinders
(CONC)—Concrete   (TURF)—Sod
(DIRT)—Dirt

The full dimensions of helipads are shown, i.e., 50X50.

RUNWAY WEIGHT BEARING CAPACITY

Runway strength data shown in this publication is derived from available information and is a realistic estimate of capability at an average level of activity. It is not intended as a maximum allowable weight or as an operating limitation. Many airport pavements are capable of supporting limited operations with gross weights of 25-50% in excess of the published figures. Permissible operating weights, insofar as runway strengths are concerned, are a matter of agreement between the owner and user. When desiring to operate into any airport at weights in excess of those published in the publication, users should contact the airport management for permission. Add 000 to figure following S, D, DT, DDT and MAX for gross weight capacity:

S—Runway weight bearing capacity for aircraft with single-wheel type landing gear, (DC-3), etc.
D—Runway weight bearing capacity for aircraft with dual-wheel type landing gear, (DC-6), etc.
DT—Runway weight bearing capacity for aircraft with dual-tandem type landing gear, (707), etc.
DDT—Runway weight bearing capacity for aircraft with double dual-tandem type landing gear, (747), etc.
Quadricycle and dual-tandem are considered virtually equal for runway weight bearing consideration, as are single-tandem and dual-wheel.
Omission of weight bearing capacity indicates information unknown.

RUNWAY LIGHTING

Lights are in operation sunset to sunrise. Lighting available by prior arrangement only or operating part of the night only and/or pilot controlled and with specific operating hours are indicated under airport remarks as footnotes. Since obstructions are usually lighted, obstruction lighting is not included in this code. Unlighted obstructions on or surrounding an airport will be noted in airport remarks.
Temporary, emergency or limited runway edge lighting such as flares, smudge pots, lanterns or portable runway lights will also be shown in airport remarks, instead of being designated by code numbers.
Types of lighting are shown with the runway or runway end they serve.

LIRL—Low intensity Runway Lights
MIRL—Medium Intensity Runway Lights
HIRL—High Intensity Runway Lights
REIL—Runway End Identifier Lights
CL—Centerline Lights
TDZ—Touchdown Zone Lights
ODALS—Omni Directional Approach Lighting System.
AF OVRN—Air Force Overrun 1000′ Standard Approach Lighting System.
LDIN—Lead-In Lighting System.
MALS—Medium Intensity Approach Lighting System.
MALSF—Medium Intensity Approach Lighting System with Sequenced Flashing Lights.
MALSR—Medium Intensity Approach Lighting System with Runway Alignment Indicator Lights.

SALS—Short Approach Lighting System.
SALSF—Short Approach Lighting System with Sequenced Flashing Lights.
SSALS—Simplified Short Approach Lighting System.
SSALF—Simplified Short Approach Lighting System with Sequenced Flashing Lights.
SSALR—Simplified Short Approach Lighting System with Runway Alignment Indicator Lights.
ALSF1—High Intensity Approach Lighting System with Sequenced Flashing Lights, Category I, Configuration.
ALSF2—High Intensity Approach Lighting System with Sequenced Flashing Lights, Category II, Configuration.
VASI—Visual Approach Slope Indicator System.

VASI approach slope angle and TCH will be shown only when slope angle exceeds 3°.

RUNWAY GRADIENT

Runway gradient will be shown only when it is 0.3 percent or more. When available the direction of slope upward will be indicated, i.e., 0.5% up NW.

RUNWAY END DATA

Lighting systems such as VASI, MALSR, REIL; obstructions; displaced thresholds will be shown on the specific runway end. "Rgt tfc"-Right traffic indicates right turns should be made on landing and takeoff for specified runway end.

⑱ AIRPORT REMARKS

"Landing Fee" indicates landing charges for private or non-revenue producing aircraft, in addition, fees may be charged for planes that remain over a couple of hours and buy no services, or at major airline terminals for all aircraft.
Obstructions—Because of space limitations only the more prominent obstacles are indicated. Natural obstruction, such as trees, clearly discernible for contact operations are not included. On the other hand, all obstructions within at least a 20:1 approach ratio are indicated.
Remarks—Data is confined to operational items affecting the status and usability of the airport.

⑲ COMMUNICATIONS

Communications will be listed in sequence in the order shown below:
Automatic Terminal Information Service (ATIS) and Private Aeronautical Stations (UNICOM) along with their frequency is shown, where available, on the line following the heading "COMMUNICATIONS".
Flight Service Station (FSS) information. The associated FSS will be shown followed by the identifier and information concerning availability of telephone service, e.g. Direct Line (DL). Local Call (LC), etc. Where the airport NOTAM File identifier is different then the associated FSS it will be shown as "NOTAM File DCA". Where the FSS is located

DIRECTORY LEGEND

on the field it will be indicated as ''on arpt'' following the identifier. Frequencies available will follow. The FSS telephone number will follow along with any significant operational information. FSS's whose name is not the same as the airport on which located will also be listed in the normal alphabetical name listing for the state in which located. Limited Remote Communication Outlet (LRCO) or Remote Communications Outlet (RCO) providing service to the airport followed by the frequency and name of the Controlling FSS.

FSS's and CS/Ts provide information on airport conditions, radio aids and other facilities, and process flight plans. Airport Advisory Service is provided at the pilot's request on 123.6 or 123.65 by FSS's located at non-tower airports or when the tower is not in operation. (See AIM, ADVISORIES AT NON TOWER AIRPORTS.)

Aviation weather briefing service is provided by FSS's and CS/T's: however, CS/T personnel are not certified weather briefers and therefore provide only factual data from weather reports and forecasts. Flight and weather briefing services are also available by calling the telephone numbers listed.

Limited Remote Communications Outlet (LRCO)—Unmanned satellite air/ground communications facility, which may be associated with a VOR. These outlets effectively extend service range of the FSS and provide greater communications reliability.

Remote Communications Outlet (RCO)—An unmanned satellite air to ground communication stations remotely controlled and providing UHF and VHF communications capability to extend the service range of an FSS.

Civil communications frequencies used in the FSS air/ground system are now operated simplex on 122.0, 122.2, 122.3, 122.4, 122.6, 123.6; emergency 121.5; plus receive-only on 122.05, 122.1, 122.15 and 123.6.

    a. 122.0 is assigned as the Enroute Flight Advisory Service channel at selected FSS's.

    b. 122.2 is assigned to all FSS's as a common enroute simplex service.

    c. 123.6 is assigned as the airport advisory channel at non-tower FSS locations, however, it is still in commission at some FSS's collocated with towers to provide part-time Airport Advisory Service.

    d. 122.1 is the primary receive-only frequency at VORs. 122.05, 122.15 and 123.6 are assigned at selected VORs meeting certain criteria.

    e. Some FSS's are assigned 50kHz channels for simplex operation in the 122-123 MHz band (e.g. 122.35). Pilots using the FSS A/G system should refer to this directory or appropriate charts to determine frequencies available at the FSS or remoted facility through which they wish to communicate.

Part time FSS hours of operation are shown in remarks under facility name.

Emergency frequency 121.5 is available at all Flight Service Stations. Towers, Approach Control and RADAR facilities, unless indicated as not available.

TERMINAL SERVICES

ATIS—A continuous broadcast of recorded non-control information in selected areas of high activity.

UNICOM—A non-government air/ground radio communications facility utilized to provide general airport advisory service.

APP CON—Approach Control. The symbol Ⓡ indicates radar approach control.

TOWER—Control tower

GND CON—Ground Control

DEP CON—Departure Control. The symbol Ⓡ indicates radar departure control.

CLNC DEL—Clearance Delivery.

PRE TAXI CLNC—Pre taxi clearance

VFR ADVSY SVC—VFR Advisory Service. Service provided by Non-Radar Approach Control.

STAGE I SVC—Radar Advisory Service for VFR aircraft

STAGE II SVC—Radar Advisory and Sequencing Service for VFR aircraft

STAGE III SVC—Radar Sequencing and Separation Service for participating VFR Aircraft within a Terminal Radar Service Area (TRSA)

TCA—Radar Sequencing and Separation Service for all aircraft in a Terminal Control Area (TCA)

TOWER, APP CON and DEP CON RADIO CALL will be the same as the airport name unless indicated otherwise.

⑳ RADIO AIDS TO NAVIGATION

The Airport/Facility Directory lists by facility name all Radio Aids to Navigation in the National Airspace System and those upon which the FAA has approved an instrument approach. Private or military Radio Aids to Navigation not in the National Airspace System are not tabulated.

All VOR, VORTAC and ILS equipment in the National Airspace System has an automatic monitoring and shutdown feature in the event of malfunction. Unmonitored as used in the publication means that FSS or tower personnel cannot observe the malfunction or shutdown signal.

NAVAID information is tabulated as indicated in the following sample:

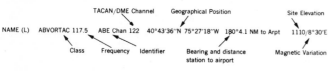

NAME (L)   ABVORTAC 117.5   ABE Chan 122   40°43'36''N 75°27'18''W   180°4.1 NM to Arpt   1110/8°30'E

VOR unusable 020°-060° beyond 26 NM below 3500'

Restrictions

NAME (L) BVORTAC 300°/36 NM

Bearing and distance from VORTAC or VOR/DME facility to airport

ASR—indicates that civil radar instrument approach minimums are published.

DIRECTORY LEGEND

RADIO CLASS DESIGNATIONS

Identification of VOR/VORTAC/TACAN Stations by Class (Operational Limitations):

Normal Usable Altitudes and Radius Distances

Class	Altitudes	Distance (miles)
(T)	12,000' and below	25
(L)	Below 18,000'	40
(H)	Below 18,000'	40
(H)	Within the Conterminous 48 States only, between 14,500' and 17,999'	100
(H)	18,000'-FL 450	130
(H)	Above FL 450	100

(H) = High   (L) = Low   (T) = Terminal

NOTE: An (H) facility is capable of providing (L) and (T) service volume and an (L) facility additionally provides (T) service volume.

The term VOR is, operationally, a general term covering the VHF omnidirectional bearing type of facility without regard to the fact that the power, the frequency-protected service volume, the equipment configuration, and operational requirements may vary between facilities at different locations.

AB	Automatic Weather Broadcast (also shown with ∎ following frequency.)
B	Scheduled Broadcast Station (broadcasts weather at 15 minutes after the hour.)
DF	Direction Finding Service.
DME	UHF standard (TACAN compatible) distance measuring equipment.
H	Non-directional radio beacon (homing), power 50 watts to less than 2,000 watts.
HH	Non-directional radio beacon (homing), power 2,000 watts or more.
H-SAB	Non-directional radio beacons providing automatic transcribed weather service.
ILS	Instrument Landing System (voice, where available, on localizer channel).
LDA	Localizer Directional Aid.
LMM	Compass locator station when installed at middle marker site.
LOM	Compass locator station when installed at outer marker site.
MH	Non-directional radio beacon (homing) power less than 50 watts.
S	Simultaneous range homing signal and/or voice.
SABH	Non-directional radio beacon not authorized for IFR or ATC. Provides automatic weather broadcasts.
SDF	Simplified Direction Facility.
TACAN	UHF navigational facility-omnidirectional course and distance information.
VOR	VHF navigational facility-omnidirectional course only.
VOR/DME	Collocated VOR navigational facility and UHF standard distance measuring equipment.
VORTAC	Collacted VOR and TACAN navigational facilities.
W	Without voice on radio facility frequency.
Z	VHF station on location marker at a LF radio facility.

㉑ COMM/NAVAID REMARKS:

Pertinent remarks concerning communications and NAVAIDS.

APPENDIX C

(Credit: *Airman's Information Manual*)

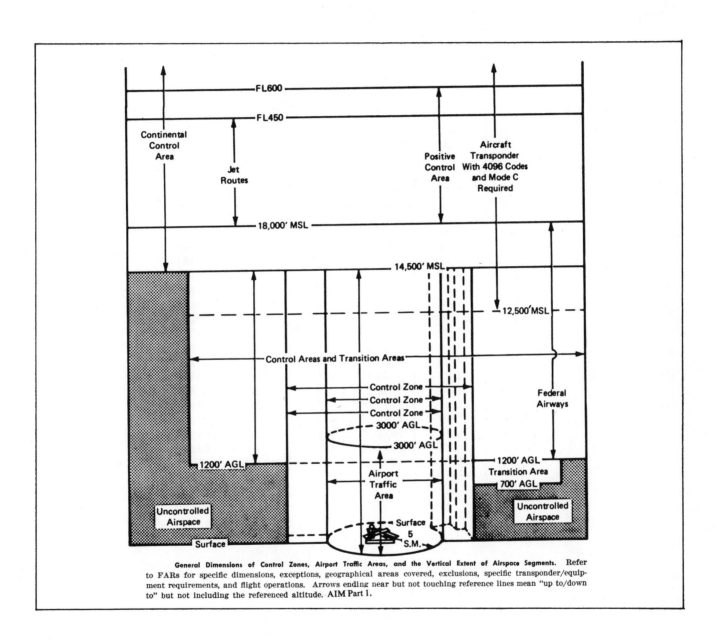

General Dimensions of Control Zones, Airport Traffic Areas, and the Vertical Extent of Airspace Segments. Refer to FARs for specific dimensions, exceptions, geographical areas covered, exclusions, specific transponder/equipment requirements, and flight operations. Arrows ending near but not touching reference lines mean "up to/down to" but not including the referenced altitude. AIM Part 1.

APPENDIX D

STANDARD ATMOSPHERE CHART

Altitude (ft)	Pressure (in Hg)	Pressure (psf)	Temp. (°C)	Temp. (°F)	Density- slugs per cubic foot
0	29.92	2116.22	15.0	59.0	.002378
1,000	28.86	2040.85	13.0	55.4	.002309
2,000	27.82	1967.68	11.C	51.9	.002242
3,000	26.82	1896.64	9.1	48.3	.002176
4,000	25.84	1827.69	7.1	44.7	.002112
5,000	24.89	1760.79	5.1	41.2	.002049
6,000	23.98	1695.89	3.1	37.6	.001988
7,000	23.09	1632.93	1.1	34.0	.001928
8,000	22.22	1571.88	-0.9	30.5	.001869
9,000	21.38	1512.70	-2.8	26.9	.001812
10,000	20.57	1455.33	-4.8	23.3	.001756
11,000	19.79	1399.73	-6.8	19.8	.001701
12,000	19.02	1345.87	-8.8	16.2	.001648
13,000	18.29	1293.70	-10.8	12.6	.001596
14,000	17.57	1243.18	-12.7	9.1	.001545
15,000	16.88	1194.27	-14.7	5.5	.001496
16,000	16.21	1146.92	-16.7	1.9	.001448
17,000	15.56	1101.11	-18.7	-1.6	.001401
18,000	14.94	1056.80	-20.7	-5.2	.001355
19,000	14.33	1013.93	-22.6	-8.8	.001310
20,000	13.74	972.49	-24.6	-12.3	.001267

BIBLIOGRAPHY

Airman's Information Manual, Washington, D. C., USGPO.

Air Traffic Service Manual 7110.65 (FAA 1976).

Aviation Weather. Washington, D.C., FAA Flight Standards Service and National Weather Service 1975.

Aviation Weather Services. FAA and National Weather Service 1975.

Civil Use of U.S. Government Instrument Approach Procedure Charts. AC 90-1A. 1968.

Flight Test Guide — Instrument Pilot, Airplane. AC 61-56A. Washington, D.C., FAA Flight Standards Service, April, 1976.

Instrument Flying. Department of the Air Force, AF Manual 51-37, 1960.

Instrument Rating Question Book. FAA-T-8080-7. FAA Aviation Standards National Field Office, 1984.

Kershner, W. K., *Advanced Pilot's Flight Manual*, 4th ed. Ames, Iowa, Iowa State University Press, 1976.

Kershner, W. K., *Student Pilot's Flight Manual*, 5th ed. Ames, Iowa, Iowa State University Press, 1979.

Liston, Joseph, *Power Plants for Aircraft*. New York, McGraw-Hill, 1953.

Perkins, C. D. and Hage, R. E., *Airplane Performance, Stability and Control*. New York, John Wiley & Sons, 1949.

National Oceanic and Atmospheric Administration (National Ocean Survey), Enroute Low Altitude Charts, Instrument Approach Charts and Standard Instrument Departures. Riverdale, MD.

Two very good instrument reference books for you to add to your library are *Weather Flying* by Robert N. Buck (The Macmillan Company, 866 Third Ave., New York, 10022) and *Instrument Flying* by Richard L. Taylor (same address). These two books give a practical look at actual coping with weather and ATC.

Pilot's Operating Handbooks, products referred to, information received from companies, or sources of information on aviation products:

Aircraft Radio Corporation (ARC), Boonton, N.J.

Bendix Corporation, Bendix Radio Division, Baltimore, Md.

Cessna Aircraft Co., Wichita, Kans. Pilot's Operating Handbooks for 150, 172, 180, 182, 210, 310.

Collins Radio Company, Cedar Rapids, Iowa.

Continental Development Corp., Ridgefield, Conn. Instantaneous Vertical Speed Indicator (IVSI).

King Radio Corp., Olathe, Kansas. (KX 145, KI 205, KAE 125, 126, 127 and 138, KCS 55A)

Lycoming Division, AVCO, Williamsport, Pa. Detail Engine Specifications.

Motorola Aviation Electronics, Culver City, Calif.

National Aeronautical Corp. (NARCO), Fort Washington, Pa. (NAV 111, DME 190 and RNAV 112)

Piper Aircraft Corp., Lock Haven, Pa. Aztec B and C, Comanche 250 and 260.

Safetech, Inc., Six Terry Drive, Newtown, PA 18940. E-6B Computer model: FDF-57-B.

INDEX